Stochastic Models

WA 1094847 3

D0588680

UNIVERSITY OF GLAMORGAN
LEARNING RESOURCES CENTRE

Pontypridd, Mid Glamorgan, CF37 1DL
Telephone: Pontypridd (0443) 480480

Books are to be returned on or before the last date below

Stochastic Models
An Algorithmic Approach

Henk C. Tijms
Vrije Universiteit, Amsterdam

Learning Resources
Centre

JOHN WILEY & SONS
Chichester • New York • Brisbane • Toronto • Singapore

1094847 3   519.2
                 TIJ

Copyright © 1994 by John Wiley & Sons Ltd,
              Baffins Lane, Chichester,
              West Sussex PO19 1UD, England

              National (01243) 779777
              International (+44) 1243 779777

Reprinted January 1995

All rights reserved.

No part of this book may be reproduced by any means,
or transmitted, or translated into a machine language
without the written permission of the publisher.

Other Wiley Editorial Offices

John Wiley & Sons, Inc., 605 Third Avenue,
New York, NY 10158–0012, USA

Jacaranda Wiley Ltd, 33 Park Road, Milton,
Queensland 4064, Australia

John Wiley & Sons (Canada) Ltd, 22 Worcester Road,
Rexdale, Ontario M9W 1L1, Canada

John Wiley & Sons (SEA) Pte Ltd, 37 Jalan Pemimpin #05-04,
Block B, Union Industrial Building, Singapore 2057

Library of Congress Cataloging-in-Publication Data

Tijms, H. C.
    Stochastic models : an algorithmic approach / Henk C. Tijms.
       p.   cm. — (Wiley series in probability and mathematical statistics)
    Includes bibliographical references.
    ISBN 0 471 94380 0;   0 471 95123 4 (pbk)
    1. Stochastic systems.   2. Algorithms.   I. Title.   II. Series.
  QA274.T47   1994
  519.2 — dc20   94-1010
                                                  CIP

British Library Cataloguing in Publication Data

A catalogue record for this book is available from the British Library

ISBN 0 471 94380 0;   0 471 95123 4 (pbk)

Typeset in 10/12pt Times by Laser Words, Madras
Printed and bound in Great Britain by
Biddles Ltd, Guildford and King's Lynn

6.3.95

Contents

Chapter 3   Markovian Decision Processes and their Applications   181

Chapter 4   Algorithmic Analysis of Queueing Models   259

Preface

OVERVIEW

The goal of this book is twofold. The first is to show the student how useful simple stochastic models can be for getting insight into the behaviour of complex stochastic systems. Though the primary goal of stochastic modelling is to provide insights and not numbers, numerical answers are often indispensable for gaining system understanding. The second goal of the book is to give the student a feeling of how numerical solutions to specific situations can be obtained. The student will see that algorithmic analysis of stochastic models is more than getting numerical answers. Its essence is to find probabilistic ideas which make the computations transparent and natural.

The above two goals are the guidelines for this book as they were for my previous Wiley book *Stochastic Modelling and Analysis*. The present book, however, is much more oriented to the student than to the researcher. An attempt has been made to write a textbook in which an easy-to-understand but rigorous presentation of the theory is given, while interweaving at the same time the theory with a wide variety of realistic examples. These examples are carefully chosen to illustrate the basic models and the associated solution methods. Students benefit best from examples and problems in order to understand and appreciate the use of stochastic models.

The integrated presentation of theory, applications and algorithms makes this book unique. In addition many interesting and thought-provoking problems are provided. The student is urged to try these problems. The best way to acquire a feeling for stochastic modelling is to go through the process of learning-by-doing.

INSTRUCTORS

This book can be used for senior undergraduate and graduate level courses for students in operations research, statistics and engineering. The prerequisites of the book are only calculus and a first course in probability. The book is organized in such a way that several one-semester or one-quarter courses on stochastic models can be given on the basis of this book. The various chapters are independent of each other to a large extent. Only the basic principles of renewal theory in Chapter 1 are necessary to understand Chapter 2 on Markov chains. Chapter 3 on

Markov decision processes and Chapter 4 on queueing models require only a basic knowledge of Markov chains.

Instructors who adopt the book for their courses can order both a solutions manual and a software package supporting this book by writing directly to the author at the address: Vrije Universiteit, Department of Econometrics, De Boelelaan 1105, 1081 HV Amsterdam, The Netherlands. The software package contains user-friendly modules for analysing queueing models, stochastic inventory models and Markov chains among others.

ACKNOWLEDGEMENTS

In writing this book I benefited from discussions with several colleagues. In particular, I would like to thank Frank van der Duyn Schouten and Paul Schweitzer. A special word of appreciation is due to my student Sindo Núñez Queija for carefully reading the manuscript and providing many helpful comments. An excellent typing job was done by Gloria Wirz-Wagenaar who patiently typed the various versions of the manuscript.

CHAPTER 1

Renewal Processes with Applications

1.0 INTRODUCTION

Renewal theory provides elegant and powerful tools for analysing stochastic processes that generate themselves from time to time. The long-run behaviour of a regenerative stochastic process can be studied in terms of its behaviour during a single regeneration cycle. In many applied probability problems, such as in inventory, queueing and reliability, regenerative stochastic processes are prevalent. Particularly useful is the renewal-reward model that arises by imposing a reward structure on a regenerative process. This simple and intuitively appealing model has numerous applications. In practical applications the Poisson process is the most frequently used renewal process. The Poisson process models not only many real-world phenomena, but allows as well for tractable analysis because of its memoryless property.

Renewal theory and the Poisson process are the subjects of Sections 1.1 and 1.2. The treatment of the Poisson process includes the compound Poisson process and the nonstationary Poisson process. Section 1.3 deals with renewal-reward processes and shows how to calculate long-run averages such as the long-run average reward per unit time and the long-run fraction of time the process spends in a given set of states. The emphasis is on basic results and how to apply them to specific problems rather than on proofs, although most proofs are given to keep the treatment self-contained. Many realistic examples illustrate the theory. The examples are taken from the fields of production/inventory, queueing and reliability. In particular, Section 1.4 is devoted to applications from reliability and maintenance, while Section 1.5 deals with inventory applications. Section 1.6 discusses the formula of Little. This formula is a kind of law of nature in operations research and relates the average queue size to the average waiting time in queueing systems. Another fundamental result that is frequently used in inventory and queueing applications is the property that Poisson arrivals see time averages. This result is discussed in some detail in Section 1.7. In production/inventory systems and queueing systems it is often important to have asymptotic estimates for the probability of stockout occurrence and the waiting-time probability. Such estimates are obtained in the final Section 1.8 by using renewal-theoretic methods. This section illustrates the simplicity of analysis to be achieved by a general renewal-theoretic approach to hard individual problems.

1.1   RENEWAL THEORY

Renewal theory concerns the study of stochastic processes counting the number of events that take place as a function of time. Here the interoccurrence times between successive events are independent and identically distributed random variables. For instance, the events could be the arrivals of customers to a waiting line or the successive replacements of light bulbs. Although renewal theory originated from the analysis of replacement problems for components such as light bulbs, the theory has many applications to quite a wide range of practical probability problems. In inventory, queueing and reliability problems the analysis is often based on an appropriate identification of embedded renewal processes for the specific problem considered. For example, in a queueing process the embedded events could be the arrivals of customers who find the system empty or in an inventory process the embedded events could be the replenishments of stock when the inventory position drops at or below the reorder point.

Formally, let $X_1, X_2, \ldots$ be a sequence of non-negative, independent random variables having a common probability distribution function

$$F(x) = P\{\dot{X}_k \leq x\}, \quad x \geq 0, \ k = 1, 2, \ldots .$$

Letting $\mu_1 = E(X_i)$, it is assumed that

$$0 < \mu_1 < \infty.$$

The random variable X_n denotes the interoccurrence time between the $(n - 1)$th and nth event in some specific probability problem. Letting

$$S_0 = 0, \quad S_n = \sum_{i=1}^{n} X_i, \quad n = 1, 2, \ldots,$$

we have that S_n is the epoch at which the nth event occurs. Define for each $t \geq 0$,

$$N(t) = \text{the largest integer } n \geq 0 \text{ for which } S_n \leq t.$$

Then the random variable $N(t)$ represents the number of events up to time t.

Definition 1.1.1   *The counting process $\{N(t), \ t \geq 0\}$ is called the renewal process generated by the interoccurrence times $X_1, X_2, \ldots$.*

It is said that a renewal occurs at time t if $S_n = t$ for some n. For each $t \geq 0$, the number of renewals up to time t is finite with probability 1. This is an immediate consequence of the strong law of large numbers stating that $S_n/n \to E(X_1)$ with probability 1 as $n \to \infty$ and thus $S_n \leq t$ only for finitely many n. We give some examples of a renewal process.

Example 1.1.1 A replacement problem

Suppose we have an infinite supply of electric bulbs, where the burning times of the bulbs are independent and identically distributed random variables. If the bulb in use fails, it is immediately replaced by a new bulb. Letting

$$X_i = \text{the burning time of the } i\text{th bulb,}$$

then

$$N(t) = \text{the total number of bulbs to be replaced up to time } t.$$

Example 1.1.2 An inventory problem

Consider a periodic-review inventory system for which the demands for a single product in the successive weeks $n = 1, 2, \ldots$ are independent and identically distributed random variables having a continuous distribution. Letting

$$X_i = \text{the demand in the } i\text{th week,}$$

then

$$1 + N(x) = \text{the number of weeks until depletion of the current inventory } x.$$

1.1.1 The renewal function

An important role in renewal theory is played by the *renewal function* $M(t)$ defined by

$$M(t) = E[N(t)], \quad t \geq 0. \tag{1.1.1}$$

Define for $n = 1, 2, \ldots$ the probability distribution function

$$F_n(t) = P\{S_n \leq t\}, \quad t \geq 0.$$

Note that $F_1(t) = F(t)$. A basic relation is

$$N(t) \geq n \text{ if and only if } S_n \leq t. \tag{1.1.2}$$

This relation implies that

$$P\{N(t) \geq n\} = F_n(t), \quad n = 1, 2, \ldots . \tag{1.1.3}$$

Lemma 1.1.1 *For any $t \geq 0$,*

$$M(t) = \sum_{n=1}^{\infty} F_n(t). \tag{1.1.4}$$

Moreover, $M(t) < \infty$ for all $t > 0$.

Proof Since $E(N) = \sum_{k=0}^{\infty} P\{N > k\}$ for any non-negative, integer-valued random variable N (see Appendix A), the relation (1.1.4) follows directly from

(1.1.3). To see that $M(t)$ is finite for each t, note that $F(0) < 1$ implies the existence of an $\alpha > 0$ such that $P\{X_n \geq \alpha\} > 0$. Let the renewal process $\{\overline{N}(t)\}$ be associated with the sequence $\{\overline{X}_n\}$, where $\overline{X}_n = 0$ if $X_n \leq \alpha$ and $\overline{X}_n = \alpha$ if $X_n > \alpha$. Then, for any $t \geq 0$, $N(t) \leq \overline{N}(t)$ with probability 1. In the renewal process $\{\overline{N}(t)\}$ the number of renewals at any time $k\alpha$ is geometrically distributed with parameter $F(\alpha)$ and thus $E[\overline{N}(t)] < \infty$ for all t.

A useful characterization of the renewal function is provided by the so-called *renewal equation*.

Theorem 1.1.2   *Assume that the probability distribution function $F(x)$ of the interoccurrence times has a density $f(x)$. Then the renewal function $M(t)$ satisfies the integral equation*

$$M(t) = F(t) + \int_0^t M(t - x) f(x) \, dx, \quad t \geq 0. \tag{1.1.5}$$

This integral equation has a unique solution that is bounded on finite intervals.

Proof   The proof of (1.1.5) is instructive. Fix $t > 0$. To compute $E[N(t)]$, we condition on the time of the first renewal and use that the process probabilistically starts over after each renewal. Under the condition that $X_1 = x$, the random variable $N(t)$ is distributed as $1 + N(t - x)$ when $0 \leq x \leq t$ and $N(t)$ is 0 otherwise. Hence, by conditioning upon X_1, we find

$$E[N(t)] = \int_0^\infty E[N(t) | X_1 = x] f(x) \, dx$$

$$= \int_0^t E[1 + N(t - x)] f(x) \, dx,$$

which gives (1.1.5). To prove that the equation (1.1.5) has a unique solution, suppose that $H(t) = F(t) + \int_0^t H(t - x) f(x) \, dx$, $t \geq 0$ for a function $H(t)$ that is bounded on finite intervals. We substitute this equation repeatedly into itself. Next we use the convolution formula

$$F_n(t) = \int_0^t F(t - x) f_{n-1}(x) \, dx.$$

This formula follows by taking $X = X_1 + \cdots + X_{n-1}$ and $Y = X_n$ in relation (A.6) in Appendix A, where $f_k(x)$ denotes the probability density of $X_1 + \cdots + X_k$. Thus

$$H(t) = \sum_{k=1}^n F_k(t) + \int_0^t H(t - x) f_n(x) \, dx, \quad n = 1, 2, \dots . \tag{1.1.6}$$

Fix now $t > 0$. Since $H(x)$ is bounded on $[0, t]$, the second term of the right side of (1.1.6) is bounded by $c F_n(t)$ for some $c > 0$. Lemma 1.1.1 implies that $F_n(t) \to 0$

as $n \to \infty$. Thus, by letting $n \to \infty$ in (1.1.6), we find $H(t) = \sum_{k=1}^{\infty} F_k(t)$ showing that $H(t) = M(t)$.

Theorem 1.1.2 allows for the following important generalization.

Theorem 1.1.3 *Assume that $F(x)$ has a probability density $f(x)$. Let $a(x)$ be a given, integrable function that is bounded on finite intervals. Suppose the function $Z(t)$, $t \geq 0$, is defined by the integral equation*

$$Z(t) = a(t) + \int_0^t Z(t - x) f(x) \, dx, \quad t \geq 0. \tag{1.1.7}$$

Then this equation has a unique solution that is bounded on finite intervals. The solution is given by

$$Z(t) = a(t) + \int_0^t a(t - x) m(x) \, dx, \quad t \geq 0, \tag{1.1.8}$$

where the renewal density $m(x)$ denotes the derivative of $M(x)$.

Proof We give only a sketch of the proof. The proof is similar to the proof of the second part of Theorem 1.1.2. Substituting the equation (1.1.7) repeatedly into itself yields

$$Z(t) = a(t) + \sum_{k=1}^{n} \int_0^t a(t - x) f_k(x) \, dx + \int_0^t Z(t - x) f_{n+1}(x) \, dx.$$

Next by letting $n \to \infty$ the desired result readily follows. It is left to the reader to verify that the various mathematical operations are allowed.

The integral equation (1.1.7) is called the *renewal equation*. This important equation arises in many applied probability problems. The renewal equation will be further studied in the next section.

Next we use the above results to study another basic quantity in renewal theory, namely the excess life.

Excess life

In many practical probability problems an important quantity is the random variable γ_t defined as the time from t until the next renewal. More precisely, γ_t is defined as

$$\gamma_t = S_{N(t)+1} - t,$$

see also Figure 1.1.1 in which a renewal epoch is denoted by $\times$. The random variable γ_t is called the excess or residual life at time t. For the replacement problem of Example 1.1.1, the random variable γ_t denotes the residual lifetime of the light bulb in use at time t.

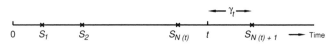

Figure 1.1.1   The excess life

Lemma 1.1.4   *For any $t \geq 0$,*

$$E(\gamma_t) = \mu_1[1 + M(t)] - t. \qquad (1.1.9)$$

Moreover, assuming that $\mu_2 = E(X_1^2)$ is finite,

$$E(\gamma_t^2) = \mu_2[1 + M(t)] - 2\mu_1 \left[t + \int_0^t M(x)\,dx \right] + t^2, \quad t \geq 0. \qquad (1.1.10)$$

Proof   Fix $t \geq 0$. To prove (1.1.9), we apply Wald's equation discussed in Appendix A. To do so, note that $N(t) \leq n-1$ if and only if $X_1 + \cdots + X_n > t$. Hence the event $\{N(t) + 1 = n\}$ is determined only by $X_1, \ldots, X_n$ and is independent of $X_{n+1}, X_{n+2}, \ldots$. Thus

$$E \left[\sum_{k=1}^{N(t)+1} X_k \right] = E(X_1)E[N(t) + 1],$$

which gives (1.1.9). The proof of (1.1.10) is only sketched. For ease assume that $F(x)$ has a density $f(x)$. By conditioning on the time of the first renewal, it follows that

$$E(\gamma_t^2) = \int_0^\infty E(\gamma_t^2 | X_1 = x) f(x)\,dx$$

$$= \int_0^t E(\gamma_{t-x}^2) f(x)\,dx + \int_t^\infty (x - t)^2 f(x)\,dx.$$

Hence, by letting $Z(t) = E(\gamma_t^2)$ and $a(t) = \int_t^\infty (x - t)^2 f(x)\,dx$, we obtain a renewal equation of the form (1.1.7). Next it is a question of tedious algebra to derive (1.1.10) from (1.1.8). The details of the derivation are omitted.

To conclude this section, the following bounds on the renewal function are given.

Lemma 1.1.5   *For any $t \geq 0$,*

$$\frac{t}{\mu_1} - 1 \leq M(t) \leq \frac{t}{\mu_1} + \frac{\mu_2}{\mu_1^2},$$

where the second inequality assumes that $\mu_2 = E(X_1^2)$ is finite.

Proof   The left inequality is an immediate consequence of (1.1.9) and the fact that $\gamma_t \geq 0$. The proof of the other inequality is demanding and lengthy. The interested reader is referred to Lorden (1970).

1.1.2 Asymptotic expansions

We first prove a law of large numbers for the renewal process $\{N(t)\}$. It will be intuitively clear that the long-run average number of renewals per unit time is equal to the inverse of the expected time between two consecutive renewals.

Lemma 1.1.6 *With probability 1,*

$$\frac{N(t)}{t} \to \frac{1}{\mu_1} \text{ as } t \to \infty.$$

Proof By the definition of $N(t)$, we have $S_{N(t)} \le t < S_{N(t)+1}$ and so

$$\frac{S_{N(t)}}{N(t)} \le \frac{t}{N(t)} < \frac{S_{N(t)+1}}{N(t)+1} \frac{N(t)+1}{N(t)}.$$

Note that, with probability 1, $N(t) \to \infty$ as $t \to \infty$, since $P\{S_n < \infty\} = 1$ for all n. Thus the desired result follows by letting $t \to \infty$ in the above inequality and using the strong law of large numbers for the sequence $\{S_n\}$. This law states that, with probability 1, $S_n/n \to \mu_1$ as $n \to \infty$.

We note that Lemma 1.1.6 is also true when $\mu_1 = \infty$. It is tempting to conclude from this Lemma that $M(t)/t \to 1/\mu_1$ as $t \to \infty$. Although this result is correct, it cannot be directly concluded from Lemma 1.1.6. The reason is that in general the random variable $N(t)/t$ is not bounded in t. For a sequence of unbounded random variables Y_n it is not necessarily true that $E(Y_n) \to c$ as $n \to \infty$ when Y_n converges to a constant c with probability 1. Consider the counterexample in which $Y_n = 0$ with probability $1 - 1/n$ and $Y_n = n$ with probability $1/n$. Then $E(Y_n) = 1$ for all n while Y_n converges to 0 with probability 1.

Theorem 1.1.7 (Elementary Renewal Theorem)

$$\lim_{t \to \infty} \frac{M(t)}{t} = \frac{1}{\mu_1}.$$

Proof The proof will be based on the relation (1.1.9) for the excess variable. By the relation (1.1.9), we have $\mu_1[1 + M(t)] - t \ge 0$ and so we obtain the inequality

$$\frac{M(t)}{t} \ge \frac{1}{\mu_1} - \frac{1}{t} \quad \text{for all } t \ge 0.$$

Also, for any constant $c > 0$,

$$\frac{M(t)}{t} \le \frac{1}{\mu(c)} + \frac{1}{t}\left(\frac{c}{\mu(c)} - 1\right) \quad \text{for all } t \ge 0,$$

where $\mu(c) = \int_0^c \{1 - F(x)\}\,dx$. To prove the second inequality, fix $c > 0$ and consider the renewal process $\{\overline{N}(t)\}$ associated with the sequence $\{\overline{X}_n\}$ defined by $\overline{X}_n = X_n$ if $X_n \leq c$ and $\overline{X}_n = c$ if $X_n > c$. Since $N(t) \leq \overline{N}(t)$, we have $M(t) \leq \overline{M}(t)$. For the renewal process $\{\overline{N}(t)\}$ the excess life $\overline{\gamma}_t$ satisfies $\overline{\gamma}_t \leq c$ for all t. By relation (A.9) in Appendix A, $E(\overline{X}_1) = \mu(c)$. Thus, by (1.1.9), $\mu(c)[\overline{M}(t) + 1] \leq t + c$. This relation in conjunction with $M(t) \leq \overline{M}(t)$ for all t yields the second inequality.

The remainder of the proof is simple. Letting $t \to \infty$ in the above two inequalities, we find that

$$\frac{1}{\mu(c)} \geq \lim_{t \to \infty} \sup \frac{M(t)}{t} \geq \lim_{t \to \infty} \inf \frac{M(t)}{t} \geq \frac{1}{\mu_1}$$

for any constant $c > 0$. Next, by letting $c \to \infty$ and noting that $\mu(c) \to \mu_1$ as $c \to \infty$, we obtain the desired result.

So far our results have not required any assumption about the distribution function $F(x)$ of the interoccurrence times. Next we state the key renewal theorem that can be used to characterize the asymptotic behaviour of the solution to the renewal equation. This famous theorem requires that the distribution function $F(x)$ is *non-arithmetic*, that is, the mass of F is not concentrated on a discrete set of points $0, \lambda, 2\lambda, \ldots$ for some $\lambda > 0$. A distribution function that has a positive density on some interval is non-arithmetic. In the discussion below we make for convenience the even stronger assumption that $F(x)$ has a probability density. To establish the limiting behaviour of the solution to the renewal equation (1.1.7), we need also to impose on the function $a(x)$ a stronger condition than integrability. It must be required that the function $a(x)$ is directly Riemann integrable. Direct Riemann integrability can be characterized in several ways. A convenient definition is the following one. A function $a(x)$ defined on $[0, \infty)$ is said to be *directly Riemann integrable* when $a(x)$ is almost everywhere continuous and $\Sigma_{n=1}^{\infty} a_n < \infty$, where a_n is the supremum of $|a(x)|$ on the interval $[n-1, n)$. A sufficient condition for a function $a(x)$ to be directly Riemann integrable is that it can be written as a finite sum of monotone, integrable functions. This condition suffices for most applications.

Theorem 1.1.8 (Key Renewal Theorem)   *Assume $F(x)$ has a probability density $f(x)$. For a given function $a(t)$ that is bounded on finite intervals, let the function $Z(t)$ be defined by the renewal equation*

$$Z(t) = a(t) + \int_0^t Z(t - x) f(x) \, dx, \quad t \geq 0.$$

Suppose that $a(t)$ is directly Riemann integrable. Then

$$\lim_{t \to \infty} Z(t) = \frac{1}{\mu_1} \int_0^{\infty} a(x) \, dx.$$

The proof of this theorem is demanding and will not be given. The interested reader is referred to Feller (1971). Next we derive a number of useful results from the Key Renewal Theorem.

Theorem 1.1.9 *Suppose $F(x)$ is non-arithmetic with a finite second moment $\mu_2 = E(X_1^2)$. Then*

(a) $\displaystyle \lim_{t \to \infty} \left[M(t) - \frac{t}{\mu_1} \right] = \frac{\mu_2}{2\mu_1^2} - 1$,

(b) $\displaystyle \lim_{t \to \infty} \left[\int_0^t M(x)\,dx - \left\{ \frac{t^2}{2\mu_1} + \left(\frac{\mu_2}{2\mu_1^2} - 1 \right) t \right\} \right] = \frac{\mu_2^2}{4\mu_1^3} - \frac{\mu_3}{6\mu_1^2}$,

provided that $\mu_3 = E(X_1^3) < \infty$.

Proof (a) The asymptotic result $M(t)/t \to 1/\mu_1$ as $t \to \infty$ suggests that, for some constant c, $M(t) \approx t/\mu_1 + c$ for t large. To determine the constant c, define the function

$$Z(t) = M(t) - \frac{t}{\mu_1}, \quad t \geq 0.$$

Assuming for ease that F has a density, we easily deduce from (1.1.5) that

$$Z(t) = a(t) + \int_0^t Z(t-x) f(x)\,dx, \quad t \geq 0$$

where

$$a(t) = F(t) - \frac{t}{\mu_1} + \frac{1}{\mu_1} \int_0^t (t-x) f(x)\,dx, \quad t \geq 0.$$

Writing

$$\int_0^t (t-x) f(x)\,dx = \int_0^\infty (t-x) f(x)\,dx - \int_t^\infty (t-x) f(x)\,dx,$$

we find

$$a(t) = -[1 - F(t)] + \frac{1}{\mu_1} \int_t^\infty (x-t) f(x)\,dx.$$

This shows that $a(t)$ is the sum of two monotone functions. Each of the two terms is integrable. This follows from $\int_0^\infty [1 - F(t)]\,dt = \mu_1$ and

$$\frac{1}{\mu_1} \int_0^\infty dt \int_t^\infty (x-t) f(x)\,dx = \frac{1}{\mu_1} \int_0^\infty f(x)\,dx \int_0^x (x-t)\,dt$$

$$= \frac{1}{\mu_1} \int_0^\infty \frac{1}{2} x^2 f(x)\,dx = \frac{\mu_2}{2\mu_1}.$$

By applying the Key Renewal Theorem the desired result now follows.

(b) The proof of (b) proceeds along the same lines. The result in part (a) of the theorem suggests that, for some constant c,

$$\int_0^t M(x)\,dx \approx \frac{t^2}{2\mu_1} + t\left(\frac{\mu_2}{2\mu_1^2} - 1\right) + c \quad \text{for } t \text{ large.}$$

To determine the constant c, define the function

$$Z(t) = \int_0^t M(x)\,dx - \left[\frac{t^2}{2\mu_1} + t\left(\frac{\mu_2}{2\mu_1^2} - 1\right)\right], \quad t \geq 0.$$

By integrating both sides of the equation (1.1.5) over t and interchanging the order of integration, we get the following renewal equation for the function $U(t) = \int_0^t M(x)\,dx$,

$$U(t) = \int_0^t F(x)\,dx + \int_0^t U(t-x)f(x)\,dx, \quad t \geq 0.$$

From this renewal equation it follows that $Z(t)$ satisfies the renewal equation

$$Z(t) = a(t) + \int_0^t Z(t-x)f(x)\,dx, \quad t \geq 0$$

for a given function $a(t)$. Using the formulas (A.9) and (A.10) in Appendix A, it is a matter of some algebra to derive that $a(t)$ can be written as

$$a(t) = \frac{\mu_2}{2\mu_1^2}\int_t^\infty \{1 - F(x)\}\,dx + \frac{1}{\mu_1}\left[t\int_t^\infty \{1 - F(x)\}\,dx - \int_t^\infty x\{1 - F(x)\}\,dx\right].$$

The function $a(t)$ is the sum of two monotone functions. Each of the two terms is integrable. Using formula (A.10) in Appendix A, we find

$$\int_0^\infty a(t)\,dt = \frac{\mu_2^2}{4\mu_1^2} - \frac{\mu_3}{6\mu_1}.$$

Next, by applying the Key Renewal Theorem, part (b) follows.

The asymptotic expansions in Theorem 1.1.9 are very useful. They are accurate for practical purposes already for moderate values of t (see also Section 1.1.3). Theorem 1.1.9 and Lemma 1.1.4 have the following corollary.

Corollary 1.1.10 *Suppose $F(x)$ is non-arithmetic. Then it holds for the excess life γ_t that*

$$\lim_{t\to\infty} E(\gamma_t) = \frac{\mu_2}{2\mu_1} \quad \text{and} \quad \lim_{t\to\infty} E(\gamma_t^2) = \frac{\mu_3}{3\mu_1}.$$

Equilibrium excess distribution

The excess life γ_t has a limiting distribution for $t \to \infty$ when $F(x)$ is non-arithmetic. This limiting distribution plays an important role in a wide variety of applied problems.

Theorem 1.1.11   *Suppose $F(x)$ is non-arithmetic. Then*

$$\lim_{t \to \infty} P\{\gamma_t \leq x\} = \frac{1}{\mu_1} \int_0^x \{1 - F(y)\} \, dy, \quad x \geq 0. \tag{1.1.11}$$

Proof   For fixed $u \geq 0$, define $Z(t) = P\{\gamma_t > u\}$, $t \geq 0$. By conditioning on the time of the first renewal, we derive a renewal equation for $Z(t)$. Since after each renewal the renewal process probabilistically starts over, it follows that

$$P\{\gamma_t > u | X_1 = x\} = \begin{cases} P\{\gamma_{t-x} > u\} & \text{if } x \leq t, \\ 0 & \text{if } t < x \leq t + u, \\ 1 & \text{if } x > t + u, \end{cases}$$

By the law of total probability,

$$P\{\gamma_t > u\} = \int_0^\infty P\{\gamma_t > u | X_1 = x\} f(x) \, dx.$$

This yields the renewal equation

$$Z(t) = 1 - F(t + u) + \int_0^t Z(t - x) f(x) \, dx, \quad t \geq 0. \tag{1.1.12}$$

The function $a(t) = 1 - F(t + u)$, $t \geq 0$, is monotone and integrable. By applying the Key Renewal Theorem it now follows that

$$\lim_{t \to \infty} Z(t) = \frac{1}{\mu_1} \int_0^\infty \{1 - F(y + u)\} \, dy = \frac{1}{\mu_1} \int_u^\infty \{1 - F(y)\} \, dy,$$

yielding the desired result by using $\int_0^\infty \{1 - F(y)\} \, dy = \mu_1$.

In many practical applications the asymptotic expansion (1.1.11) gives a useful approximation to the distribution of γ_t already for moderate values of t. The limiting distribution of the excess life is called the *equilibrium excess distribution* and has applications in a wide variety of contexts. The equilibrium excess distribution can be given the following interpretation. Suppose that an outside person observes the state of the process at an arbitrarily chosen point in time when the process has been in operation since a very long time. Assuming that the outside person has no information about the past history of the process, the best prediction the person can give about the residual life of the item in use is according to the equilibrium excess distribution.

Remark 1.1.1 Waiting-time paradox

We all have experienced long waits at a bus stop when buses depart irregularly and we arrive at the bus stop at random. A theoretical explanation of this phenomenon is provided by the expression for $\lim_{t \to \infty} E(\gamma_t)$. It is convenient to rewrite this expression in terms of the coefficient of variation (= ratio of standard deviation and mean) of the interdeparture times X_i. Denote by

$$c_X^2 = \frac{\sigma^2(X_1)}{E^2(X_1)}$$

the squared coefficient of variation of the X_i's. Then, by Corollary 1.1.10,

$$\lim_{t \to \infty} E(\gamma_t) = \frac{1}{2}(1 + c_X^2)\mu_1, \tag{1.1.13}$$

where μ_1 is the mean interdeparture time. This representation shows that

$$\lim_{t \to \infty} E(\gamma_t) = \begin{cases} < \mu_1 & \text{if } c_X^2 < 1, \\ > \mu_1 & \text{if } c_X^2 > 1, \end{cases}$$

Thus the mean waiting time for the next bus depends on the regularity of the bus service and increases with the coefficient of variation of the interdeparture times. If we arrive at the bus stop at random, then for highly irregular service ($c_X^2 > 1$) the mean waiting time for the next bus is even larger than the mean interdeparture time. This surprising result is sometimes called the *waiting-time paradox*. A heuristic explanation is as follows. For highly irregular bus service most realizations of the interdeparture times will be short and very long realizations occur only occasionally. Thus it is more likely to hit a long interdeparture time than a short one when arriving at the bus stop at random. To illustrate this, consider the extreme situation in which the interdeparture time is 0 minutes with probability 9/10 and is 10 minutes with probability 1/10. Then the mean interdeparture time is 1 minute, but your mean waiting time for the next bus is 5 minutes when you arrive at the bus stop at random.

Another illustration of the importance of the concept of excess life is given in the following example.

Example 1.1.3 Approximations to an (s, S) inventory system

Suppose a periodic-review inventory system for which the demands $X_1, X_2, \ldots$ for a single product in the successive weeks $1, 2, \ldots$ are independent random variables having a common probability distribution function $F(x)$ with finite mean μ and finite standard deviation σ. Any demand exceeding the current inventory is backlogged until inventory becomes available by the arrival of a replenishment order. The inventory position is reviewed at the beginning of each week and is controlled by an (s, S) rule with $0 \leq s < S$. Under this control rule a replenishment order of size $S - x$ is placed when at a review the inventory level x is

Figure 1.1.2 The inventory process modelled as a renewal process

below the reorder point s; otherwise, no ordering is done. We assume instantaneous delivery of every replenishment order.

We are interested in the average frequency of ordering and the mean order size. Since the inventory process starts from scratch on each occasion the inventory position is ordered up to level S, the operating characteristics can be calculated by using a renewal model in which the weekly demand sizes $X_1, X_2, \ldots$ represent the interoccurrence times of renewals. The number of weeks between two consecutive orderings equals the number of weeks needed for a cumulative demand larger than $S - s$. The order size is the sum of $S - s$ and the undershoot of the reorder point s at the epoch of ordering (cf. Figure 1.1.2 in which a renewal occurrence is denoted by an $\times$). Denote by $\{N(t)\}$ the renewal process associated with the weekly demands $X_1, X_2, \ldots$. Then the number of weeks needed for a cumulative demand exceeding $S - s$ is given by $1 + N(S - s)$. The undershoot of the reorder point s is just the excess life γ_{S-s} of the renewal process. Hence

$$E[\text{number of weeks between two replenishment orders}] = 1 + M(S - s)$$

and

$$E[\text{order size}] = S - s + E(\gamma_{S-s}).$$

Assuming that the distribution function $F(x)$ of the weekly demand is non-arithmetic, the following two approximations are obtained from Theorem 1.1.9 and Corollary 1.1.10,

$$E[\text{number of weeks between two replenishments}] \approx \frac{S - s}{\mu} + \frac{\sigma^2 + \mu^2}{2\mu^2}$$

and

$$E[\text{order size}] \approx S - s + \frac{\sigma^2 + \mu^2}{2\mu},$$

provided that $S - s$ is sufficiently large relatively to the mean μ of the weekly demand. The symbol $\approx$ stands for 'approximately equal'.

To conclude this section, we state a Central Limit Theorem for the renewal process $\{N(t)\}$.

Theorem 1.1.12 (Central Limit Theorem)   *Assuming that $\sigma^2 = \sigma^2(X_1)$ is finite,*

$$\lim_{t \to \infty} P\left\{ \frac{N(t) - t/\mu_1}{\sigma\sqrt{t/\mu_1^3}} \leq x \right\} = \frac{1}{\sqrt{2\pi}} \int_{-\infty}^{x} e^{-(1/2)y^2} dy, \quad x \geq 0.$$

In other words, $N(t)$ is asymptotically normal distributed with mean t/μ_1 and variance $\sigma^2 t/\mu_1^3$. The proof of the Theorem is based on two basic results. The first result is that $N(t) \geq n$ if and only if $S_n \leq t$. The other result used is the Central Limit Theorem for the sequence $\{S_n\}$. This theorem states that $(S_n - n\mu_1)/\sigma\sqrt{n}$ converges in distribution to the standard normal distribution as $n \to \infty$. The remainder of the proof is technical and is not given. The interested reader is referred to Karlin and Taylor (1975).

1.1.3 Computation of the renewal function

A question of practical interest is how to compute the renewal function $M(t)$. The asymptotic expansion in Theorem 1.1.9 is very useful for computational purposes. Denoting by $c_X^2 = \sigma^2(X_1)/E^2(X_1)$ the squared coefficient of variation of the interoccurrence times, the asymptotic expansion states that

$$M(t) \approx \frac{t}{\mu_1} + \frac{1}{2}(c_X^2 - 1) \quad \text{for } t \text{ large enough.} \tag{1.1.14}$$

As will be seen below, the asymptotic expansion is exact for all $t \geq 0$ when $c_X^2 = 1$. It turns out that the asymptotic expansion is accurate enough for practical purposes already for moderate values of t provided that c_X^2 is not too large or too close to 0. Numerical investigations indicate that for practical purposes the asymptotic expansion for $M(t)$ can be used for $t \geq t_0$, where

$$t_0 = \begin{cases} \frac{3}{2}c_X^2\mu_1 & \text{if } c_X^2 > 1, \\ \mu_1 & \text{if } 0.2 < c_X^2 \leq 1, \\ \frac{1}{2c_X}\mu_1 & \text{if } 0 < c_X^2 \leq 0.2. \end{cases}$$

The asymptotic expansion deteriorates as $c_X^2 \to 0$ (note that $c_X^2 = 0$ corresponds to deterministic interoccurrence times). The relative error percentage of the approximation (1.1.14) is typically below 5% for $t \geq t_0$ and is usually less than 2%.

Next we discuss the computation of $M(t)$ for smaller values of t.

Exact computation

An exact computation of $M(t)$ for all $t \geq 0$ is tractable only for special cases of the distribution function $F(x)$. An analytical expression for $M(t)$ can be given when $F(x)$ has the density

$$f(x) = p_1\lambda_1 e^{-\lambda_1 x} + p_2\lambda_2 e^{-\lambda_2 x}, \quad x \geq 0,$$

where $\lambda_1, \lambda_2 > 0$ and $p_1 + p_2 = 1$. Then the renewal function $M(t)$ equals

$$M(t) = \frac{t}{\mu_1} + \frac{1}{2}(c_X^2 - 1)\left[1 - e^{-(p_1\lambda_2 + p_2\lambda_1)t}\right], \quad t \geq 0, \tag{1.1.15}$$

where $\mu_1 = p_1/\lambda_1 + p_2/\lambda_2$. This result can be verified by direct substitution into the renewal equation (1.1.5). In particular, for exponentially distributed interoccurrence times with density

$$f(x) = \lambda e^{-\lambda x}, \quad x \geq 0,$$

we obtain from (1.1.15) together with $c_X^2 = 1$ that

$$M(t) = \lambda t, \quad t \geq 0. \tag{1.1.16}$$

Next consider the case in which the interoccurrence times X_i are gamma (α, λ) distributed, that is,

$$F(x) = \frac{1}{\Gamma(\alpha)} \int_0^x e^{-\lambda y} \lambda^\alpha y^{\alpha-1} \, dy, \quad x \geq 0.$$

Then a computationally tractable approach can be based on

$$M(t) = \sum_{n=1}^{\infty} F_n(t), \quad t \geq 0. \tag{1.1.17}$$

Recall that $F_n(t)$ denotes the distribution function of $X_1 + \cdots + X_n$. Since the sum of two independent random variables having a gamma (α, λ) and a gamma (β, λ) distribution is gamma $(\alpha + \beta, \lambda)$ distributed, it follows that, for any $n \geq 1$,

$$\cdot F_n(t) = \frac{1}{\Gamma(n\alpha)} \int_0^x e^{-\lambda y} \lambda^{n\alpha} y^{n\alpha-1} \, dy, \quad x \geq 0. \tag{1.1.18}$$

This result explains why (1.1.17) is useful for the numerical calculation of $M(t)$. Fast and reliable codes to evaluate numerically the incomplete gamma integral are readily available, see e.g. Press *et al.* (1986). In numerical computations the infinite series (1.1.17) should be truncated. This is not difficult to do in view of the inequality

$$\sum_{n=k+1}^{\infty} F_n(t) \leq \frac{F_k(t)F(t)}{1 - F(t)}.$$

To see this inequality, note that, for any $n > k$,

$$P\{X_1 + \cdots + X_n \leq t\} \leq P\{X_1 + \cdots + X_k \leq t\}P\{X_{k+1} + \cdots + X_n \leq t\}$$

$$\leq P\{X_1 + \cdots + X_k \leq t\}P\{X_{k+1} \leq t\} \cdots P\{X_n \leq t\},$$

and so $F_n(t) \leq F_k(t)[F(t)]^{n-k}$ for all $n > k$ and $t \geq 0$. Thus, for fixed t, the evaluation of the terms in the series (1.1.17) is stopped as soon as $F_k(t)F(t)/[1 - F(t)] \leq \varepsilon$ for some prespecified $\varepsilon > 0$ (e.g. $\varepsilon = 10^{-5}$).

Discretization method

The renewal equation (1.1.5) for $M(t)$ is a special case of an integral equation which is known in numerical analysis as a Volterra integral equation of the second kind. Many numerical methods have been proposed to solve such equations. Unfortunately, these methods typically suffer from the accumulation of roundoff errors when t gets larger. However, using basic concepts from the theory of Riemann–Stieltjes integration, a simple and direct solution method with good convergence properties can be given for the renewal equation (1.1.5). This method discretizes the time and computes recursively the renewal function on a grid of points.

For fixed $t > 0$, let the time interval $[0, t]$ be partitioned according to $0 = t_0 < t_1 < t_2 \cdots < t_n = t$, where $t_i = ih$ for a given grid size $h > 0$. Put, for abbreviation,

$$M_i = M(ih), \quad F_i = F\left(\left(i - \frac{1}{2}\right)h\right) \text{ and } A_i = F(ih), \quad 1 \le i \le n.$$

The recursion scheme for computing the M_i's is as follows:

$$M_i = \frac{1}{1 - F_1}\left[A_i + \sum_{j=1}^{i-1}(M_j - M_{j-1})F_{i-j+1} - M_{i-1}F_1\right], \quad 1 \le i \le n,$$

starting with $M_0 = 0$. This recursion scheme is a minor modification of the Riemann–Stieltjes method proposed in Xie (1989) (the original method uses F_i instead of A_i). The recursion scheme is easy to program and gives surprisingly accurate results. It is remarkable how well the recursion scheme is able to resist the accumulation of roundoff errors as t gets larger. How to choose the grid size $h > 0$ depends not only on the desired accuracy in the answers, but also on the shape of the distribution function $F(x)$ and the length of the time interval $[0, t]$. The usual way to find out whether the answers are accurate enough is to do the computations for both a grid size h and a grid size $h/2$. In many cases of practical interest a four-digit accuracy is obtained with a grid size h in the range of 0.05–0.01. The discretization method has been applied to compute the exact values of the renewal function for the Weibull distribution. In Table 1.1.1 some results are given where a grid size of $h = 0.02$ is used for the case of $c_X^2 = 0.25$ and a grid size of $h = 0.01$ for the case of $c_X^2 = 2$. In both cases the normalization $\mu_1 = 1$ is used for the mean interoccurrence time.

Finally, it is noted that the discretization algorithm can also be used to solve an integral equation of the type (1.1.7). The only change is to replace $A_i = F(ih)$ by $A_i = a(ih) + a(0)F(ih)$.

Gamma approximation

In view of the fact that the asymptotic expansion (1.1.14) uses only the first two moments of the interoccurrence times, it is tempting to calculate a two-moment

Table 1.1.1   Renewal function for the Weibull distribution

	$c_X^2 = 0.25$				$c_X^2 = 2$		
t	app1	app2	exact	asympt	t	app1	app2	exact	asympt
---	---	---	---	---	---	---	---	---	---
0.1	0.0061	0.0061	0.0061	−0.275	0.2	0.4286	0.3993	0.3841	0.700
0.2	0.0260	0.0261	0.0261	−0.175	0.5	0.8455	0.8171	0.7785	1.000
0.4	0.1071	0.1086	0.1087	0.025	1.0	1.429	1.418	1.357	1.500
0.6	0.2360	0.2421	0.2422	0.225	1.5	1.965	1.966	1.901	2.000
0.8	0.4012	0.4133	0.4141	0.425	2.0	2.482	2.490	2.428	2.500
1.0	0.5913	0.6069	0.6091	0.625	2.5	2.900	3.001	2.947	3.000
1.2	0.7965	0.8093	0.8143	0.825	3.0	3.494	3.505	3.460	3.500
1.5	1.114	1.113	1.124	1.125	3.5	3.996	4.006	3.969	4.000
2.0	1.631	1.611	1.627	1.625	5.0	5.498	5.503	5.485	5.500
2.5	2.130	2.118	2.125	2.125	7.5	7.999	7.998	7.995	8.000

approximation for $M(t)$. In a two-moment approximation we replace the original distribution function $F(x)$ by a more tractable distribution function $\overline{F}(x)$ having the same first two moments as $F(x)$. For smaller values of t it is hazardous to compute $M(t)$ from a two-moment approximation. To illustrate this, consider a distribution function $F(x)$ with mean $\mu = 1$ and a squared coefficient of variation of $c_X^2 = 0.5$. Then the renewal function $M(t)$ has the respective values 0.0176 and 0.0318 for $t = 0.1$ and the respective values 0.2838 and 0.3044 for $t = 0.5$ when $F(x)$ is, respectively, a gamma distribution and a Weibull distribution. Obviously, the behaviour of $F(t)$ near the origin becomes essential for the computation of $M(t)$ when t is small.

The idea is to compute exactly the first few terms in the series (1.1.17) and to approximate the other terms using a two-moments match. This suggest the following two approximations:

$$M_{\text{app1}}(t) = F(t) + \sum_{n=2}^{\infty} \overline{F}_n(t), \quad t \geq 0$$

and

$$M_{\text{app2}}(t) = F(t) + F_2(t) + \sum_{n=3}^{\infty} \overline{F}_n(t), \quad t \geq 0.$$

Here $\overline{F}_n(t)$ is the distribution function of $\overline{X}_1 + \cdots + \overline{X}_n$, where the random variables $\overline{X}_1, \ldots, \overline{X}_n$ are independent and have a common gamma (α, λ) distribution. The parameters α and λ are chosen such that the first two moments of the original interoccurrence times X_i are matched by the first two moments of the gamma (α, λ) distribution. The distribution function $\overline{F}_n(t)$ is given by the right side of (1.1.18). The computation of the $\overline{F}_n(t)$'s as well as the truncation of the infinite series have been discussed above. The second-order approximation $M_{\text{app2}}(t)$ requires the computation of $F_2(t)$. In general numerical integration must be used to compute

$$F_2(t) = \int_0^t F(t-x)f(x)\,dx.$$

A convenient method to calculate this finite integral is the Gauss–Legendre integration method, see e.g. Press *et al.* (1986).

The above two approximations were proposed in Smeitink and Dekker (1990). Numerical investigations indicate that these approximations yield quick and useful estimates for $M(t)$ provided that c_X^2 is not too large. In Table 1.1.1 the approximations $M_{app1}(t)$ and $M_{app2}(t)$ are given for Weibull distributions with $c_X^2 = 0.25$ and $c_X^2 = 2$. In addition this table contains the exact value of $M(t)$ and the asymptotic estimate (1.1.14).

1.2 POISSON PROCESS AND EXTENSIONS

The Poisson process is the most important case of a renewal process. It is both mathematically tractable and practically useful. The Poisson process plays a key role in the analysis of many applied probability problems. In this section we first discuss thoroughly a number of basic properties of the Poisson process. Next we consider several extensions of the Poisson process such as the compound Poisson process and the nonstationary Poisson process.

1.2.1 Poisson Process

The Poisson process is an extremely useful renewal process for modelling purposes in numerous practical applications, e.g. to model arrival processes for queueing systems or demand processes for inventory systems. It is empirically found that in a wide variety of circumstances the arising stochastic processes can quite well be represented by a Poisson process. An intuitive explanation of this fact can be based on the phenomenon that in the situation of many individual events, each having a small probability of occurrence, the actual number of events occurring approximately follows a Poisson process.

There are several equivalent definitions of the Poisson process. We give the following definition.

Definition 1.2.1   *A renewal process* $\{N(t), \ t \geq 0\}$ *with interoccurrence times* $X_1, X_2, \ldots$ *between successive events is called a Poisson process if the interoccurrence times have the exponential probability distribution function*

$$F(t) = 1 - e^{-\lambda t}, \quad t \geq 0.$$

In the remainder of this section we use for the Poisson process $\{N(t)\}$ the terminology 'arrivals' instead of 'renewals'. The parameter λ is called the arrival rate of the process.

We shall first characterize the distribution of the counting variable $N(t)$, $t \geq 0$. For that purpose we define for $k = 1, 2, \ldots$ the random variable S_k by

$$S_k = \sum_{i=1}^{k} X_i.$$

In other words, S_k is the epoch at which the kth arrival occurs. In Appendix A it is shown that the sum of k independent random variables having a common exponential distribution has an Erlang-k distribution. Thus the random variable S_k has the probability distribution function

$$F_k(t) = 1 - \sum_{j=0}^{k-1} e^{-\lambda t} \frac{(\lambda t)^j}{j!}, \quad t \geq 0 \tag{1.2.1}$$

with probability density

$$f_k(t) = \lambda^k \frac{t^{k-1}}{(k-1)!} e^{-\lambda t}, \quad t \geq 0. \tag{1.2.2}$$

Incidentally, the relations (1.2.1) and (1.2.2) imply the useful identity

$$\int_0^t \lambda^k \frac{x^{k-1}}{(k-1)!} e^{-\lambda x} \, dx = 1 - \sum_{j=0}^{k-1} e^{-\lambda t} \frac{(\lambda t)^j}{j!}, \quad t \geq 0, k = 1, 2, \ldots. \tag{1.2.3}$$

The next theorem shows that the random variable $N(t)$ representing the total number of arrivals up to time t has a Poisson distribution with mean λt.

Theorem 1.2.1 *For any $t \geq 0$,*

$$P\{N(t) = k\} = e^{-\lambda t} \frac{(\lambda t)^k}{k!}, \quad k = 0, 1, \ldots.$$

Proof Fix $t \geq 0$. Since $N(t) \geq k$ if and only if $S_k \leq t$, we have

$$P\{N(t) \geq k\} = P\{S_k \leq t\}.$$

Next the desired result follows from $P\{N(t) = k\} = P\{N(t) \geq k\} - P\{N(t) \geq k+1\}$ and (1.2.1). ∎

Remark 1.2.1 Computation of Poisson probabilities

In many practical applications the calculation of (cumulative) Poisson probabilities is required. For fixed t, the probabilities $p_j(t) = e^{-\lambda t} (\lambda t)^j / j!$ can be calculated recursively from

$$p_0(t) = e^{-\lambda t} \quad \text{and} \quad p_j(t) = \frac{\lambda t}{j} p_{j-1}(t), \quad j = 1, 2, \ldots.$$

However, exponent underflow may occur in the calculation of $p_0(t)$ when λt is very large. There is a simple trick to avoid the exponent underflow. Define

$$q_j(t) = \ln[p_j(t)], \quad j = 0, 1, \ldots .$$

The recursion scheme

$$q_0(t) = -\lambda t \quad \text{and} \quad q_j(t) = \ln\left(\frac{\lambda t}{j}\right) + q_{j-1}(t), \quad j = 1, 2, \ldots$$

offers no numerical difficulties at all. The desired $p_j(t)$ is calculated from $q_j(t)$ using

$$p_j(t) = \begin{cases} e^{q_j(t)} & \text{if } q_j(t) \geq -M, \\ 0 & \text{if } q_j(t) < -M, \end{cases}$$

where M is a large number (e.g. $M = 500$) such that your computer code gives no underflow for e^{-x} when $x \leq M$. The trick of working with logarithms is one of the most useful tricks to avoid underflow in numerical analysis. Logarithms enable us to reduce the manipulation with extremely large (small) numbers to the manipulation with moderately sized numbers.

The memoryless property of the Poisson process

Next we derive a memoryless property that is peculiar to the Poisson process. Recall the definition of the excess life,

$$\gamma_t = \text{the waiting time from epoch } t \text{ until the next arrival.}$$

Theorem 1.2.2   *For any $t \geq 0$,*

$$P\{\gamma_t \leq x\} = 1 - e^{-\lambda x}, \quad x \geq 0.$$

Proof   The starting point is the renewal equation (1.1.12) for $P\{\gamma_t > u\}$. Applying Theorem 1.1.3, the solution of this renewal equation is given by

$$P\{\gamma_t > u\} = 1 - F(t + u) + \int_0^t \{1 - F(t + u - x)\}m(x)\,dx,$$

where $m(x)$ is the derivative of the renewal function $M(x)$. Since $M(x) = E[N(x)]$, we have by Theorem 1.2.1 that $M(x) = \lambda x$ for $x \geq 0$ and so $m(x) = \lambda$ for $x > 0$. Substituting $F(y) = 1 - e^{-\lambda y}$ and $m(x) = \lambda$ into the above equation yields the desired result.

The theorem states that at each point in time the waiting time until the next arrival has the same exponential distribution as the original interarrival time, regardless of how long ago the last arrival occurred. The Poisson process is the only renewal process having this memoryless property. How much time is elapsed since the last arrival gives no information about how long to wait until the next arrival.

This remarkable property does not hold for general arrival processes (e.g. consider the case of constant interarrival times). The lack of memory of the Poisson process explains the mathematical tractability of the process. In specific applications the analysis does not require a state variable keeping track of the time elapsed since the last arrival. The memoryless property of the Poisson process is of course closely related to the lack of memory of the exponential distribution.

In view of the lack of memory of the Poisson process, it will be intuitively clear that the Poisson process has the following properties:

(A) *Independent increments: the number of arrivals occurring in two disjoint intervals of time are independent.*

(B) *Stationary increments: the number of arrivals occurring in a given time interval depends only on the length of the interval.*

A formal proof of these properties will not be given here (see Exercise 1.12). Theorem 1.2.1 and the properties (A) and (B) imply that, for any $t, s > 0$,

$$P\{N(t + s) - N(t) = k\} = e^{-\lambda s} \frac{(\lambda s)^k}{k!}, \quad k = 0, 1, \ldots, \qquad (1.2.4)$$

regardless of t. From (1.2.4) we obtain the properties:

(C) *The probability of one arrival occurring in a time interval of length Δt is $\lambda \Delta t + o(\Delta t)$ for $\Delta t \to 0$.*

(D) *The probability of two or more arrivals occurring in a time interval of length Δt is $o(\Delta t)$ for $\Delta t \to 0$.*

Here the standard symbol $o(h)$ denotes some unspecified function $g(h)$ with the property that $g(h)/h \to 0$ as $h \to 0$, that is $o(h)$ is negligibly small compared with h as $h \to 0$. The sum and the product of two $o(h)$ functions is again $o(h)$. The property (D) states that the possibility of two or more arrivals in a very small time interval has a negligibly small probability. The properties (C) and (D) are an immediate consequence of (1.2.4) and the series expansion

$$e^{-\Delta t} = 1 - \Delta t + \frac{(\Delta t)^2}{2!} - \frac{(\Delta t)^3}{3!} + \cdots = 1 - \Delta t + o(\Delta t) \quad \text{as } \Delta t \to 0.$$

As said before, several equivalent definitions of the Poisson process can be given. The Poisson process could be alternatively defined by taking (A), (B), (C) and (D) as postulates.

Merging and splitting of Poisson processes

Many applications involve the merging of independent Poisson processes or the splitting of the events of a Poisson process in different categories. The next theorem shows that these situations again lead to Poisson processes.

Theorem 1.2.3   (a) *Suppose that* $\{N_1(t), \ t \geq 0\}$ *and* $\{N_2(t), \ t \geq 0\}$ *are independent Poisson processes with respective rates* λ_1 *and* λ_2, *where the process* $\{N_i(t)\}$ *corresponds to type-i arrivals. Let* $N(t) = N_1(t) + N_2(t), t \geq 0$. *Then the merged process* $\{N(t), \ t \geq 0\}$ *is a Poisson process with rate* $\lambda = \lambda_1 + \lambda_2$. *Denoting by* Z_k *the interarrival time between the* $(k-1)$*th and kth arrival in the merged process and letting* $I_k = i$ *if the kth arrival in the merged process is a type-i arrival, then for any* $k = 1, 2, \ldots,$

$$P\{I_k = i | Z_k = t\} = \frac{\lambda_i}{\lambda_1 + \lambda_2}, \quad i = 1, 2, \tag{1.2.5}$$

independently of t.

(b) *Let* $\{N(t), \ t \geq 0\}$ *be a Poisson process with rate* λ. *Suppose that each arrival of the process is classified as being a type-1 arrival or type-2 arrival with respective probabilities* p_1 *and* p_2, *independently of all other arrivals. Let* $N_i(t)$ *be the number of type-i arrivals up to time* $t (i = 1, 2)$. *Then* $\{N_1(t)\}$ *and* $\{N_2(t)\}$ *are two independent Poisson processes having respective rates* λp_1 *and* λp_2.

Proof   We give only a sketch of the proof using the properties (A), (B), (C) and (D).

(a) It will be obvious that the process $\{N(t)\}$ satisfies the properties (A) and (B). To verify property (C), note that

$P\{\text{one arrival in } (t, t + \Delta t]\}$

$$= \sum_{i=1}^{2} P\{\text{one arrival of type } i \text{ and no arrival of the other type in } (t, t + \Delta t]\}$$

$$= [\lambda_1 \Delta t + o(\Delta t)][1 - \lambda_2 \Delta t + o(\Delta t)] + [\lambda_2 \Delta t + o(\Delta t)][1 - \lambda_1 \Delta t + o(\Delta t)]$$

$$= (\lambda_1 + \lambda_2)\Delta t + o(\Delta t) \quad \text{as } \Delta t \to 0.$$

Property (D) follows by noting that

$$P\{\text{no arrival in } (t, t + \Delta t]\} = [1 - \lambda_1 \Delta t + o(\Delta t)][1 - \lambda_2 \Delta t + o(\Delta t)]$$

$$= 1 - (\lambda_1 + \lambda_2)\Delta t + o(\Delta t) \quad \text{as } \Delta t \to 0.$$

To prove the other assertion in part (a), denote by the generic variable Y_i the interarrival time in the process $\{N_i(t)\}$. Then

$$P\{Z_k > t, I_k = 1\} = P\{Y_2 > Y_1 > t\} = \int_t^{\infty} P\{Y_2 > Y_1 > t | Y_1 = x\}\lambda_1 e^{-\lambda_1 x} \, dx$$

$$= \int_t^{\infty} e^{-\lambda_2 x}\lambda_1 e^{-\lambda_1 x} \, dx = \frac{\lambda_1}{\lambda_1 + \lambda_2}e^{-(\lambda_1 + \lambda_2)t}.$$

By taking $t = 0$ in this relation, we find $P\{I_k = 1\} = \lambda_1/(\lambda_1 + \lambda_2)$. Since $\{N(t)\}$ is a Poisson process with rate $\lambda_1 + \lambda_2$, we have $P\{Z_k > t\} = \exp[-(\lambda_1 + \lambda_2)t]$.

Hence

$$P\{I_k = 1, Z_k > t\} = P\{I_k = 1\}P\{Z_k > t\},$$

showing that $P(I_k = 1 | Z_k = t) = \lambda_1/(\lambda_1 + \lambda_2)$ independently of t.

(b) Obviously, the process $\{N_i(t)\}$ satisfies the properties (A), (B) and (D). To verify property (C), note that

$$P\{\text{one arrival of type } i \text{ in } (t, t + \Delta t]\} = (\lambda \Delta t) p_i + o(\Delta t)$$

$$= (\lambda p_i)\Delta t + o(\Delta t).$$

It remains to prove that the processes $\{N_1(t)\}$ and $\{N_2(t)\}$ are independent. Fix $t > 0$. Then, by conditioning,

$$P\{N_1(t) = k, \ N_2(t) = m\}$$

$$= \sum_{n=0}^{\infty} P\{N_1(t) = k, N_2(t) = m | N(t) = n\}P\{N(t) = n\}$$

$$= P\{N_1(t) = k, N_2(t) = m | N(t) = k + m\}P\{N(t) = k + m\}$$

$$= \binom{k+m}{k} p_1^k p_2^m e^{-\lambda t} \frac{(\lambda t)^{k+m}}{(k + m)!}$$

$$= e^{-\lambda p_1 t} \frac{(\lambda p_1 t)^k}{k!} e^{-\lambda p_2 t} \frac{(\lambda p_2 t)^m}{m!},$$

showing that $P\{N_1(t) = k, \ N_2(t) = m\} = P\{N_1(t) = k\}P\{N_2(t) = m\}$.

The remarkable result (1.2.5) states that the next arrival is of type i with probability $\lambda_i/(\lambda_1 + \lambda_2)$ regardless of how long it takes until the next arrival. This result is characteristic for competing Poisson processes which are independent of each other.

Poisson process and the uniform distribution

In any small time interval of the same length the occurrence of a Poisson arrival is equally likely. In other words, Poisson arrivals occur completely random in time. To make this statement more precise, we relate the Poisson process to the uniform distribution.

Lemma 1.2.4 *For any $t > 0$ and $n = 1, 2, \ldots,$*

$$P\{S_k \le x | N(t) = n\} = \sum_{j=k}^{n} \binom{n}{j} \left(\frac{x}{t}\right)^j \left(1 - \frac{x}{t}\right)^{n-j}, \qquad 0 \le x \le t \text{ and } 1 \le k \le n,$$

$$(1.2.6)$$

and

$$E(S_k | N(t) = n) = \frac{kt}{n + 1}, \qquad 1 \le k \le n.$$

Proof   Using that the Poisson process has independent and stationary increments, we have

$$P\{S_k \le x | N(t) = n\} = \frac{P\{S_k \le x, N(t) = n\}}{P\{N(t) = n\}}$$

$$= \frac{P\{N(x) \ge k, N(t) = n\}}{P\{N(t) = n\}}$$

$$= \frac{\displaystyle\sum_{j=k}^{n} P\{N(x) = j, N(t) - N(x) = n - j\}}{P\{N(t) = n\}}$$

$$= \frac{1}{e^{-\lambda t} (\lambda t)^n / n!} \sum_{j=k}^{n} e^{-\lambda x} \frac{(\lambda x)^j}{j!} e^{-\lambda(t-x)} \frac{[\lambda(t-x)]^{n-j}}{(n-j)!}$$

$$= \sum_{j=k}^{n} \binom{n}{j} \left(\frac{x}{t}\right)^j \left(1 - \frac{x}{t}\right)^{n-j},$$

proving the first assertion. The second assertion is an immediate consequence of the first assertion and the identity

$$\frac{(p+q+1)!}{p! q!} \int_0^1 y^p (1 - y)^q \, dy = 1, \quad p, q = 0, 1, \dots$$

for the beta distribution.

The right side of (1.2.6) can also be given the following interpretation. Let $U_1, \dots, U_n$ be n independent random variables that are uniformly distributed over the interval $(0, t)$. Then the right side of (1.2.6) represents also the probability that the kth smallest among $U_1, \dots, U_n$ is smaller than or equal to x. More generally, we have:

Theorem 1.2.5   *For any $t > 0$ and $n = 1, 2, \dots$,*

$$P\{S_1 \le x_1, \dots, S_n \le x_n | N(t) = n\} = P\{U_{(1)} \le x_1, \dots, U_{(n)} \le x_n\},$$

where $U_{(k)}$ denotes the kth smallest among n independent random variables $U_1, \dots, U_n$ that are uniformly distributed over the interval $(0, t)$.

The proof of this theorem proceeds along the same lines as that of Lemma 1.2.4. In other words, given the occurrence of n arrivals in $(0, t)$, the n arrival epochs are statistically indistinguishable from n independent observations taken from the uniform distribution on $(0, t)$. Thus Poisson arrivals occur completely random in time.

Example 1.2.1 An occupancy problem

Ships enter the North Sea according to a Poisson process $\{N(t)\}$ with intensity λ. The amounts of time the ships stay on the North Sea are independent random variables with a common probability distribution function $G(x)$. What is the probability distribution of the number of ships on the North Sea at t time units hence? Assume for ease that there are no ships present at the present time 0.

Denoting by $A(t)$ the number of ships present at time t, the answer to the question is

$$P\{A(t) = k\} = e^{-\lambda p_t} \frac{(\lambda p_t)^k}{k!} \quad \text{for } k = 0, 1, \ldots, \text{ where } p_t = \int_0^t \{1 - G(x)\}\,dx.$$

(1.2.7)

The result is obtained as follows. Fix $t > 0$. By conditioning on the number of arrivals up to time t, it follows that

$$P\{A(t) = k\} = \sum_{n=k}^{\infty} P\{A(t) = k \mid N(t) = n\} e^{-\lambda t} \frac{(\lambda t)^n}{n!}.$$

For fixed n, denote by $U_{(i)}$ for $i = 1, \ldots, n$ the ith smallest among n independent random variables $U_1, \ldots, U_n$ that are uniformly distributed on $(0, t)$. Then, by Theorem 1.2.5,

$$P\{A(t) = k \mid N(t) = n\} = P\{\text{exactly } k \text{ of the } U_{(1)} + G_1, \ldots, U_{(n)} + G_n \text{ exceed } t\},$$

where G_i denotes the amount of time the ith arriving ship stays on the North Sea. Some reflections show that

$$P\{\text{exactly } k \text{ of the } U_{(1)} + G_1, \ldots, U_{(n)} + G_n \text{ exceed } t\}$$

$$= P\{\text{exactly } k \text{ of the } U_1 + G_1, \ldots, U_n + G_n \text{ exceed } t\}.$$

Thus

$$P\{A(t) = k \mid N(t) = n\} = \binom{n}{k} p^k (1 - p)^{n-k}$$

where $p = P\{U_1 + G_1 > t\}$. By conditioning, $p = (1/t)\int_0^t \{1 - G(x)\}\,dx$. Hence

$$P\{A(t) = k\} = \sum_{n=k}^{\infty} \binom{n}{k} p^k (1 - p)^{n-k} e^{-\lambda t} \frac{(\lambda t)^n}{n!}.$$

Next we easily obtain the desired result (1.2.7).

A theoretical explanation of the occurrence of Poisson processes

We conclude this section with a theoretical explanation of the occurrence in practice of processes closely approximating to Poisson processes. For that purpose, consider n independent renewal processes $\{N_k(t), t \geq 0\}$, $k = 1, \ldots, n$. Define the superposition process $\{N(t), t \geq 0\}$ by $N(t) = \sum_{k=1}^{n} N_k(t)$. For example $N_k(t)$ is the number of ships from some harbour k that arrive during $(0, t]$ in the harbour of Rotterdam, and $N(t)$ is the sum of these arrivals up to time t. For the renewal process $\{N_k(t)\}$, denote by F_k the probability distribution function of the interoccurrence time between two consecutive arrivals and denote by α_k the first moment of F_k. Assuming that at the present moment the processes are already in progress for a very long time and that no information is known about the histories of the processes up to now, we define W_k as the waiting time until the next renewal in the process $\{N_k(t)\}$. Also, let $W = \min(W_1, \ldots, W_n)$ be the waiting time until the next event in the superposition process $\{N(t)\}$. Suppose now that the number n of renewal processes gets very large and that in each individual renewal process the renewals occur very rarely so that the pooled effect is due to many small causes. More precisely, it is supposed that the arrival rates $1/\alpha_k$ get very small such that

$$\frac{1}{\alpha_1} + \cdots + \frac{1}{\alpha_n} \to \lambda \quad \text{as } n \to \infty.$$

Then it can heuristically be argued that

$$P\{W > u\} \approx e^{-\lambda u} \quad \text{for all } u \geq 0,$$

supporting the claim that the superposition process $\{N(t)\}$ is approximately a Poisson process. To obtain this result, note first that

$$P\{W > u\} = P\{W_1 > u\} \ldots P\{W_n > u\},$$

since $W_1, \ldots, W_n$ are independent. Since the processes are assumed to be in progress for a very long time, we have by (1.1.11) that

$$P\{W_k \leq u\} \approx \frac{1}{\alpha_k} \int_0^u \{1 - F_k(x)\}\, dx.$$

Now fix u. For n sufficiently large, the average interarrival times α_k will be very large and so, for any k, $F_k(x) \approx 0$ for $0 \leq x \leq u$. Thus, for n large enough, $P\{W_k \leq u\} \approx u/\alpha_k$ and so

$$P\{W > u\} \approx \left(1 - \frac{u}{\alpha_1}\right) \ldots \left(1 - \frac{u}{\alpha_n}\right) \approx \left\{1 - \left(\frac{u}{\alpha_1} + \cdots + \frac{u}{\alpha_n}\right)\right\}$$

$$\approx \exp[-u(1/\alpha_1 + \cdots + 1/\alpha_n)],$$

which yields the desired result. This limiting result provides a theoretical explanation of the occurrence of Poisson processes in a wide variety of circumstances. For example, the number of calls received at a large telephone exchange is the superposition of the individual calls of many subscribers each calling infrequently. Thus the overall number of calls can be expected to closely be a Poisson process. Similarly, a Poisson demand process for a given product can be expected if the demands are the superposition of the individual requests of many customers each asking infrequently for that product.

1.2.2 Compound Poisson process

A compound Poisson process generalizes the Poisson process by allowing jumps that are not necessarily of unit magnitude.

Definition 1.2.2   *A stochastic process* $\{X(t),\ t \geq 0\}$ *is said to be a compound Poisson process if it can be represented by*

$$X(t) = \sum_{i=1}^{N(t)} D_i, \quad t \geq 0,$$

where $\{N(t),\ t \geq 0\}$ *is a Poisson process with rate* λ *and* $D_1, D_2, \ldots$ *are independent and identically distributed random variables which are also independent of the process* $\{N(t)\}$.

Compound Poisson processes arise in a variety of contexts. As an example, consider an insurance company at which claims arrive according to a Poisson process and the claim sizes are independent and identically distributed random variables, which are also independent of the arrival process. Then the cumulative amount claimed up to time t is a compound Poisson variable. Also, the compound Poisson process has applications in inventory theory. Suppose customers asking for a given product arrive according to a Poisson process. The demands of the customers are independent and identically distributed random variables, which are also independent of the arrival process. Then the cumulative demand up to time t is a compound Poisson variable.

The mean and variance of the compound Poisson variable $X(t)$ are given by

$$E[X(t)] = \lambda t E(D_1) \quad \text{and} \quad \sigma^2[X(t)] = \lambda t E(D_1^2), \quad t \geq 0. \tag{1.2.8}$$

These formulas follow directly from the formulas (A.11) and (A.12) in Appendix A and the fact that both the mean and the variance of the Poisson variable $N(t)$ are equal to λt.

An efficient algorithm for the probability distribution of the compound Poisson variable can be given when the D_i's are non-negative random variables having a discrete distribution

$$\phi_j = P\{D_1 = j\}, \quad j = 0, 1, \ldots .$$

Before doing so, we give some preliminaries. The generating function of a discrete random variable N with probability distribution $\{p_k,\ k = 0, 1, \ldots\}$ is defined as

$$P(z) = \sum_{k=0}^{\infty} p_k z^k, \quad |z| \leq 1. \tag{1.2.9}$$

Note the probabilistic interpretation

$$P(z) = E(z^N), \quad |z| \leq 1.$$

The probability distribution $\{p_k, \ k = 0, 1, \dots\}$ can be recovered analytically from the compressed function $P(z)$ by

$$p_k = \frac{1}{k!} \frac{d^k P(z)}{dz^k}\bigg|_{z=0}, \quad k = 0, 1, \dots \ . \tag{1.2.10}$$

For fixed $t \geq 0$, define now

$$r_k = P\{X(t) = k\}, \quad k = 0, 1, \dots \ .$$

For ease of notation the dependence of the r_k's on t is suppressed. A recursion scheme for the computation of the probabilities r_k is given in the next theorem. This recursion is sometimes called Adelson's recursion scheme, named after Adelson (1966).

Theorem 1.2.6 *For any $t \geq 0$,*

$$r_0 = e^{-\lambda t(1-\phi_0)},$$

$$r_k = \frac{\lambda t}{k} \sum_{j=0}^{k-1}(k-j)\phi_{k-j}r_j, \quad k = 1, 2, \dots \ . \tag{1.2.11}$$

Proof Fix $t \geq 0$. By conditioning on the number of arrivals up to time t,

$$r_k = \sum_{n=0}^{\infty} P\{X(t) = k | N(t) = n\}P\{N(t) = n\}$$

$$= \sum_{n=0}^{\infty} P\{D_0 + \cdots + D_n = k\}e^{-\lambda t}\frac{(\lambda t)^n}{n!}, \quad k = 0, 1, \dots$$

using the convention that $D_0 = 0$. Define the generating functions

$$R(z) = \sum_{k=0}^{\infty} r_k z^k \quad \text{and} \quad Q(z) = \sum_{k=0}^{\infty} \phi_k z^k, \quad |z| \leq 1.$$

Since the D_i's are independent of each other, we have by (1.2.9) that

$$E(z^{D_0+\cdots D_n}) = E(z^{D_0})\dots E(z^{D_n}) = [Q(z)]^n, \quad n \geq 0.$$

Thus

$$R(z) = \sum_{k=0}^{\infty} r_k z^k = \sum_{n=0}^{\infty} e^{-\lambda t}\frac{(\lambda t)^n}{n!} \sum_{k=0}^{\infty} P\{D_0 + \cdots + D_n = k\}z^k$$

$$= \sum_{n=0}^{\infty} e^{-\lambda t}\frac{(\lambda t)^n}{n!}[Q(z)]^n,$$

which yields

$$R(z) = e^{-\lambda t[1-Q(z)]}, \quad |z| \le 1. \tag{1.2.12}$$

It is not possible to obtain (1.2.11) directly from (1.2.10) and (1.2.12). The following intermediate step is needed. By differentiation

$$R'(z) = \lambda t \, Q'(z) R(z), \quad |z| \le 1.$$

By applying the Leibniz rule for the differentiation of a product, we find

$$R^{(k)}(0) = \lambda t \sum_{j=0}^{k-1} \binom{k-1}{j} Q^{(k-j)}(0) R^{(j)}(0), \quad k = 1, 2, \dots .$$

Next, using (1.2.10), we get the recursion scheme (1.2.11).

The recursion scheme for the r_k's is easy to program and is numerically stable. Note that for the special case of $\phi_1 = 1$ the recursion (1.2.11) reduces to the well-known recursion scheme for computing Poisson probabilities.

Remark 1.2.2 Fast Fourier Transform method

The total number of additions and multiplications required by the recursion scheme (1.2.11) is of the order N^2 when the probabilities r_k must be computed for $k = 0, 1, \dots, N$. In some cases the generating function $R(z) = e^{-\lambda t[1-Q(z)]}$ allows for a simple analytical expression, e.g. $R(z) = e^{-\lambda t(1-z)/(1-pz)}$ when $\phi_j = (1-p)p^j$, $j \ge 0$, for some $0 < p < 1$. In those cases the (discrete) Fast Fourier Transform method (FFT method) may be preferred to the algorithm (1.2.11).

It will be instructive to know how the FFT method computes the r_k's from $R(z)$. Choose an integer N such that the tail probability $\sum_{k=N}^{\infty} r_k$ is negligibly small and so

$$R(z) \approx \sum_{n=0}^{N-1} r_n z^n, \quad |z| \le 1. \tag{1.2.13}$$

The FFT method requires that $N = 2^m$ for some integer m. Define now the complex numbers

$$f_k = R(e^{2\pi i k/N}), \quad k = 0, 1, \dots, N-1, \tag{1.2.14}$$

where i is the complex number with $i^2 = -1$. The numerical computation of the f_k's from (1.2.14) is not demanding when $R(z)$ allows for a simple analytical expression. By taking $z = e^{2\pi i k/N}$ in (1.2.13), it follows that

$$f_k \approx \sum_{n=0}^{N-1} r_n e^{2\pi i k n/N}, \quad k = 0, 1, \dots, N-1.$$

The FFT method recovers the r_n's from the f_k's by using the inverse formula

$$r_n \approx \frac{1}{N} \sum_{k=0}^{N-1} f_k e^{-2\pi i nk/N}, \quad n = 0, 1, \ldots, N-1. \qquad (1.2.15)$$

This inversion formula follows from the fact that the matrix W_N whose (n, k)th element equals $e^{2\pi i nk/N}$ can be shown to have $(1/N)\overline{W}_N$ as inverse matrix, where the elements of $\overline{W}_N$ are the complex conjugates of the elements of W_N. The vector of r_n's is obtained from (1.2.15) by multiplying the vector of f_k's by a matrix whose (n, k)th element is the constant $w = e^{-2\pi i/N}$ to the power $n \times k$. The matrix multiplications in (1.2.15) would normally require N^2 multiplications. However, the FFT method performs the multiplications in an extremely fast and ingenious way that requires only $N \log_2 N$ multiplications instead of N^2. The key to the method is the simple observation that the discrete Fourier transform of length N can be written as the sum of two discrete Fourier transforms, each of length $N/2$. It holds that

$$\sum_{k=0}^{N-1} f_k e^{-2\pi i kn/N} = \sum_{k=0}^{N/2-1} f_{2k} e^{-2\pi i n(2k)/N} + \sum_{k=0}^{N/2-1} f_{2k+1} e^{-2\pi i n(2k+1)/N}$$

$$= \sum_{k=0}^{N/2-1} f_{2k} e^{-2\pi i nk/(N/2)} + w^n \sum_{k=0}^{N/2-1} f_{2k+1} e^{-2\pi i nk/(N/2)},$$

where $w = e^{-2\pi i/N}$. In other words, the problem of computing the discrete inverse Fourier transform of N points can be reduced to the problem of computing the transform of its $N/2$ even-numbered points and its $N/2$ odd-numbered points. This beautiful trick can be applied recursively. It will now be clear why N should be equal to 2^m for some integer m. Computer codes for the FFT method are widely available, see e.g. Press *et al.* (1986).

The FFT method represents a breakthrough in numerical analyses. It has numerous applications. Another applied probability problem for which the FFT method may be useful is the calculation of the convolution of two discrete probability distributions.

1.2.3 Nonstationary Poisson process

The nonstationary Poisson process is another useful stochastic process for counting the events that occur over time. It generalizes the Poisson process by allowing that the arrival rate need not be constant. Nonstationary Poisson processes are used to model arrival processes where the arrival rate fluctuates significantly over time.

Definition 1.2.3 *A counting process* $\{N(t), t \geq 0\}$ *is said to be a nonstationary Poisson process with intensity function* $\lambda(t)$, $t \geq 0$, *if it satisfies the following properties:*

(a) $N(0) = 0$.

(b) The process $\{N(t)\}$ has independent increments.
(c) $P\{N(t + \Delta t) - N(t) = 1\} = \lambda(t)\Delta t + o(\Delta t)$ as $\Delta t \to 0$.
(d) $P\{N(t + \Delta t) - N(t) \geq 2\} = o(\Delta t)$ as $\Delta t \to 0$.

The next theorem proves that the total number of arrivals in a given time interval is Poisson distributed.

Theorem 1.2.7   *For any $t, s \geq 0$,*

$$P\{N(t + s) - N(t) = k\} = e^{-[M(t+s)-M(t)]} \frac{[M(t + s) - M(t)]^k}{k!}, \quad k = 0, 1, \ldots,$$

(1.2.16)

where $M(x) = \int_0^x \lambda(y)\,dy, \ x \geq 0$.

Proof   The proof is instructive. Fix $t \geq 0$. Put for abbreviation

$$p_k(s) = P\{N(t + s) - N(t) = k\}, \quad k = 0, 1, \ldots .$$

Since the probability of two or more arrivals in a small time interval is negligibly small and using that the process $\{N(t)\}$ has independent increments, it follows by conditioning on the number of arrivals up to time $t + s$ that

$$p_k(s + \Delta s) = p_{k-1}(s)[\lambda(t + s)\Delta s + o(\Delta s)] + p_k(s)[1 - \lambda(t + s)\Delta s + o(\Delta s)]$$

$$\text{as } \Delta s \to 0.$$

Subtracting $p_k(s)$ from both sides of this equation and dividing by Δs, we obtain

$$p_k'(s) = -\lambda(t + s)[p_k(s) - p_{k-1}(s)], \quad k \geq 1.$$

For $k = 0$, we have $p_0'(s) = -\lambda(t + s)p_0(s)$. From the latter differential equation together with $p_0(0) = 1$, it follows that

$$p_0(s) = \exp\left(-\int_0^s \lambda(t + u)\,du\right) = \exp\left(-[M(t + s) - M(t)]\right).$$

The relation (1.2.16) next follows by induction from the above differential equation for $p_k(s)$. The details are left to the reader.

The next example illustrates the usefulness of nonstationary Poisson processes.

Example 1.2.2 Replacement with minimal repair

A machine has a stochastic lifetime with a continuous distribution. The machine is replaced by a new one at fixed times $T, 2T, \ldots$, whereas a minimal repair is done at each failure occurring between two planned replacements. A minimal repair brings the machine into the condition it had just before the failure. It is assumed that each minimal repair takes a negligible time. What is the probability distribution of the total number of minimal repairs between two planned replacements?

Denote by $F(x)$ and $f(x)$ the probability distribution function and the probability density of the lifetime of the machine. Also, let $r(t) = f(t)/[1 - F(t)]$ denote the failure rate function of the machine. Then the answer to the above question is

$P\{\text{there are } k \text{ minimal repairs between two planned replacements}\}$

$$= e^{-M(T)} \frac{[M(T)]^k}{k!}, \quad k = 0, 1, \ldots,$$

where $M(T) = \int_0^T r(t)\, dt$. This result is an immediate consequence of Theorem 1.2.7. The process counting the number of minimal repairs between two planned replacements satisfies the properties (a), (b), (c) and (d) of Definition 1.2.3. In particular, the probability of one failure of the machine in a small time interval $(t, t + \Delta t]$ is equal to $r(t)\Delta t + o(\Delta t)$ as shown in Appendix B.

1.3   RENEWAL-REWARD PROCESSES

The renewal-reward model provides an extremely useful tool for analysing numerous applied probability models. The model is simple and intuitively appealing. Many stochastic processes have the property of regenerating themselves at certain points in time so that the behaviour of the process after the regeneration epoch is a probabilistic replica of the behaviour starting at time zero and is independent of the behaviour before the regeneration epoch.

A formal definition of a regenerative process is as follows.

Definition 1.3.1   *A stochastic process* $\{X(t), \ t \in T\}$ *with state space X and time-index set T is said to be regenerative if there exists a (random) epoch* S_1 *such that:*

(a) $\{X(t + S_1), \ t \in T\}$ *is independent of* $\{X(t), \ 0 \leq t < S_1\}$,
(b) $\{X(t + S_1), \ t \in T\}$ *has the same distribution as* $\{X(t), \ t \in T\}$.

It is assumed that the index set T is either the interval $T = [0, \infty)$ or the countable set $T = \{0, 1, \ldots\}$. In the former case we have a continuous-time regenerative process and in the other case a discrete-time regenerative process. The state space X of the process $\{X(t)\}$ is assumed to be a subset of some Euclidean space and may be a discrete set or a continuous set.

The existence of the regeneration epoch S_1 implies the existence of further regeneration epochs $S_2, S_3, \ldots$ having the same property as S_1. Intuitively speaking, a regenerative process can be split into independent and identically distributed renewal cycles. A cycle is defined as the time interval between two consecutive regeneration epochs. Examples of regenerative processes are:

(i) The continuous-time process $\{X(t), \ t \geq 0\}$ with $X(t)$ denoting the number of customers present at time t in a single-server queue in which the customers arrive according to a renewal process and the service times are independent and identically distributed random variables. The regeneration epochs $S_1, S_2, \ldots$ are the epochs at which an arriving customer finds the system empty.

(ii) The discrete-time process $\{I_n, \ n = 0, 1, \ldots\}$ with I_n denoting the inventory level at the beginning of the nth week in the (s, S) inventory model dealt with in Example 1.1.3. The regeneration epochs are the beginnings of the weeks in which the inventory level is ordered up to the level S.

Let us define the random variables $L_n = S_n - S_{n-1}, n = 1, 2, \ldots$, where $S_0 = 0$ by convention. In other words,

$$L_n = \text{the length of the } n\text{th renewal cycle}, \quad n = 1, 2, \ldots .$$

Note that the cycle length can only take on the discrete values $0, 1, \ldots$ when the index set $T = \{0, 1, \ldots\}$. In the following it is assumed that

$$0 < E(L_1) < \infty.$$

In many practical situations a reward structure is imposed on the regenerative process $\{X(t), \ t \in T\}$. The reward structure usually consists of reward rates that are earned continuously over time and lump rewards that are only earned at certain state transitions. Denote by

$$R_n = \text{the total reward earned in the } n\text{th renewal cycle}, \quad n = 1, 2, \ldots .$$

It is assumed that $R_1, R_2, \ldots$ are independent and identically distributed random variables. Note that R_n typically depends on L_n. In case R_n can take on both positive and negative values, it is assumed that $E(|R_1|) < \infty$. Define now, for any $t \in T$,

$$R(t) = \text{the cumulative reward earned up to time } t.$$

The process $\{R(t), \ t \geq 0\}$ is called a *renewal-reward process*. We are now ready to prove a theorem of utmost importance.

Theorem 1.3.1 (renewal-reward theorem)

$$\lim_{t \to \infty} \frac{R(t)}{t} = \frac{E(R_1)}{E(L_1)} \quad \text{with probability 1.}$$

In other words, the long-run average reward per unit time is equal to the expected reward earned during one cycle divided by the expected length of one cycle.

Proof For ease, let us assume that the rewards are non-negative. Define $N(t)$ as the number of renewal cycles that are completed up to time t. Then, for any $t > 0$,

$$\sum_{i=1}^{N(t)} R_i \leq R(t) \leq \sum_{i=1}^{N(t)+1} R_i,$$

and so

$$\frac{\sum_{i=1}^{N(t)} R_i}{N(t)} \frac{N(t)}{t} \leq \frac{R(t)}{t} \leq \frac{\sum_{i=1}^{N(t)+1} R_i}{N(t)+1} \frac{N(t)+1}{t}.$$

Note that $N(t) \to \infty$ with probability 1 as $t \to \infty$. By the strong law of large numbers for a sequence of independent and identically distributed random variables,

$$\lim_{n \to \infty} \frac{1}{n} \sum_{i=1}^{n} R_i = E(R_1) \quad \text{with probability 1.}$$

By the strong law of large numbers for a renewal process,

$$\lim_{t \to \infty} \frac{N(t)}{t} = \frac{1}{E(L_1)} \quad \text{with probability 1.}$$

Together the above relations imply the assertion of the theorem. In case the rewards are not non-negative, then the proof of the theorem follows by treating separately the positive and negative parts of the rewards.

In a natural way Theorem 1.3.1 relates the behaviour of the renewal-reward process over time to the behaviour of the process over a single renewal cycle. It is noteworthy that the outcome of the long-run average actual reward per unit time can be predicted with probability 1. If we are going to run the process over an infinitely long period of time, then we can say beforehand that in the long run the average *actual* reward per unit time will be equal to the constant $E(R_1)/E(L_1)$ with probability 1. This is a much stronger and more useful statement than the statement that the long-run *expected* average reward per unit time equals $E(R_1)/E(L_1)$.

Remark 1.3.1 Expected-value version of the renewal-reward theorem

It will be no surprise that the result

$$\lim_{t \to \infty} \frac{E[R(t)]}{t} = \frac{E(R_1)}{E(L_1)}$$

holds. This result follows directly from Theorem 1.3.1 and the bounded convergence theorem when $|R(t)/t| \leq M$ for all t for some $M > 0$, but the proof is rather intricate when $R(t)/t$ is not bounded. The expected-value version of the renewal-reward theorem will occasionally be used in theoretical considerations.

An important application of the renewal-reward theorem is the characterization of the long-run fraction of time the regenerative process $\{X(t), \ t \in T\}$ spends in some set B of states. For a given set B of states, define for any $t \in T$ the indicator variable

$$I_B(t) = \begin{cases} 1 & \text{if } X(t) \in B, \\ 0 & \text{if } X(t) \notin B. \end{cases}$$

Also, define the random variable

$T_B = $ the amount of time the process spends in the set B of states during one cycle.

Obviously, $T_B = \int_0^{S_1} I_B(u)\,du$ for a continuous-time process $\{X(t),\ t \geq 0\}$, while for a discrete-time process $\{X(n),\ n = 0, 1, \ldots\}$ the random variable T_B denotes the number of indices k with $0 \leq k < S_1$, such that $X(k) \in B$.

The following theorem is an immediate consequence of the renewal-reward theorem.

Theorem 1.3.2   *For a continuous-time regenerative process* $\{X(t),\ t \geq 0\}$

$$\lim_{t \to \infty} \frac{1}{t} \int_0^t I_B(u)\,du = \frac{E(T_B)}{E(L_1)} \quad \text{with probability 1,}$$

Similarly, for a discrete-time regenerative process $\{X(n),\ n = 0, 1, \ldots\}$

$$\lim_{n \to \infty} \frac{1}{n} \sum_{k=0}^{n} I_B(k) = \frac{E(T_B)}{E(L_1)} \quad \text{with probability 1.}$$

Proof   The long-run average reward per unit time gives the long-run fraction of time the process $\{X(t)\}$ is in the set B when it is assumed that a reward at rate 1 is earned when the process is in the set B. For this reward structure the expected reward earned during one cycle equals the expected amount of time the process spends in the set B during one cycle. Applying the renewal-reward theorem, the result next follows.

In words, Theorem 1.3.2 states that the long-run fraction of time the process $\{X(t),\ t \in T\}$ will spend in the set B of states equals with probability 1 the expected amount of time the process spends in the set B during one cycle divided by the expected length of one cycle. This result will often be utilized in this book.

For completeness we also state the following theorem.

Theorem 1.3.3   *For the regenerative process* $\{X(t),\ t \in T\}$,

$$\lim_{t \to \infty} P\{X(t) \in B\} = \frac{E(T_B)}{E(L_1)}$$

provided that the probability distribution of the cycle length has a continuous part in the continuous-time case and is aperiodic in the discrete-time case.

A distribution function is said to have a continuous part if it has a positive density on some interval. A discrete distribution $\{a_j,\ j = 0, 1, \ldots\}$ is said to be aperiodic if the greatest common divisor of the indices $j \geq 1$ for which $a_j > 0$ is equal to 1.

The proof of Theorem 1.3.3 requires deep mathematics and is beyond the scope of this book. The interested reader is referred to Miller (1972). It is remarkable that the proof for the time-average limit is much simpler than the proof for the ordinary limit. This is all the more striking when we take into account that the time-average limit is in general much more useful for practical purposes than the ordinary limit. Another advantage of the time-average limit is that it is easier to understand than the ordinary limit. In interpreting the ordinary limit one should be quite careful. The ordinary limit represents the probability that an outside person will find the process in some state of the set B when inspecting the process at an arbitrary point in time after the process has been in operation since a very long time. It is essential for this interpretation that the outside person has no information about the past of the process when inspecting the process. How much more concrete is the interpretation of the time-average limit as the long-run fraction of time the process will spend in the set B of states.

As another application of the renewal-reward theorem we derive a theoretically useful representation for the long-run average value of a regenerative stochastic process. Suppose that $\{X(t)\}$ is a regenerative process that takes on non-negative real values. For example, $X(t)$ is the number of customers present at time t in a queueing system. In many applications it is possible to calculate the long-run fraction of time the process $\{X(t)\}$ takes on a value larger than y for any y. Then the long-run average value of the process $\{X(t)\}$ can be computed by using this distribution. The next theorem is only stated for a continuous-time process, but it applies equally well to a discrete-time process.

Theorem 1.3.4   *Let $\{X(t), \ t \geq 0\}$ be a continuous-time regenerative process whose state space is a subset of the non-negative reals. Then*

$$\lim_{t \to \infty} \frac{1}{t} \int_0^t X(u)\, du = \int_0^\infty \overline{P}(y)\, dy \quad \text{with probability 1,}$$

where $\overline{P}(y)$ represents the long-run fraction of time the process $\{X(t)\}$ takes on a value larger than y.

Proof   Denote by S_1 the first regeneration epoch of the process $\{X(t)\}$. Assuming that a reward at rate x is incurred whenever the process $\{X(t)\}$ is in state x, it follows from the renewal-reward theorem that

$$\lim_{t \to \infty} \frac{1}{t} \int_0^t X(u)\, du = \frac{E\left[\int_0^{S_1} X(u)\, du\right]}{E(S_1)} \quad \text{with probability 1.}$$

For fixed $y \geq 0$, define the indicator variable $I_y(t)$ by

$$I_y(t) = \begin{cases} 1 & \text{if } X(t) > y, \\ 0 & \text{otherwise.} \end{cases}$$

Since $\overline{P}(y) = \lim_{t \to \infty} (1/t) \int_0^t I_y(u)\, du$ with probability 1, we also have by the renewal-reward theorem that

$$\overline{P}(y) = \frac{E\left[\int_0^{S_1} I_y(u)\, du\right]}{E(S_1)} \quad \text{with probability 1.}$$

The desired result now follows from

$$\int_{0}^{\infty} \overline{P}(y)\,dy = \frac{1}{E(S_1)} \int_{0}^{\infty} E\left[\int_{0}^{S_1} I_y(u)\,du\right]\,dy$$

$$= \frac{1}{E(S_1)} E\left[\int_{0}^{S_1} \left\{\int_{0}^{\infty} I_y(u)\,dy\right\}\,du\right]$$

$$= \frac{1}{E(S_1)} E\left[\int_{0}^{S_1} X(u)\,du\right],$$

where we used the relation $\int_{0}^{\infty} I_y(t)\,dy = X(t)$ and the interchange of the order of integration is justified by the fact that the terms involved are non-negative.

Before giving a number of applications of the renewal-reward theorem, we state a central limit theorem for renewal-reward processes.

Theorem 1.3.5 *Assuming that $R(t) \geq 0$ and that $E(L_1^2) < \infty$ and $E(R_1^2) < \infty$,*

$$\lim_{t \to \infty} P\left\{\frac{R(t) - gt}{v\sqrt{t/\mu}} \leq x\right\} = \frac{1}{\sqrt{2\pi}} \int_{-\infty}^{x} e^{-(1/2)y^2}\,dy, \quad x \geq 0,$$

where $\mu = E(L_1)$, $g = E(R_1)/E(L_1)$ and $v^2 = E(R_1 - gL_1)^2$.

A proof of this theorem can be found in Wolff (1989). In applying this theorem, the difficulty is usually to find the constant v. In specific applications one might use simulation to find v.

Next we give a number of examples in which the renewal-reward theorem is applied.

Example 1.3.1 A control-limit rule for a deteriorating item

Consider an item that deteriorates by incurring an amount of damage each period. The damages incurred in the successive periods are independent random variables having a common exponential density $f(x) = \alpha e^{-\alpha x}$, $x \geq 0$. The successive damages accumulate. At the end of each period the item is inspected. The item has to be repaired at a high cost of c_0 when the inspection reveals an accumulated damage larger than a given value z_0. However, a preventive repair at a lower cost of $c_1 > 0$ is possible when the accumulated damage at inspection is at or below z_0. Each repair takes a negligible time and after a repair the item is as good as new. The following control-limit rule is used. The item is repaired at each inspection that reveals an accumulated damage larger than the repair limit z with $0 < z \leq z_0$. How to choose the repair limit z such that the long-run average cost per period is minimal?

Under a control rule with a given repair limit z, the stochastic process $\{X(t),\ t \geq 0\}$ describing the accumulated damage generates itself each time a repair is completed. The length of a cycle is given by the number of periods needed for a cumulative damage exceeding z. Thus, denoting by $M(x)$ the renewal function associated with the damage density $f(x)$, it follows that

$$\text{the expected length of a cycle} = 1 + M(z).$$

The repair cost incurred at the end of a cycle is equal to c_0 when the excess in damage of the repair level z is larger than $z_0 - z$ and is equal to c_1 otherwise. Denoting this excess by γ_z, it follows that

$$\text{the expected cost incurred per cycle} = c_0 P\{\gamma_z > z_0 - z\} + c_1 P\{\gamma_z \le z_0 - z\}.$$

Since the damage density $f(x)$ is exponential with mean $1/\alpha$, we have by relation (1.1.16) and Theorem 1.2.2 that $M(z) = \alpha z$ and $P\{\gamma_z > x\} = \mathrm{e}^{-\alpha x}$ for all z. Hence, under the control rule with repair limit z,

$$\text{the long-run average cost per period} = \frac{c_0 P\{\gamma_z > z_0 - z\} + c_1 P\{\gamma_z \le z_0 - z\}}{1 + M(z)}$$

$$= \frac{(c_0 - c_1)\mathrm{e}^{-\alpha(z_0 - z)} + c_1}{1 + \alpha z}$$

with probability 1. It is a matter of simple analysis to verify that the average cost per period is minimal for the unique solution z of the equation

$$\alpha z \mathrm{e}^{-\alpha(z_0 - z)} = \frac{c_1}{c_0 - c_1},$$

provided $\alpha z_0 > c_1/(c_0 - c_1)$; otherwise the optimal value of z equals z_0. This equation has to be solved numerically. Bisection is a safe and fast method to find the root of the equation once one has located an interval for the root.

Example 1.3.2 A stochastic clearing system

In a communication system messages requiring transmission arrive according to a Poisson process with rate λ. The messages are temporarily stored in a buffer having ample capacity. Every T time units the buffer is cleared from all messages present. The buffer is empty at time $t = 0$. A fixed cost of $K > 0$ is incurred for each clearing of the buffer. Also, there is a holding cost of $h > 0$ for each time unit a message has to wait in the buffer. What is the value of T for which the long-run average cost per unit time is minimal?

We first derive an expression for the average cost per unit time for a given value of the control parameter T. To do so, observe that the stochastic process describing the number of messages in the system regenerates itself each time the buffer is cleared from all messages present. This fact uses the lack of memory of the Poisson arrival process so that at any clearing epoch it is not relevant how long ago the last message arrived. Since a cycle is the time interval between two clearings of the buffer, we have

$$\text{the expected length of one cycle} = T.$$

To find the expected holding cost incurred during $(0, T]$, the following trick is used. This trick can be typically used when a Poisson process is involved. Define the function $\alpha(t)$ by

$$\alpha(t) = \text{the expected holding costs incurred during } (0, t], \quad 0 < t < T.$$

Then, by conditioning on what may happen in the first Δt time units,

$$\alpha(t + \Delta t) = \lambda \Delta t \times ht + \alpha(t) + o(\Delta t) \quad \text{as } \Delta t \to 0.$$

This implies $\alpha'(t) = h\lambda t$ and so $\alpha(t) = \frac{1}{2}h\lambda t^2$, $0 < t < T$. Hence

$$E[\text{holding cost incurred up to time } T] = \frac{1}{2}h\lambda T^2. \tag{1.3.1}$$

This shows that the expected cost in one cycle is $K + \frac{1}{2}h\lambda T^2$ and so

$$\text{the long-run average cost per unit time} = \frac{1}{T}\left(K + \frac{1}{2}h\lambda T^2\right)$$

with probability 1. The average cost per unit time is minimal for

$$T^* = \sqrt{\frac{2K}{h\lambda}}.$$

Example 1.3.3 A reliability system with redundancies

An electronic system consists of a number of independent and identical components hooked up in parallel. The lifetime of each component has an exponential distribution with mean $1/\mu$. The system is operative only if m or more components are operating. The non-failed units remain in operation when the system as a whole is in a non-operative state. The system availability is increased by periodic maintenance and by putting r redundant components into operation in addition to the minimum number m of components required. Under the periodic maintenance the system is inspected every T time units, where at inspection the failed components are repaired. The repair time is negligible and each repaired component is as good as new. The periodic inspections provide the only repair opportunities. The following costs are involved. For each component there is a depreciation cost of $I > 0$ per unit time. A fixed cost of $K > 0$ is made for each inspection and there is a repair cost of $R > 0$ for each failed component. How to choose the number r of redundant components and the time T between two consecutive inspections such that the long-run average cost per unit time is minimal subject to the requirement that the probability of system failure between two inspections is no more than a prespecified value α?

We first derive the performance measures for given values of the parameters r and T. The stochastic process describing the number of operating components is regenerative. Using the lack of memory of the exponential lifetimes of the components, it follows that the process regenerates itself after each inspection. Since a cycle is the time interval between two inspections, we have

$$E(\text{length of one cycle}) = T.$$

Further, using that a given component fails within a time T with probability $1 - e^{-\mu T}$, it follows that

$$P\{\text{the system as a whole fails between two inspections}\}$$

$$= \sum_{k=r+1}^{m+r} \binom{m+r}{k} (1 - e^{-\mu T})^k e^{-\mu T(m+r-k)}$$

and

$$E(\text{number of components that fail between two inspections}) = (m+r)(1 - e^{-\mu T}).$$

Hence

$$E(\text{total costs in one cycle}) = (m+r)I \times T + K + (m+r)(1 - e^{-\mu T})R$$

and so

the long-run average cost per unit time

$$= \frac{1}{T}[(m+r)I \times T + K + (m+r)(1 - e^{-\mu T})R].$$

The optimal values of the parameters r and T are found from the minimization problem:

$$\text{Minimize}_{r,T} \frac{1}{T}\left[(m+r)I \times T + K + (m+r)(1 - e^{-\mu T})R\right]$$

$$\text{subject to} \sum_{k=r+1}^{m+r} \binom{m+r}{k}(1 - e^{-\mu T})^k e^{-\mu T(m+r-k)} \leq \alpha.$$

Using the Lagrange method this problem can be numerically solved.

Next we give a more elaborate application of the renewal-reward theorem.

Example 1.3.4 The N-policy for starting up and shutting down a production process

Consider a production facility at which production orders arrive according to a Poisson process with rate λ. The production times $\tau_1, \tau_2, \ldots$ of the orders are independent random variables having a common probability distribution function F with finite first two moments. Also, the production process is independent of the arrival process. The facility can only work on one order at a time. It is assumed that $E(\tau_1) < 1/\lambda$; that is, the average production time per order is less than the mean interarrival time between two consecutive orders. The facility operates only intermittently and is shut down when no orders are present any more. A fixed setup cost of $K > 0$ is incurred each time the facility is re-opened. Also, a holding cost $h > 0$ per unit time is incurred for each order present. The facility is only turned on when enough orders have accumulated. The so-called N-policy reactivates the facility as soon as N orders are present. For ease we assume that it takes a zero setup time to restart production. How to choose the value of the control parameter N such that the long-run average cost per unit time is minimal?

To analyse this problem, we first observe that for a given N-policy the stochastic process describing jointly the number of orders present and the status of the facility (on or off) regenerates itself each time the facility is turned on. Define a cycle as the time elapsed between two consecutive reactivations of the facility. Clearly, each cycle consists of a busy period B with production and an idle period I with no production. We deal separately with the idle and the busy period. Using the memoryless property of the Poisson process, the length of the idle period is the sum of N exponential random variables each having mean $1/\lambda$. Hence

$$E(I) = \frac{N}{\lambda} \quad \text{and} \quad E(\text{holding cost incurred during } I) = h \left(\frac{N-1}{\lambda} + \cdots + \frac{1}{\lambda} \right).$$

To deal with the busy period, we define for $n = 1, 2, \ldots$ the quantities

$t_n = $ the expected time until the facility becomes empty given that at epoch 0 a production starts with n orders present,

and

$h_n = $ the expected holding costs incurred until the facility becomes empty given that at epoch 0 a production starts with n orders present.

These quantities are independent of the control rule considered. In particular,

$$E(B) = t_N \quad \text{and} \quad E(\text{holding cost incurred during } B) = h_N.$$

Now, by the renewal-reward theorem,

$$\text{the long-run average cost per unit time} = \frac{\left(\frac{1}{2} h/\lambda \right) N(N-1) + K + h_N}{N/\lambda + t_N}$$

with probability 1. To find the functions t_n and h_n, define

$a_j = $ the probability that j orders arrive during the production time of a single order.

This definition uses the lack of memory of the Poisson arrival process. Assume for ease that the production time has a probability density $f(x)$. By conditioning on the production time and noting that the number of orders arriving in a fixed time y is Poisson distributed with mean λy, it follows that

$$a_j = \int_0^\infty e^{-\lambda y} \frac{(\lambda y)^j}{j!} f(y)\, dy, \quad j = 0, 1, \ldots .$$

It is readily verified that

$$\sum_{j=1}^\infty j a_j = \lambda E(\tau_1) \quad \text{and} \quad \sum_{j=1}^\infty j^2 a_j = \lambda^2 E(\tau_1^2) + \lambda E(\tau_1). \tag{1.3.2}$$

We now derive recursion relations for the quantities t_n and h_n. Suppose that at epoch 0 a production starts with n orders present. If the number of new orders arriving during the production time of the first order is j, then the time to empty the system equals the first production time plus the time to empty the system starting with $n - 1 + j$ orders present. Thus

$$t_n = E(\tau_1) + \sum_{j=0}^\infty t_{n-1+j} a_j, \quad n = 1, 2, \ldots,$$

where $t_0 = 0$. Similarly, we derive a recursion relation for the h_n's. Therefore note that relation (1.3.1) implies that the expected holding costs incurred during the first production equals $hn E(\tau_1) + \frac{1}{2} h\lambda E(\tau_1^2)$. Thus

$$h_n = hn E(\tau_1) + \frac{1}{2} h\lambda E(\tau_1^2) + \sum_{j=0}^\infty h_{n-1+j} a_j, \quad n = 1, 2, \ldots,$$

where $h_0 = 0$. In a moment it will be shown that t_n is linear in n and h_n is quadratic in n. Substituting these functional forms in the above recursion relations and using (1.3.2), we find after some algebra that for $n = 1, 2, \ldots,$

$$t_n = \frac{n E(\tau_1)}{1 - \lambda E(\tau_1)}, \tag{1.3.3}$$

$$h_n = \frac{h}{1 - \lambda E(\tau_1)} \left[\frac{1}{2} n(n-1) E(\tau_1) + n E(\tau_1) + \frac{\lambda n E(\tau_1^2)}{2\{1 - \lambda E(\tau_1)\}} \right]. \tag{1.3.4}$$

To verify that t_n is linear in n and h_n is quadratic in n the following arguments are used. First observe that t_n and h_n do not depend on the specific order in which the production orders are coped with during the production process. Imagine now the following production discipline. The n initial orders $O_1, \ldots, O_n$ (say) are separated. Order O_1 is produced first, after which all orders (if any) are produced that have arrived during the production time of O_1, and this way of production is continued until the facility is free of all orders but $O_2, \ldots, O_n$. Next this procedure is repeated with order O_2, etc. Thus we find that $t_n = nt_1$, proving that t_n is linear in n. Using the same arguments and noting that $h_1 + h(n-k)t_1$ gives the expected holding cost incurred during the time to free the system of order O_k and its direct descendants until only the orders $O_{k+1}, \ldots, O_n$ are left, it follows that

$$h_n = \sum_{k=1}^{n} \{h_1 + h(n-k)t_1\} = nh_1 + \frac{1}{2}hn(n-1)t_1.$$

Combining the above results we find for an N-policy that, with probability 1,

the long-run average cost per unit time

$$= \frac{\lambda(1-\rho)K}{N} + h\left\{\rho + \frac{\lambda^2 E(\tau_1^2)}{2(1-\rho)} + \frac{N-1}{2}\right\}, \qquad (1.3.5)$$

where $\rho = \lambda E(\tau_1)$. It is noteworthy that this expression needs from the production time only the first two moments. Also note that, by putting $K = 0$ and $h = 1$ in (1.3.5), we find that, with probability 1,

the long-run average number of orders in the system $= \rho + \dfrac{\lambda^2 E(\tau_1^2)}{2(1-\rho)} + \dfrac{N-1}{2}.$

For the special case of $N = 1$ this formula reduces to the famous Pollaczek–Khintchine formula for the average number of customers in the system for the standard single-server queue with Poisson arrivals and general service times. In addition to these results, it is interesting to note that

the long-run fraction of time the facility is on $= \dfrac{E(B)}{E(I) + E(B)} = \rho.$

This relation follows from the renewal-reward theorem by assuming that the system earns a reward at rate 1 when the facility is on.

The optimal value of N can be obtained by differentiating the right side of (1.3.5) in which we take N as a continuous variable. Since the average cost is convex in N, it follows that the average cost is minimal for one of the two integers nearest to

$$N = \sqrt{\frac{2\lambda(1-\rho)K}{h}}.$$

1.4   RELIABILITY APPLICATIONS

Renewal theory has many applications in the field of reliability. This section gives a number of representative applications.

Example 1.4.1 *Age replacement and block replacement*

Age replacement and block replacement are two basic rules for controlling items (e.g. light bulbs or machines) that are subject to stochastic breakdowns.

Let us first discuss the age-replacement rule. This rule prescribes to replace the item by a new one upon failure or upon reaching the critical age T, whicheyer occurs first. The age T is the control parameter. It is assumed that the lifetimes of the items are independent random variables $X_1, X_2, \ldots$ having a common probability distribution function $F(x)$ with probability density $f(x)$. A cost of $c_p > 0$ is incurred for each planned (preventive) replacement and a cost of c_f for each failure replacement, where $c_f > c_p$. The stochastic process describing the age of the item in use is regenerative. The regeneration epochs are the epochs at which a new item is installed. The cycle length is distributed as $\min(T, X_1)$. By conditioning on X_1, we find

$$E(\text{length of one cycle}) = \int_0^T x f(x) \, dx + \int_T^\infty T f(x) \, dx$$

$$= \int_0^T \{1 - F(x)\} \, dx.$$

Similarly, we find

$$E(\text{costs incurred during one cycle}) = c_f F(T) + c_p \{1 - F(T)\}.$$

Hence, by the renewal-reward theorem,

$$\text{the long-run average cost per unit time} = \frac{c_p + (c_f - c_p) F(T)}{\int_0^T \{1 - F(x)\} \, dx}$$

with probability 1. Assume now that the failure rate function $r(x)$ defined by

$$r(x) = \frac{f(x)}{1 - F(x)}, \quad x \geq 0,$$

is continuous and strictly increasing to infinity. Recall that $r(x)\Delta x + o(\Delta x)$ represents the probability that an item of age x will fail in the next time Δx for Δx small. By putting the derivative of the average cost function equal to zero, it is readily verified that the minimizing value of T is the unique solution to the equation

$$r(T) \int_0^T \{1 - F(x)\} \, dx - F(T) = \frac{c_p}{c_f - c_p}.$$

In general a numerical procedure should be used to solve this equation.

Denote by $g(T)$ the long-run average cost per unit time for the age-replacement rule with limit T and let T^* be the optimal value of T. It is a matter of simple algebra to derive from the above equations that

$$g(T^*) = (c_f - c_p) r(T^*). \tag{1.4.1}$$

An interesting heuristic derivation of this relation is by marginal analysis. Suppose that for an age-replacement rule with limit T a preventive replacement is delayed by

ΔT time units when the item has reached the critical age T. This delay causes extra replacement costs of $c_f - c_p$ when the item fails in $(T, T + \Delta T]$. The probability of such a failure is $r(T)\Delta T + o(\Delta T)$. Thus the expected extra replacement costs when delaying the preventive replacement are equal to $(c_f - c_p)r(T)\Delta T + o(\Delta T)$. In other words, delaying the preventive replacement for a very small period of time causes an extra cost at rate $(c_f - c_p)r(T)$. It is intuitively reasonable that this cost rate should be equal to the minimal average cost rate $g(T^*)$ when $T = T^*$, in accordance with (1.4.1).

Let us next consider the block-replacement rule. Under this rule the item is replaced by a new one upon failure and upon scheduled times $T, 2T, \ldots$. There is always a replacement at the scheduled times regardless of the age of the item in use. It is noted that block replacement is usually applied when a group of items has to be maintained. The cost structure is the same as in the age-replacement model. The stochastic process describing the age of the item in use is regenerative. The times $0, T, 2T, \ldots$ for scheduled replacements are taken as regeneration epochs. Then the length of one cycle is T. Further

$$E(\text{costs incurred during one cycle}) = c_p + c_f M(T),$$

where $M(x)$ denotes the renewal function associated with the lifetime distribution function $F(x)$. This result follows by noting that the number of renewals up to time T in the renewal process generated by the lifetimes $X_1, X_2, \ldots$ is nothing else than the number of failure replacements up to time T. Hence, for the block-replacement rule with parameter T,

$$\text{the long-run average cost per unit time} = \frac{1}{T}\{c_p + c_f M(T)\}$$

with probability 1. The optimal value of T has to be found by a numerical procedure.

A basic question in reliability problems is how to calculate the probability distribution of the time until the first failure. The next example deals with this question and shows that the time until the first failure is often approximately exponentially distributed.

Example 1.4.2 A reliability problem with periodic inspections

High reliability of an electronic system is often achieved by employing redundant components and having periodic inspections. Let us consider a reliability system with two identical units, where one unit is in full operation and the other unit is in warm standby. The operating unit has a constant failure rate of λ_0 and the unit in standby has a constant failure rate of λ_1, where $0 \le \lambda_1 < \lambda_0$. Upon failure of the operating unit the standby unit is put into full operation provided the standby is not in the failure state. Failed units are replaced only at the scheduled times $T, 2T, \ldots$ when the system is inspected. The time to replace any failed unit is negligible. A system failure occurs if both units are down. In designing highly reliable systems a key measure of system performance is the probability distribution of the time until the first system failure. This probability distribution will be derived under the assumption that $(\lambda_0 + \lambda_1)T$ is very small. This assumption will be satisfied when a high reliability is pursued. Further it is supposed that both units are in a good condition at time $t = 0$. Note that after each inspection both units can be considered as good as new.

To find the distribution of the time until the first system failure, we first compute the probability q defined by

$$q = P\{\text{system failure occurs between two inspections}\}.$$

To do so, observe that a constant failure rate λ for the lifetime of a unit implies that the lifetime has an exponential distribution with mean $1/\lambda$. Using that the minimum of two independent exponentials with respective means $1/\lambda_0$ and $1/\lambda_1$ is exponentially distributed with mean $1/(\lambda_0 + \lambda_1)$, we find by conditioning on the epoch of the first failure of a unit that

$$q = \int_0^T \{1 - e^{-\lambda_0(T-x)}\}(\lambda_0 + \lambda_1)e^{-(\lambda_0+\lambda_1)x}\, dx$$

$$= 1 - \frac{(\lambda_0 + \lambda_1)}{\lambda_1}e^{-\lambda_0 T} + \frac{\lambda_0}{\lambda_1}e^{-(\lambda_0+\lambda_1)T}.$$

The assumption that $(\lambda_0 + \lambda_1)T$ is very small in conjunction with the asymptotic expansion $e^{-x} \approx 1 - x + x^2/2$ for $x \to 0$ enables us to approximate the probability q by $q \approx \tfrac{1}{2}\lambda_0(\lambda_0+\lambda_1)T^2$, showing that q will be typically close to 0. Define now

$$U = \text{the time until the first system failure.}$$

Since the process describing the state of the two units regenerates itself every T time units, it follows that

$$P\{U > nT\} = (1 - q)^n, \quad n = 0, 1, \dots .$$

Assuming that the failure probability q is close to 0, the asymptotic expansions $(1 - q)^n \approx 1 - nq$ and $e^{-nq} \approx 1 - nq$ apply. Thus we find that

$$P\{U > t\} \approx e^{-tq/T}, \quad t \geq 0. \tag{1.4.2}$$

In other words, the time until the first system failure is approximately exponentially distributed.

The characterization (1.4.2) holds generally for reliability models in which the occurrence of a system failure is a rare event. Loosely formulated, the following result holds. Let $\{X(t)\}$ be a regenerative process having a set B of (bad) states such that the probability q that the process visits the set B during a given cycle is very small. Denote by the random variable U the time until the process visits the set B for the first time. Assuming that the cycle length has a finite and positive mean $E(T)$, it holds that $P\{U > t\} \approx e^{-tq/E(T)}$ for $t \geq 0$; see Keilson (1979) or Solovjez (1971) for a proof. The result that the time until the first occurrence of a rare event in a regenerative process is approximately exponentially distributed is very useful. It gives not only quantitative insight, but it also indicates that the computation of the mean of the first-passage time suffices. Besides applications to reliability models, the result has applications to finite-buffer communication (or production) systems for which the occurrence of buffer overflow is a rare event.

The next example deals with the calculation of the probability distribution of the availability of a reliability system during a given time period.

Example 1.4.3 The 1-out-of-2 reliability model with repair

The 1-out-of-2 reliability model deals with a repairable system that has one operating unit and one cold standby unit as protection against failures. The lifetime of an operating unit has a general probability distribution function $F_L(x)$ having density $f_L(x)$ with mean μ_L. If the operating unit fails, it is replaced immediately by the standby unit if available. The failed unit is sent to a repair facility and immediately enters repair if the facility is idle. Only one unit can be in repair at a time. The repair time of a failed unit has a general probability distribution function $G_R(x)$ with mean μ_R. It is assumed that $\mu_R \ll \mu_L$. The operating times and repair times are mutually independent. The system is down when both units are broken down and is up otherwise.

We are interested in the probability distribution function

$$A(x, t_0) = \lim_{t \to \infty} P\{\text{the total up-time in } (t, t + t_0] \text{ is } \leq x\}$$

for a prespecified interval of length t_0. In other words, the performance measure is the probability distribution function of total amount of time the system is available during a time interval of given length t_0 when the system has reached statistical equilibrium.

An approximate analysis will be given. The analysis is based on the following ideas:

1. Compute the means of the up- and down-periods.
2. Approximate the stochastic process of the up- and down-periods by an alternating renewal process in which both the up-periods and the down-periods are independent, exponential random variables and the up-periods are independent of the down-periods.

In view of the assumption $\mu_R \ll \mu_L$ the occurrence of a system failure is a rare event. This justifies the approximate step of assuming an exponential distribution for the up-period. A similar justification for approximating the distribution of the down-time by an exponential distribution cannot be given. However, in view of the fact that the up-time will typically dominate the down-time, it is reasonable to expect that the distributional form of the down-time has only a minor effect on the accuracy of the approximation.

Before working out the above two steps, let us give some results for an alternating renewal process.

Alternating renewal process

An alternating renewal process is a two-state process alternating between an on-state and an off-state, where both the on-times and the off-times are independent and identically distributed random variables. In addition we suppose that the sequences of on-times and off-times are mutually independent. Assuming that an on-time starts at epoch 0, define for any $s > 0$ the probability $P_{on}(s)$ by

$$P_{on}(s) = P\{\text{the process is in the on-state at time } s\}, \quad s \geq 0$$

and the random variable $U(s)$ by

$$U(s) = \text{the total amount of time the process is in the on-state during } [0, s].$$

The following result holds.

Theorem 1.4.1 *Suppose that the on-times and off-times have exponential distributions with respective means $1/\alpha$ and $1/\beta$. Then*

$$P_{on}(s) = \frac{\beta}{\alpha + \beta} + \frac{\alpha}{\alpha + \beta} e^{-(\alpha+\beta)s}, \quad s \geq 0 \tag{1.4.3}$$

and

$$P\{U(s) \leq x\} = \sum_{n=0}^{\infty} e^{-\beta(s-x)} \left[\frac{\beta(s-x)}{n!} \right]^n \left[1 - \sum_{k=0}^{n} e^{-\alpha x} \frac{(\alpha x)^k}{k!} \right], \quad 0 \leq x < s. \tag{1.4.4}$$

The distribution function $P\{U(s) \leq x\}$ has a mass of $e^{-\alpha s}$ at $x = s$.

Proof Let $P_{off}(s) = P\{\text{the process is in the off-state at time } s\}$. By considering what may happen in the time interval $(s, s+\Delta s]$ with Δs small, it is straightforward to derive the linear differential equation

$$P'_{on}(s) = \beta P_{off}(s) - \alpha P_{on}(s), \quad s > 0.$$

Since $P_{off}(s) = 1 - P_{on}(s)$, we find $P'_{on}(s) = -(\alpha + \beta)P_{on}(s) + \beta$, $s > 0$. The solution of this differential equation is given by (1.4.3).

The proof of (1.4.4) is more complicated. Let us first note that the random variable $U(s)$ is equal to s if and only if the first on-time exceeds s. Hence the distribution function $P\{U(s) \leq x\}$ has mass $e^{-\alpha s}$ at $x = s$. Fix now $0 \leq x < s$. By conditioning on the lengths of the first on-time and the first off-time, we obtain

$$P\{U(s) \leq x\} = \int_0^x \alpha e^{-\alpha y} \, dy \int_0^\infty P\{U(s - y - u) \leq x - y\} \beta e^{-\beta u} \, du.$$

Observing that $P\{U(s - y - u) \le x - y\} = 1$ if $s - y - u \le x - y$, this equation can be simplified to

$$P\{U(s) \le x\} = e^{-\beta(s-x)}(1 - e^{-\alpha x}) + \int_0^x \alpha e^{-\alpha y}\, dy$$

$$\times \int_0^{s-x} P\{U(s - y - u) \le x - y\}\beta e^{-\beta u}\, du.$$

Substituting this equation repeatedly into itself leads to the desired result (1.4.4). We omit the tedious details.

Corollary 1.4.2   *For any $t_0 > 0$,*

$$\lim_{t \to \infty} P\{U(t + t_0) - U(t) \le x\}$$

$$= \frac{\beta}{\alpha + \beta} \sum_{n=0}^{\infty} e^{-\beta(t_0-x)} \frac{[\beta(t_0 - x)]^n}{n!} \left[1 - \sum_{k=0}^{n} e^{-\alpha x} \frac{(\alpha x)^k}{k!}\right]$$

$$+ \frac{\alpha}{\alpha + \beta} \sum_{j=0}^{\infty} e^{-\beta(t_0-x)} \frac{[\beta(t_0 - x)]^j}{j!} \left[1 - \sum_{k=0}^{j-1} e^{-\alpha x} \frac{(\alpha x)^k}{k!}\right], \quad 0 \le x < t_0. \ (1.4.5)$$

Proof   Using Theorem 1.4.1, we have

$$\lim_{t \to \infty} P\{U(t + t_0) - U(t) \le x\} = \frac{\beta}{\alpha + \beta} P\{U(t_0) \le x\}$$

$$+ \frac{\alpha}{\alpha + \beta} \left[\int_0^{t_0-x} P\{U(t_0 - y) \le x\}\beta e^{-\beta y}\, dy + \int_{t_0-x}^{\infty} \beta e^{-\beta y}\, dy\right].$$

Next it is a matter of algebra to obtain the desired result.

Approximate analysis for the reliability problem of Example 1.4.3

The system alternates between the up-state and the down-state. With the possible exception of the first up-period, the up-periods start when a unit is put into operation while the other unit enters repair. The system regenerates itself at the beginning of the up-periods. For ease we assume that epoch 0 is such a regeneration epoch. Let the generic variables τ_{up} and τ_{down} denote the lengths of an up-period and a down-period. Denote by the sequences $\{L_i, i \ge 1\}$ and $\{R_i, i \ge 1\}$ the successive operating times and the successive repair times. Then

$$E(\tau_{\text{up}}) = E\left[\sum_{i=1}^{N} L_i\right],$$

where $N = \min\{n \ge 1 | R_n > L_n\}$. It is readily seen that the event $\{N = n\}$ is independent of $L_{n+1}, L_{n+2}, \ldots$ for any $n \ge 1$. Thus, by Wald's equation, $E(\tau_{\text{up}}) =$

$E(N)\mu_L$. Define the probability q by

$$q = P\{R > L\}$$

where the generic variables L and R denote the operating time and the repair time of a unit. Since $P\{N = n\} = (1 - q)^{n-1}q$, $n \geq 1$, we find

$$E(\tau_{\text{up}}) = \frac{\mu_L}{q}. \tag{1.4.6}$$

By conditioning on the lifetime, we have

$$q = \int_0^\infty \{1 - G_R(x)\} f_L(x) \, dx. \tag{1.4.7}$$

To find $E(\tau_{\text{down}})$, note that $E(\tau_{\text{down}}) = E(R - L | R > L)$. Using the formula (A.9) in Appendix A, we find

$$
\begin{aligned}
E(\tau_{\text{down}}) &= \int_0^\infty P\{R - L > t | R > L\} \, dt \\
&= \frac{1}{q} \int_0^\infty \left[\int_0^\infty \{1 - G_R(x + t)\} f_L(x) \, dx \right] dt \\
&= \frac{1}{q} \int_0^\infty f_L(x) \left[\int_x^\infty \{1 - G_R(u)\} \, du \right] dx,
\end{aligned}
$$

where the latter equality uses an interchange of the order of integration. Interchanging again the order of integration, we next find that

$$E(\tau_{\text{down}}) = \frac{1}{q} \int_0^\infty \{1 - G_R(u)\} F_L(u) \, du. \tag{1.4.8}$$

We are now in a position to calculate an approximation for the probability distribution function of the total up-time in a time interval of given length t_0 when the system has reached statistical equilibrium. An approximation to the desired probability $A(x; t_0)$ is obtained by applying formula (1.4.5) in which $1/\alpha$ and $1/\beta$ are replaced by $E(\tau_{\text{up}})$ and $E(\tau_{\text{down}})$ respectively. The numerical evaluation of the right side of (1.4.5) is easy, since the infinite series converges rapidly and involves only Poisson probabilities. Numerical integration is required to calculate the integrals for $E(\tau_{\text{up}})$ and $E(\tau_{\text{down}})$.

It remains to investigate the quality of the approximation for the probabilities $A(x; t_0)$. Several assumptions has been made to get the approximation. The most serious weakness of the approximation is the assumption that the off-time is approximately exponentially distributed. Nevertheless it turns out that the approximation performs very well for practical purposes. Denoting by D_x the probability that the fraction of time the system is unavailable in the time interval of length t_0 is more than $x\%$, we give in Table 1.4.1 the approximate and exact values of D_x for

Table 1.4.1 The unavailability probabilities

		$c_L^2 = 0.5$				$c_L^2 = 1$			
		D_0	D_2	D_5	D_{10}	D_0	D_2	D_5	D_{10}
$c_R^2 = 0$	app	0.044	0.030	0.016	0.006	0.117	0.086	0.054	0.024
	sim	0.043	0.033	0.020	0.005	0.108	0.091	0.066	0.027
$c_R^2 = 0.5$	app	0.051	0.040	0.028	0.015	0.117	0.095	0.068	0.040
	sim	0.050	0.040	0.029	0.016	0.109	0.092	0.070	0.042
$c_R^2 = 1$	app	0.056	0.047	0.036	0.024	0.117	0.099	0.077	0.050
	sim	0.055	0.047	0.036	0.024	0.110	0.094	0.074	0.050
$c_R^2 = 4$	app	0.076	0.071	0.063	0.053	0.117	0.108	0.096	0.079
	sim	0.075	0.069	0.061	0.050	0.112	0.101	0.089	0.072

several values of x. Note that $D_x = A(1 - t_0 x / 100; t_0)$. The exact values of D_x are obtained by computer simulation. The length of the simulation run has been taken long enough to ensure that the half-width of the 95% confidence interval for the simulated probability is no more than 0.001. It is assumed that the lifetime L of a unit has a Weibull distribution with mean $E(L) = 1$ and the repair time R of a unit has a gamma distribution with mean $E(R) = 0.125$. The squared coefficients of variation of the lifetime and the repair time are varied as $c_L^2 = 0.5$, 1 and $c_R^2 = 0$, 0.5, 1, 4. For the length of the interval we have taken $t_0 = 1$.

1.5 INVENTORY APPLICATIONS

This section discusses a number of basic stochastic inventory models for long-term control. It is assumed that the parameters of the models do not appreciably change with time. The situation of stochastic demand and positive replenishment lead times will be considered. In this situation stockouts cannot always be avoided. A stockout occurs when the demand during the lead time exceeds the stock on hand. Two basic questions to answer are:

1. When to order? (the reorder point).
2. How much to order? (the order quantity).

In order to provide adequate service to the customers, the reorder point will be typically larger than the expected demand during the replenishment lead time. The amount by which the reorder point is set above the expected lead-time demand is called the safety stock. This stock provides a buffer against random fluctuations in the lead-time demand. Increasing the safety stock decreases the probability of a stockout but increases the inventory level. A balance has to be found between the service provided to the customers and the costs of holding inventories. The order quantity and the reorder point are often determined separately in practice. The determination of the order quantity is usually based on holding and replenishment cost considerations only, while the determination of the reorder point is based on the service level requirement. Empirical studies indicate that the approach of determining sequentially the order quantity and the reorder point works excellently in most practical situations.

In probabilistic inventory models, it is important to specify what happens to demand that occurs when the system is out of stock. Two cases are considered:

(a) **Backlog case.** Any demand occurring when the system is out of stock is backlogged and is filled as soon as an adequate-sized replenishment arrives.

(b) **Lost-sales case.** Any demand when out of stock is lost.

The following inventory concepts will be used in the following discussion.

1. **On-hand stock.** This is stock that is physically on the shelves. The on-hand stock is always non-negative.

2. **Net stock** = (on-hand stock) − (amount backlogged). This quantity is negative only if a backlog exists. In the lost-sales model the net stock coincides with the on-hand stock

3. **Inventory position** = (net stock) + (on-order stock). The on-order stock is stock which has been already requisitioned but not yet received.

An inventory control rule should be based on the inventory position and not on the net stock. If net stock was used in deciding on when to order, we might unnecessarily place another order today while a large shipment was due in tomorrow.

The inventory control models considered in this section are the continuous-review (s, Q) model with small individual demands, the periodic-review (R, S) model, the periodic-review (R, s, S) model and the continuous-review (s, S) model with compound Poisson demand. A probabilistic analysis of these models will be given. In particular, the service measure of the long-run fraction of demand satisfied directly from stock on hand will be considered. In general it is not possible to obtain tractable exact results (except for the continuous-review (s, Q) model with Poisson demands, cf. Exercise 1.23 in Chapter 1 and Exercise 2.25 in Chapter 2). In order to obtain implementable results we have to compromise between mathematical and practical standpoints. The analysis will therefore be approximative, but it will be backed up by sound probabilistic principles from renewal theory. The heuristic analysis leads not only to practically useful results, but it is also quite flexible. Essentially the same analysis applies to each of the models considered in this section. The heuristic analysis is first given for the backlog case, but needs only minor modifications for the lost-sales case.

1.5.1 The continuous-review (s, Q) inventory model

The assumptions of this frequently used inventory model are as follows:

1. Continuous review of the inventory, that is, the stock status is continuously monitored and is updated each time a transaction occurs.

2. The individual demand transactions are small so that the inventory level can be treated as a continuous variable.

3. A replenishment order of size Q is placed each time the inventory position drops to the reorder point s.

4. The lead time of any replenishment order is a positive constant L.

5. The demands in disjoint time intervals can be treated as independent random variables.

These assumptions, in one way or another, represent approximations to a real situation. Nevertheless, the model and its heuristic solution below have proved to be extremely useful in practice. The heuristic solution can be extended to the case of stochastic lead times. In the heuristic analysis the random demand process has to be specified only through the random variable

$$\xi_L = \text{the total demand in the replenishment lead time.}$$

It is assumed that the demand variable ξ_L has a continuous distribution with probability density $f_L(x)$. The mean and the standard deviation of the leadtime demand are denoted by μ_L and σ_L. It will be seen that we need in fact only these two system characteristics for the heuristic solution. An appropriate measure for service provided to the customers is the long-run fraction of demand satisfied directly from stock on hand. This service measure is more useful for managerial purposes than the probability of not running out of a stock during a replenishment lead time.

For a given (s, Q) policy with $s \geq 0$, define

$$\beta(s, Q) = \text{the long-run fraction of demand satisfied directly from stock on hand.}$$

Let us first consider the backlog case. To derive an expression for the service level $\beta(s, Q)$ of an (s, Q) policy, we define a cycle as the time elapsed between two consecutive epochs at which a replenishment order is received. Although we cannot apply the standard theory of regenerative processes, the following basic formula applies:

the long-run fraction of demand not met directly from stock on hand

$$= \frac{E[\text{amount of demand that goes short in one cycle}]}{E[\text{total demand in one cycle}]}.$$

In the backlog case any demand is ultimately satisfied. Hence the average demand per cycle must be equal to the average amount received per cycle. This gives

$$E[\text{total demand in one cycle}] = Q.$$

Define now the basic quantities

$$B_1 = E[\text{shortage present at the beginning of a cycle}]$$

and
$$B_2 = E[\text{shortage present at the end of a cycle}].$$

In other words, B_1 is the expected shortage just after a replenishment order has been received and B_2 is the expected shortage just prior to the arrival of a replenishment order. Then

$$\beta(s, Q) = 1 - \frac{(B_2 - B_1)}{Q}. \tag{1.5.1}$$

To find expressions for B_1 and B_2, we need the key result:

$$\text{the net stock just before a replenishment comes in} = s - \xi_L. \tag{1.5.2}$$

The proof of this result is simple. Tag one of the replenishment orders. In view of the assumption of constant lead times, replenishment orders will be received in the same order as they were placed. Hence any stock that was still on order just before the placing of the tagged order has arrived when the tagged order comes in. The inventory position was s at the moment the tagged order was placed. Noting that excess demand is backlogged, it now follows that the net stock just prior to the arrival of the tagged order equals s minus the total demand in the lead time of the tagged order. This verifies the relation (1.5.2). Since the size of each replenishment order equals Q, it follows from (1.5.2) that

$$\text{the net stock just after a replenishment order has been received} = s - \xi_L + Q. \tag{1.5.3}$$

Let x^+ be a shorthand notation for $\max(x, 0)$. Using (1.5.2) and (1.5.3), we now find

$$B_1 = E[(\xi_L - s - Q)^+] = \int_{s+Q}^{\infty} (x - s - Q) f_L(x) \, dx$$

and

$$B_2 = E[(\xi_L - s)^+] = \int_{s}^{\infty} (x - s) f_L(x) \, dx.$$

Hence the service level $\beta(s, Q)$ can be expressed as

$$\beta(s, Q) = 1 - \frac{1}{Q} \left[\int_{s}^{\infty} (x - s) f_L(x) \, dx - \int_{s+Q}^{\infty} (x - s - Q) f_L(x) \, dx \right]. \tag{1.5.4}$$

Incidentally, the results (1.5.2) and (1.5.3) can also be used to establish the first-order approximation $\frac{1}{2}(I_1 + I_2)$ for the average stock on hand, where $I_1 = E[(s + Q - \xi_L)^+]$ and $I_2 = E[(s - \xi_L)^+]$. This approximation for the average stock on hand is practically useful only when the service level of the (s, Q) policy is sufficiently high.

Minimization of costs subject to a service level constraint

Suppose that we wish to choose the parameters s and Q such that the long-run average holding and replenishment costs per unit time are minimized subject to the service level constraint that the long-run fraction of demand satisfied directly from stock on hand is at least β , where β is a prespecified number. There is a fixed setup cost $K > 0$ for each replenishment order and a holding cost of $h > 0$ per unit time for each unit of on-hand inventory. Theoretically s and Q should be determined simultaneously by minimizing an expression for the long-run average cost per unit time subject to the constraint $\beta(s, Q) \geq \beta$. However, in practice one uses a much simpler approach. Empirical studies indicate that in most practical situations no serious cost penalty is created when the following sequential approach is used.

Sequential approach

1. Calculate first the order quantity from the economic order quantity (EOQ) formula

$$Q = \sqrt{\frac{2K\mu_1}{h}} \tag{1.5.5}$$

where μ_1 is the average demand per unit time.

2. Determine next the reorder point s by solving the equation

$$\int_s^\infty (x - s) f_L(x)\, dx - \int_{s+Q}^\infty (x - s - Q) f_L(x)\, dx = (1 - \beta)Q. \tag{1.5.6}$$

It turns out that the sequential approach gives no serious deviation from the theoretically minimal costs as long as the economic order quantity Q is larger than the standard deviation σ_L of the lead time demand.

A warning with respect to formula (1.5.6) should be made. Many texts omit the second integral in (1.5.6). This integral represents the term B_1 which is defined as the expected shortage present just after the arrival of a replenishment order. Intuitively, it seems reasonable to neglect the term B_1. However, numerical experiments indicate that for the computation of the reorder point it is hazardous to neglect this term when σ_L/μ_L is not small (say, $\sigma_L/\mu_L > 0.5$) or β is not close to 1 (say, $\beta < 0.9$). In general it is no simple task to solve numerically the equation (1.5.6). The reorder point s can be found by using method of bisection, but each step of this method requires the numerical evaluation of two integrals. In many practical situations, however, the lead time demand can be modelled by a gamma distribution. Then equation (1.5.6) can be routinely solved, since the integrals in the left side of (1.5.6) reduce to incomplete gamma integrals for which fast codes are widely available, see e.g. Press *et al.* (1986). An even simpler numerical procedure for solving (1.5.6) can be given when the lead-time demand is approximately normally distributed.

Normal demand

In practice it is often reasonable to model the lead-time demand by a normal distribution. If the demand comes from a large number of independent sources, a justification for the use of the normal distribution is provided by the central limit theorem. Assume now that the lead-time demand is normally distributed with mean μ_L and standard deviation σ_L. This assumption requires that σ_L/μ_L is not too large (say, $\sigma_L/\mu_L \leq 0.5$); otherwise, there would be a significant probability of negative demand. To simplify equation (1.5.6), we first note that for normally distributed demand

$$\int_a^\infty (x-a) f_L(x)\, dx = E[(\xi_L - a)^+] = \sigma_L E\left[\left\{\frac{(\xi_L - \mu_L)}{\sigma_L} - \frac{(a - \mu_L)}{\sigma_L}\right\}^+\right]$$

$$= \sigma_L I\left(\frac{a - \mu_L}{\sigma_L}\right),$$

where

$$I(z) = \frac{1}{\sqrt{2\pi}} \int_z^\infty (x-z) e^{-\frac{1}{2}x^2}\, dx.$$

The function $I(z)$ is called the normal loss integral and can be computed from

$$I(z) = \phi(z) - z\{1 - \Phi(z)\},$$

where $\Phi(z)$ is the standard normal probability distribution function and $\phi(z)$ is its probability density. Hence for normal demand equation (1.5.6) can be simplified to

$$\sigma_L I\left(\frac{s - \mu_L}{\sigma_L}\right) - \sigma_L I\left(\frac{s + Q - \mu_L}{\sigma_L}\right) = (1 - \beta)Q. \qquad (1.5.7)$$

For practical purposes this equation can be further simplified. In case the required service level β is sufficiently high (say, $\beta \geq 0.9$), numerical studies indicate that the second term in the left side of equation (1.5.7) can be ignored. Thus, using the representation

$$s = \mu_L + k\sigma_L,$$

it suffices to solve the simple equation

$$\sigma_L I(k) = (1 - \beta)Q. \qquad (1.5.8)$$

The factor k is called the safety factor, since $k\sigma_L$ represents the safety stock. The safety stock is in general defined as the expected net stock just prior to the arrival of a replenishment order.

Equation (1.5.8) is one of the most famous formulas in inventory theory. It is very easy to solve numerically. For hand calculations one might wish to use a table for $I(z)$ in order to find k. In software packages for scientific inventory calculations the safety factor k is usually computed by using a rational approximation to the inverse function $I^{-1}(y)$; see also Section 1.5.5.

Illustration

Suppose that the demand process is described by a renewal process. Each demand is for one unit and the interoccurrence time A between two consecutive demands is gamma distributed with mean $E(A) = 0.02$ and coeffient of variation c_A. The replenishment lead time $L = 1$. Since many demands will occur during a lead time, it is reasonable to approximate the lead-time demand by a normal distribution. Using Theorem 1.1.9 and Exercise 1.5 for asymptotic estimates of the first two moments of the number of renewals in the lead time L, we approximate the mean and the standard deviation of the lead-time demand by

$$\mu_L \approx \frac{L}{v_1} + \frac{v_2}{2v_1^2} - 1 \quad \text{and} \quad \sigma_L^2 \approx \frac{(v_2 - v_1^2)}{v_1^3}L + \frac{5v_2^2}{4v_1^4} - \frac{2v_3}{3v_1^3} - \frac{v_2}{2v_1^2},$$

where $v_k = E(A^k)$ is the kth moment of the interoccurrence time A.

Suppose now that the order quantity $Q = 50$ (obtained from the EOQ formula or from other considerations). The goal is to find the reorder point s such that the long-run fraction of demand satisfied directly from stock on hand equals $\beta = 0.99$. Applying formula (1.5.8) for the respective cases of $c_A^2 = 0.5$, 1 and 2, we find the values $s = 55$, 58 and 61 for the reorder point. Since the calculation of the reorder point involves several approximate steps, we use computer simulation to find the actual service levels. In each example we simulate 200,000 customer demands and we find for the reorder point in the respective cases of $c_A^2 = 0.5$, 1 and 2 the actual service levels 0.992 (± 0.001), 0.990 (± 0.001) and 0.991 (± 0.002), where the numbers within brackets indicate the 95% confidence intervals.

Stochastic lead times

The assumption of constant lead times was only used to establish the relation (1.5.2). In the proof of this relation it was essential that replenishment orders are received in the same order as they are placed. Thus for the case of stochastic lead times relation (1.5.2) holds approximately when the probability of orders crossing in time is negligible. Then the results (1.5.4), (1.5.6) and (1.5.8) remain approximately valid with the proper interpretation of $f_L(x)$, μ_L and σ_L.

Lost-sales case

An exact analysis for the lost-sales case is even harder than for the backlog case. However, the heuristic analysis from the previous section needs only slight modifications. The result (1.5.2) was crucial in the analysis of the backlog model. What is the corresponding result for the lost-sales model? It is tempting to state that the net stock just before a replenishment order arrives is exactly equal to $(s - \xi_L)^+$. However, this need not hold when two or more orders can be simultaneously outstanding. Nevertheless, it is reasonable to take $(s - \xi_L)^+$ as an approximation for the net stock just before the arrival of an order. Thus

$$E[\text{demand that goes short in one cycle}] \approx E[(\xi_L - s)^+].$$

Hence for the fraction of demand satisfied directly from stock on hand we find the expression

$$\beta(s, Q) \approx 1 - \frac{E[(\xi_L - s)^+]}{E[\text{total demand in one cycle}]}.$$

To find $E[\text{total demand in one cycle}]$, note that the total demand in one cycle is the sum of the total demand satisfied in one cycle and the total demand lost in one cycle. Thus

$$E[\text{total demand in one cycle}] \approx Q + E[(\xi_L - s)^+].$$

Hence for the lost-sale case we find for the fraction of demand satisfied directly from stock on hand the expression

$$\beta(s, Q) \approx 1 - \frac{\int_s^\infty (x - s) f_L(x) \, dx}{Q + \int_s^\infty (x - s) f_L(x) \, dx}.$$

In particular, formula (1.5.6) should be modified for the lost-sales case as

$$\int_s^\infty (x - s) f_L(x) \, dx = \frac{1 - \beta}{\beta} Q. \tag{1.5.9}$$

For the special case of normal demand this formula reduces to

$$\sigma_L I(k) = \frac{1 - \beta}{\beta} Q, \tag{1.5.10}$$

where k is related to the reorder point s by $s = \mu_L + k\sigma_L$.

1.5.2 The periodic-review (R, S) inventory model

In this model the inventory position is reviewed only periodically. A replenishment order can only be placed at the review instants. The specific assumptions of the periodic-review (R, S) model are as follows:

1. The inventory position is reviewed every R time periods, where R is a given positive integer.

2. At each review the inventory position is ordered up to the level S, where the order-up-to level S is positive.

3. The lead time of a replenishment order is L periods, where L is a given non-negative integer.

4. The demands in the time periods $t = 1, 2, \ldots$ are independent random variables having a common probability density $f_1(x)$ with mean μ_1 and standard deviation σ_1.

The analysis of this model is very similar to that of the continuous-review (s, Q) model. The heuristic analysis is again based on the study of the inventory process during a single cycle. As before a cycle is defined as the time interval between two consecutive epochs at which a replenishment order is received. Let the random variable T_k be defined as

$$T_k = \text{the total demand in } k \text{ periods.}$$

The probability density of T_k is denoted by $f_k(x)$. The mean and standard deviation of $f_k(x)$ are given by $\mu_k = k\mu_1$ and $\sigma_k = \sqrt{k}\sigma_1$. For a given (R, S) policy with $S > 0$, let

$\beta(R, S) = \text{the long-run fraction of demand satisfied directly from stock on hand.}$

We first analyse the backlog case. To do so, we need the key result that the net stock just before a replenishment order arrives is distributed as $S - T_{R+L}$. To derive this result, tag one of the replenishment orders. The inventory position just prior to the placing of the tagged order is $S - V$, where the random variable V is the cumulative demand since the previous review. Note that V is distributed as T_R. The assumption of constant lead times implies that any stock that was on order prior to the placing of the tagged order has arrived when the tagged order comes in. Thus, using the assumption of backlogging of excess demand, the net stock just before the tagged order comes in is $S - V - W$, where the random variable W is the cumulative demand in the replenishment lead time of the tagged order. Note that W is distributed as T_L and that V and W are independent of each other. Consequently $V + W$ is distributed as T_{R+L}. This proves that the net stock just before an order arrives is distributed as $S - T_{R+L}$. Further, since the size of any replenishment order is distributed as T_R, we have that the net stock just after the arrival of a replenishment order is distributed as $S - T_L$. We can now conclude that

$$E[\text{shortage present at the beginning of a cycle}] = E[(T_L - S)^+]$$

and

$$E[\text{shortage present at the end of a cycle}] = E[(T_{R+L} - S)^+]$$

Hence we find that

$$\beta(R, S) = 1 - \frac{E[\text{amount of demand that goes short in one cycle}]}{E[\text{total demand in one cycle}]}$$

$$= 1 - \frac{E[(T_{R+L} - S)^+] - E[(T_L - S)^+]}{R\mu_1},$$

using that $E[\text{total demand in one cycle}] = E[\text{order size}] = R\mu_1$. This leads to the expression

$$\beta(R, S) = 1 - \frac{1}{R\mu_1}\left[\int_S^\infty (x - S)f_{R+L}(x)\,dx - \int_S^\infty (x - S)f_L(x)\,dx\right].$$

If we wish to determine the order-up-to level S such that the long-run fraction of demand satisfied directly from stock on hand is at least β, then we have to 'solve the equation

$$\int_S^\infty (x - S) f_{R+L}(x) \, dx - \int_S^\infty (x - S) f_L(x) \, dx = (1 - \beta) R \mu_1. \qquad (1.5.11)$$

In case $\sigma_{R+L}/\mu_{R+L} \leq 0.5$ and $\beta \geq 0.9$, it turns out that the second term in the left side of (1.5.11) can be ignored for practical computations. Thus for the particular case of normal demand the equation (1.5.11) simplifies to

$$\sigma_{R+L} I(k) = (1 - \beta) R \mu_1 \qquad (1.5.12)$$

when we use the representation

$$S = \mu_{R+L} + k \sigma_{R+L}.$$

The function $I(k)$ denotes the normal loss integral introduced before. The resemblance between formula (1.5.12) for the periodic-review (R, S) model and formula (1.5.8) for the continuous-review (s, Q) model is striking. The only difference is that in the periodic-review model the safety stock $k \sigma_{R+L}$ should provide protection against fluctuations in the demand during the review time plus lead time rather than against fluctuations in the lead-time demand only as in the continuous-review model.

Stochastic lead times

The above results obtained for the case of constant lead times can also be used for the case of a stochastic lead time L provided that the probability of replenishment orders crossing in time is zero or negligible. In the formulas the densities $f_{R+L}(x)$ and $f_L(x)$ should then be replaced by the densities $\eta(x)$ and $\xi(x)$ of the random variables η and ξ defined as

$$\eta = \text{the total demand in the review time plus lead time}$$

$$\xi = \text{the total demand in the lead time.}$$

The quantities μ_{R+L} and σ_{R+L} should be replaced by $E(\eta)$ and $\sigma(\eta)$. Obviously,

$$E(\eta) = R\mu_1 + E(\xi) \quad \text{and} \quad \sigma(\eta) = \sqrt{R\sigma_1^2 + \sigma^2(\xi)}. \qquad (1.5.13)$$

Using the formulas (A.11) and (A.12) in Appendix A, we have

$$E(\xi) = E(L)\mu_1 \quad \text{and} \quad \sigma^2(\xi) = E(L)\sigma_1^2 + \sigma^2(L)\mu_1^2. \qquad (1.5.14)$$

Lost-sales case

In accordance with the formulas (1.5.9) and (1.5.10), we modify the formulas (1.5.11) and (1.5.12) as follows for the lost-sales case:

$$\int_S^\infty (x - S) f_{R+L}(x) \, dx = \frac{(1 - \beta)}{\beta} R\mu_1$$

and

$$\sigma_{R+L} I(k) = \frac{(1-\beta)}{\beta} R\mu_1.$$

These formulas should be used only when β is close to 1. The modification is in agreement with the empirical finding that the lost-sales model typically requires a smaller reorder point than the backlog model in order to achieve the same service level.

1.5.3 The periodic-review (R, s, S) inventory model

The periodic-review (R, s, S) inventory model differs from the periodic-review (R, S) model only in the fact that a replenishment order is not placed at every review. A replenishment order is placed at a review only when the inventory position is at or below the reorder level $s (\geq 0)$; otherwise, no replenishment order is placed. Each replenishment order raises the inventory position to S. The other assumptions of the model are the same as in the periodic-review (R, S) model. Also, in the analysis below the notation is the same as used in the previous section. In the following it is assumed that

$$S - s \gg \mu_R,$$

that is, the difference between the order-up-to level S and the reorder level s should be sufficiently large relative to the mean review-time demand. How large $S - s$ should be compared with μ_R depends to some extent on the coefficient of variation of the review-time demand; cf. relation (1.1.14) in Section 1.1.3. In most practical situations $S - s \geq 1.5\mu_R$ should suffice. The above assumption is essential for the approximate analysis in order to apply a basic result from renewal theory. The approximate analysis for the periodic-review (R, s, S) inventory model is more subtle than the analysis for the previous models, since we have to incorporate the distribution of the undershoot of the reorder point s when a replenishment order is placed. The service level $\beta(R, s, S)$ of an (R, s, S) policy is defined by

$\beta(R, s, S) = $ the long-run fraction of demand satisfied directly from stock on hand.

We first analyse the backlog model. For that we need some preliminary results. Define the random variable U by

$U = $ the undershoot of the reorder point s when a replenishment order is triggered.

In view of the assumption $S - s \gg \mu_R$, we approximate the probability distribution function and the mean of U by

$$P\{U \leq x\} \approx \frac{1}{\mu_R} \int_0^x \{1 - F_R(y)\} \, dy \quad \text{for } x \geq 0 \qquad (1.5.15)$$

and

$$E(U) \approx \frac{\sigma_R^2 + \mu_R^2}{2\mu_R},\tag{1.5.16}$$

where $F_R(x)$ is the probability distribution function of the review-time demand and μ_R and σ_R denote the mean and standard deviation of the review-time demand. These approximations are justified by Corollary 1.1.10 and Theorem 1.1.11 in Section 1, since the random variable U can be seen as the excess variable in a renewal process in which the (continuous) interoccurrence times are distributed as the review-time demands. The analysis below will be greatly facilitated by the fact that the distribution of the undershoot variable U is approximately independent of the values of s and S provided that $S - s$ is sufficiently large relative to the mean review-time demand.

As before, define the random variables ξ and η by

ξ = the total demand in the lead time of a replenishment order

η = the total demand in a review time plus lead time.

The next result is essential for our analysis.

Theorem 1.5.1   (a) *The net stock just before a replenishment order arrives is distributed as $s - U - \xi$, while the net stock just after the arrival of a replenishment order is distributed as $S - \xi$.*

(b) *The probability density $h(x)$ of $P\{U + \xi \leq x\}$ is approximately given by*

$$h(x) \approx \frac{1}{\mu_R}[P\{\xi \leq x\} - P\{\eta \leq x\}] \quad for \ x \geq 0.$$

Proof   (a) Tag a replenishment order. The inventory position just prior to the placing of the tagged order is distributed as $s - U$. Since the replenishment orders are received in the same order as they are placed, the net stock just prior to the arrival of the tagged order is distributed as $s - U - \xi$. Note that U and ξ are independent of each other, since ξ is the total demand in the lead time of the tagged order and U is determined by demand that occurred before the order was placed. The result that the net stock just after the arrival of the tagged order is distributed as $S - \xi$ follows from the observation that the inventory position just after the placing of the tagged order is S.

(b) Denote by $f_U(x)$ the density of U. Using (1.5.15) and the convolution formula (A.6) in Appendix A, we find

$$P\{U + \xi \leq x\} = \int_0^x P\{\xi \leq x - y\}f_U(y)\,dy$$

$$\approx \frac{1}{\mu_R}\int_0^x P\{\xi \leq x - y\}\{1 - F_R(y)\}\,dy$$

$$= \frac{1}{\mu_R}\int_0^x P\{\xi \leq z\}\,dz - \frac{1}{\mu_R}\int_0^x P\{\xi \leq z\}F_R(x - z)\,dz.$$

Thus

$$h(x) \approx \frac{1}{\mu_R} P\{\xi \le z\} - \frac{1}{\mu_R} \int_0^x P\{\xi \le z\} f_R(x - z) \, dz - \frac{1}{\mu_R} P\{\xi \le z\} F_R(0).$$

The third term on the right side of this relation vanishes since $F_R(0) = 0$. Write the integral $\int_0^x P\{\xi \le z\} f_R(x - z) \, dz$ as $\int_0^x P\{\xi \le x - y\} f_R(y) \, dy$. Thus the integral can be interpreted as $P\{\xi + Y \le x\}$, where Y is distributed as the review-time demand and Y is independent of ξ. This gives the desired result.

Two remarks are in order. First, the function approximating the density $h(x)$ is a true probability density. Second, Theorem 1.5.1 remains valid for the case of stochastic lead times provided that the probability of replenishment orders crossing in time is zero. In the remainder of the discussion the case of stochastic lead times will be considered, where it is assumed that the probability of orders crossing in time is zero or negligible. The heuristic solution will involve the probability densities

$$\xi(x) = \text{the probability density of the total demand in a lead time}$$

and

$$\eta(x) = \text{the probability density of the total demand in a review time plus lead time.}$$

Note that $\xi(x) = f_L(x)$ and $\eta(x) = f_{R+L}(x)$ for the case of a constant lead time L, where $f_k(x)$ is the density of the total demand in k periods. The mean and the standard deviation of each of the densities $\eta(x)$ and $\xi(x)$ are given by (1.5.13) and (1.5.14). In practical applications it usually suffices to know only these moments. A gamma or a normal distribution is then fitted to the first two moments. To conclude the preparatory analysis, we state the following lemma.

Lemma 1.5.2 *Assuming that the density $f_1(x)$ of the one-period demand has a finite third moment, then*

$$\int_s^\infty (x - s) h(x) \, dx \approx \frac{1}{2\mu_R} \left\{ \int_s^\infty (x - s)^2 \eta(x) \, dx - \int_s^\infty (x - s)^2 \xi(x) \, dx \right\}.$$

Proof The assumption of the lemma implies that $E(U + \xi)^3 < \infty$ and so $x^2 h(x) \to 0$ as $x \to \infty$. Hence, using partial integration,

$$\int_s^\infty (x - s) h(x) \, dx = \frac{1}{2} \int_s^\infty h(x) \, d(x - s)^2 = -\frac{1}{2} \int_s^\infty (x - s)^2 h'(x) \, dx.$$

Substituting $h'(x) \approx (1/\mu_R)[\xi(x) - \eta(x)]$ yields the desired result.

Approximations

Define a cycle again as the time elapsed between two consecutive epochs at which replenishment orders are received. Then, using Theorem 1.5.1(a), it follows that

$$\beta(R, s, S) = 1 - \frac{E[\text{amount of demand that goes short in one cycle}]}{E[\text{total demand in one cycle}]}$$

$$= 1 - \frac{E[(U + \xi - s)^+] - E[(\xi - S)^+]}{E[\text{order size}]}.$$

By (1.5.16), we have

$$E[\text{order size}] \approx S - s + \frac{(\sigma_R^2 + \mu_R^2)}{2\mu_R}$$

Hence, using Theorem 1.5.1(b) and Lemma 1.5.2, we find for the fraction of demand not satisfied directly from stock on hand the expression

$$1 - \beta(R, s, S)$$

$$\approx \frac{(2\mu_R)^{-1} \left[\int_s^\infty (x - s)^2 \eta(x)\, dx - \int_s^\infty (x - s)^2 \xi(x)\, dx \right] - \int_S^\infty (x - S)\xi(x)\, dx}{S - s + \frac{1}{2}(\sigma_R^2 + \mu_R^2)/\mu_R}.$$

Minimizing costs subject to a service level constraint

Suppose we wish to find s and S in order to minimize the average ordering and holding cost subject to the requirement that the fraction of demand satisfied directly from stock on hand must be at least β, where $0 < \beta < 1$ is a prespecified number. There is a fixed ordering cost of $K > 0$ for each replenishment order and there is a holding cost of $h > 0$ per unit time for each unit of on-hand inventory. Instead of taking s and S as variables, we take s and $S - s$ as variables. The difference $S - s$ is related to the order quantity. Empirical investigations indicate that in practical situations the optimal value of $S - s$ is fairly independent of the required service level β provided that β is not too small. This is in support of a heuristic approach in which $S - s$ and s are determined sequentially. The determination of $S - s$ is based on cost considerations, while the reorder point s is determined on the basis of the service level requirement. It turns out that for the choice

$$S - s = \sqrt{\frac{2K\mu_1}{h}}$$

the average holding and replenishment costs are often quite close to the theoretically minimum average costs. The sequential approach will now be given under the assumption that the required service level β is close to 1 (say, $\beta \geq 0.90$) and the lead-time demand ξ is not highly variable. Then numerical investigations indicate that the third integral in the above expression for $\beta(R, s, S)$ can be neglected. This third integral gives the expected shortage just after the arrival of a replenishment order. The sequential approach now proceeds as follows.

Sequential approach

1. Determine $S - s$ according to the above EOQ formula (or on the basis of other considerations).
2. Calculate the reorder point s as the solution of the equation

$$\int_s^\infty (x - s)^2 \eta(x)\, dx - \int_s^\infty (x - s)^2 \xi(x)\, dx = (1 - \beta)\gamma, \qquad (1.5.17)$$

where the constant γ is given by

$$\gamma = 2\mu_R\{S - s + \tfrac{1}{2}(\sigma_R^2 + \mu_R^2)/\mu_R\}.$$

Equation (1.5.17) can be numerically solved by bisection. In many practical inventory situations gamma densities can be fitted to the demand densities $\eta(x)$ and $\xi(x)$ by matching the first two moments. Then the integrals in (1.5.17) reduce to incomplete gamma integrals for which fast codes are widely available. An even simpler numerical procedure can be given when the demand variables η and ξ can be modelled by normal distributions.

Normal demand

In many cases of practical interest the demand densities $\eta(x)$ and $\xi(x)$ can be approximated by normal densities. A necessary condition for this approximation is

$$\frac{\sigma(\eta)}{E(\eta)} \le 0.5 \quad \text{and} \quad \frac{\sigma(\xi)}{E(\xi)} \le 0.5,$$

since otherwise the probability of negative demand is not negligible. The first two moments of the densities $\eta(x)$ and $\xi(x)$ are given by (1.5.13) and (1.5.14) in terms of the first two moments of the one-period demand and the replenishment lead time.

The integrals in formula (1.5.17) can be considerably simplified for the case of normal demand. For any $N(\mu, \sigma^2)$ distributed random variable X with density $g(x)$, we have

$$\int_a^\infty (x - a)^2 g(x)\, dx = E[\{(X - a)^+\}^2]$$

$$= \sigma^2 E\left[\left\{\left(\frac{(X - \mu)}{\sigma} - \frac{(a - \mu)}{\sigma}\right)^+\right\}^2\right]$$

$$= \sigma^2 J\left(\frac{a - \mu}{\sigma}\right),$$

where the function $J(z)$ is defined by

$$J(z) = \frac{1}{\sqrt{2\pi}} \int_z^\infty (x - z)^2 e^{-(1/2)x^2}\, dx.$$

The function $J(z)$ is very easy to evaluate; see Section 1.5.5. Hence for normal demand equation (1.5.17) can be written as

$$\sigma^2(\eta) J \left(\frac{s - E(\eta)}{\sigma(\eta)} \right) - \sigma^2(\xi) J \left(\frac{s - E(\xi)}{\sigma(\xi)} \right) = (1 - \beta)\gamma. \qquad (1.5.18)$$

This equation can be rapidly solved by bisection since every function evaluation of $J(z)$ is very cheap. A further simplification of equation (1.5.18) is possible when the review time R is not too small compared with the expected lead time (say, $R \geq \frac{1}{2}E(L)$). Then the second term in the left side of (1.5.18) can also be neglected. Using the representation

$$s = E(\eta) + k\sigma(\eta)$$

we then obtain the simplified equation

$$\sigma^2(\eta) J(k) = (1 - \beta)\gamma. \qquad (1.5.19)$$

This equation has the advantage that it can be numerically solved by a single-pass calculation using a rational approximation to the inverse function $J^{-1}(y)$; see Section 1.5.5.

Lost-sales case

In the lost-sales case the long-run fraction of demand satisfied directly from stock on hand is approximated by

$$\beta(R, s, S) \approx 1 - \frac{E[(U + \xi - s)^+]}{E[(U + \xi - s)^+] + S - s + \frac{1}{2}(\sigma_R^2 + \mu_R^2)/\mu_R}$$

for a given (R, s, S) policy. This approximation is only justified when the service level of the (R, s, S) policy is close to 1. In that case the average order size in the lost-sales model is nearly the same as in the backlog model. The denominator of the above ratio is next motivated by splitting the total demand in one cycle into the demand lost in one cycle and the demand satisfied in one cycle, and noting that the average demand satisfied in one cycle equals the average order size. Thus for the lost-sales case the formulas (1.5.17) and (1.5.19) should be modified as

$$\int_s^\infty (x - s)^2 \eta(x)\, dx - \int_s^\infty (x - s)^2 \xi(x)\, dx = \frac{1 - \beta}{\beta}\gamma$$

and

$$\sigma^2(\eta) J(k) = \frac{1 - \beta}{\beta}\gamma.$$

Numerical results

Let us assume that the one-period demand has a negative binomial distribution

$$p_j = \binom{r+j-1}{j} p^r (1-p)^j, \quad j = 0, 1, \ldots .$$

The parameters r and p of this distribution are uniquely determined by $p = \mu_1/\sigma_1^2$ and $r = p\mu_1/(1-p)$ provided that $\sigma_1^2/\mu_1 > 1$. In all examples we take

$$\sigma_1^2/\mu_1 = 3, \quad \mu_1 = 16, \quad R = 1 \quad \text{and} \quad S - s = 34,$$

where the value $S - s = 34$ corresponds to the economic order quantity for a fixed ordering cost of $K = 36$ and a linear holding cost of $h = 1$. Two cases are considered for the replenishment lead time:

Case (a) $P\{L = 2\} = 1$ $[E(L) = 2, \ \sigma^2(L) = 0]$,
Case (b) $P\{L = 1\} = \frac{1}{4}, \ P\{L = 2\} = \frac{1}{2}, \ P\{L = 3\} = \frac{1}{4}$ $[E(L) = 2,$
$\sigma^2(L) = \frac{1}{2}]$.

For these two respective cases the squared coefficients of variation of the demand variables ξ and η are given by

$$c_\xi^2 = 0.094, \quad c_\eta^2 = 0.063 \quad \text{for case (a),}$$

$$c_\xi^2 = 0.219, \quad c_\eta^2 = 0.118 \quad \text{for case (b).}$$

The required service level β is varied as $\beta = 0.90, 0.95$ and 0.99. In Table 1.5.1 three approximations 'gamma, normal and true' are given for the reorder point s in the backlog case. The gamma approximation corresponds to the solution of equation (1.5.17) in which gamma densities are fitted to $\eta(x)$ and $\xi(x)$, the normal approximation corresponds to the solution of the simplified equation (1.5.19) and the true approximation corresponds to the solution of the discrete version of the equation (1.5.17) in which the true discrete distributions of η and ξ are used.

Table 1.5.1 Approximate reorder points and the actual service levels

	$\beta = 0.90$		$\beta = 0.95$		$\beta = 0.99$	
Case (a)	(s, S)	$\beta\ (s, S)$	(s, S)	$\beta\ (s, S)$	(s, S)	$\beta\ (s, S)$
gamma	(42,76)	0.900	(49,83)	0.954	(61,95)	0.991
normal	(43,77)	0.909	(48,82)	0.948	(59,93)	0.988
true	(42,76)	0.900	(48,82)	0.948	(60,94)	0.989
Case (b)						
gamma	(47,81)	0.903	(55,89)	0.952	(73,107)	0.993
normal	(48,82)	0.910	(54,88)	0.947	(67,101)	0.986
true	(47,81)	0.902	(55,89)	0.951	(70,104)	0.990

The approximate values of s obtained by using a continuous demand distribution are rounded to the nearest integer. The table also gives the exact value of the service level $\beta(s, S) = \beta(R, s, S)$ for the various (R, s, S) policies when the demand follows a negative binomial distribution. These exact values are computed by using Markov-chain methods (constant lead times) and computer simulation (stochastic lead times). The numerical investigations indicate that the approximations perform very satisfactorily for practical purposes. It also appears that the use of only the first two moments of the demand variables η and ξ is justified as long their coefficients of variation are not too large (say, $0 \le c_\eta^2$, $c_\xi^2 \le 0.25$). For erratic demand the tail behaviour of the demand densities $\eta(x)$ and $\xi(x)$ becomes more essential, particularly when the required service level β is very close to 1. Additional information about the distributional form of the demand densities should then guide the choice of a fitting distribution. Fortunately, in many cases of practical interest the normal or gamma distribution can be used to model the demand distribution.

1.5.4 The continuous-review (s, S) inventory model

The assumptions of this model are as follows:

1. Continuous review of the inventory status.
2. The demand process is a compound Poisson process. That is, customers arrive according to a Poisson process and the demands of the customers are independent and identically distributed random variables.
3. A replenishment order is placed to raise the inventory position to the order-up-to level S each time the inventory position falls at or below the reorder point s; otherwise, no ordering is done.
4. The lead times of the replenishment orders are positive and independent of each other.

Further, it is assumed that $S - s$ is sufficiently large relative to the mean demand of an individual customer. The customer demands are supposed to be non-unit sized. In case the replenishment lead time L is stochastic, we make the assumption that for the relevant (s, S) policies the probability of replenishment orders crossing in time is negligible.

The heuristic analysis of the continuous-review (s, S) inventory model is identical to that of the periodic-review (R, s, S) model with $R = 1$. The role of the review-time demand is now played by the demand of an individual customer. In fact the review time is now the time between two consecutive arrivals of customers and is continuously distributed rather than fixed. Thus the heuristic formulas for the continuous-review (s, S) model follow by putting $R = 1$ in the formulas for the periodic-review (R, s, S) model and adapting in these formulas the expressions for $E(\eta)$, $\sigma(\eta)$, $E(\xi)$ and $\sigma(\xi)$. These expressions become

$$E(\eta) = \mu_1 + E(\xi), \qquad \sigma(\eta) = \sqrt{\sigma_1^2 + \sigma^2(\xi)}$$

and

$$E(\xi) = \lambda E(L)\mu_1, \quad \sigma^2(\xi) = \lambda E(L)(\sigma_1^2 + \mu_1^2) + \lambda^2 \mu_1^2 \sigma^2(L),$$

where λ denotes the average arrival rate of customers and μ_1 and σ_1 are the mean and standard deviation of the demand of an individual customer. The expressions for $E(\xi)$ and $\sigma^2(\xi) = E(\xi^2) - E^2(\xi)$ are easily derived from the formula (1.2.8) in Section 1.2.2.

Just as in the periodic-review model, the approximations for the continuous-review (s, S) model show an excellent performance. However, it should be pointed out that it is hazardous to use the simplified normal approximation (1.5.19). The reason for this is that this approximation requires that the mean customer demand is not too small relative to the mean lead-time demand. This requirement will often not be satisfied.

1.5.5 Rational approximations for inventory calculations

This section summarizes a number of useful facts for the standard normal distribution. The standard normal probability density and the standard normal probability distribution function are denoted by

$$\phi(x) = \frac{1}{\sqrt{2\pi}}e^{-(1/2)x^2} \quad \text{and} \quad \Phi(x) = \frac{1}{\sqrt{2\pi}}\int_{-\infty}^{x} e^{-(1/2)y^2} dy.$$

Computation of Φ (x)

The computation of $\Phi(x)$ looks like a formidable job, but offers no difficulty in practice. This function can be computed in any desired accuracy using a rational approximation (a rational function is the ratio of two polynomials). An approximation that is accurate enough for most practical purposes is the following one:

$$\Phi(x) = 1 - \tfrac{1}{2}(1 + d_1 x + d_2 x^2 + d_3 x^3 + d_4 x^4 + d_5 x^5 + d_6 x^6)^{-16} + \varepsilon(x), \quad x \geq 0,$$

where $|\varepsilon(x)| < 1.5 \times 10^{-7}$ and

$$\begin{aligned}
d_1 &= 0.049\ 867\ 3470 & d_4 &= 0.000\ 038\ 0036 \\
d_2 &= 0.021\ 141\ 0061 & d_5 &= 0.000\ 048\ 8906 \\
d_3 &= 0.003\ 277\ 6263 & d_6 &= 0.000\ 005\ 3830.
\end{aligned}$$

This formula applies only for $x \geq 0$. However, using the relation

$$\Phi(x) = 1 - \Phi(-x) \quad \text{for } x < 0,$$

the formula can also be used to compute $\Phi(x)$ for $x < 0$.

Computation of the inverse of Φ (x)

A frequently occurring problem is to solve the equation

$$\Phi(k) = \alpha,$$

where α is a given number between 0 and 1. The percentile k can be found in a single-pass calculation using a rational approximation to the inverse function $\Phi^{-1}(\alpha)$. A useful approximation is

$$k = w - \frac{c_0 + c_1 w + c_2 w^2}{1 + d_1 w + d_2 w^2 + d_3 w^3} + \varepsilon(\alpha), \quad 0.5 \le \alpha < 1,$$

where $|\varepsilon(\alpha)| < 4.5 \times 10^{-4}$ and

$$w = \sqrt{-\ln(1-\alpha)^2}$$
$$c_0 = 2.515\ 517 \qquad d_1 = 1.432\ 788$$
$$c_1 = 0.802\ 853 \qquad d_2 = 0.189\ 269$$
$$c_2 = 0.010\ 328 \qquad d_3 = 0.001\ 308.$$

If $0 < \alpha < 0.5$, the solution to the equation $\Phi(k) = \alpha$ can be found by applying the above inversion formula to the equation $\Phi(-k) = 1 - \alpha$.

Normal loss function I(z)

The normal loss function $I(z)$ is defined by

$$I(z) = \frac{1}{\sqrt{2\pi}} \int_z^\infty (x - z) e^{-\frac{1}{2}x^2} \, dx.$$

This integral plays an important role in inventory theory. The function $I(z)$ can be computed from

$$I(z) = \phi(z) - z\{1 - \Phi(z)\} \quad \text{for all } z.$$

Computation of the inverse of I(z)

A basic problem in inventory theory is to solve the equation

$$I(k) = \beta,$$

where β is a given number between 0 and 1. In practical situations the solution k will satisfy $-4 \le k \le 4$. Then the solution k can be obtained in a single-pass calculation using the rational approximation

$$k = \frac{a_0 + a_1 v + a_2 v^2 + a_3 v^3}{b_0 + b_1 v + b_2 v^2 + b_3 v^3 + b_4 v^4} + \varepsilon(\beta),$$

where $|\varepsilon(\beta)| < 2 \times 10^{-4}$ and

$$v = \sqrt{\ln(25/\beta^2)} \qquad b_0 = 1.000\ 0000$$
$$a_0 = -5.392\ 5569 \qquad b_1 = -7.249\ 6485 \times 10^{-1}$$
$$a_1 = 5.621\ 1054 \qquad b_2 = 5.073\ 266\ 22 \times 10^{-1}$$
$$a_2 = -3.883\ 6830 \qquad b_3 = 6.691\ 368\ 68 \times 10^{-2}$$
$$a_3 = 1.089\ 7299 \qquad b_4 = -3.291\ 291\ 14 \times 10^{-3}.$$

Normal function J(z)

The function $J(z)$ is defined by

$$J(z) = \frac{1}{\sqrt{2\pi}} \int_z^\infty (x - z)^2 e^{-\frac{1}{2}x^2} \, dx.$$

The function $J(z)$ can be computed from

$$J(z) = (1 + z^2)\{1 - \Phi(z)\} - z\phi(z) \quad \text{for all } z.$$

Certain inventory applications require the solution of the equation $J(k) = \beta$ with β a given positive number. Instead of using bisection in conjunction with the above formula for $J(z)$, the equation $J(k) = \beta$ may be solved by a single-pass calculation. In case the solution k satisfies $-4 \leq k \leq 4$, the following rational approximation can be used:

$$k = \frac{f_0 + f_1 v + f_2 v^2 + f_3 v^3}{g_0 + g_1 v + g_2 v^2 + g_3 v^3} + \varepsilon(\beta),$$

where $|\varepsilon(\beta)| \leq 2.3 \times 10^{-4}$. For $0 \leq \beta \leq 0.5$,

$$
\begin{aligned}
v &= \sqrt{\ln(1/\beta^2)} \\
f_0 &= -4.188\ 4136 \times 10^{-1} & g_0 &= 1 \\
f_1 &= -2.554\ 6970 \times 10^{-1} & g_1 &= 2.134\ 0807 \times 10^{-1} \\
f_2 &= 5.189\ 1032 \times 10^{-1} & g_2 &= 4.439\ 9342 \times 10^{-2} \\
f_3 &= 0 & g_3 &= -2.639\ 7875 \times 10^{-3},
\end{aligned}
$$

while for $\beta > 0.5$,

$$
\begin{aligned}
v &= \beta \\
f_0 &= 1.125\ 9464 & g_0 &= 1 \\
f_1 &= -1.319\ 0021 & g_1 &= 2.836\ 7383 \\
f_2 &= -1.809\ 6435 & g_2 &= 6.559\ 3780 \times 10^{-1} \\
f_3 &= -1.165\ 0097 \times 10^{-1} & g_3 &= 8.220\ 4352 \times 10^{-3}.
\end{aligned}
$$

The rational approximations for $\Phi(x)$ and $\Phi^{-1}(x)$ are taken from Abramowitz and Stegun (1965) and the rational approximations for $I^{-1}(x)$ and $J^{-1}(x)$ are taken from Schneider (1979).

1.6 LITTLE'S FORMULA

In Example 1.3.5 we dealt with a production process that is turned off when no orders are present and is turned on as soon as N orders have accumulated. The production orders arrive according to a Poisson process with rate λ and the production time of each order has a general distribution with mean μ and standard deviation σ. Denoting by L the long-run average number of orders in the system, it was shown in Example 1.3.4 that

$$L = \rho + \frac{\lambda^2(\sigma^2 + \mu^2)}{2(1 - \rho)} + \frac{N - 1}{2},$$

where $\rho = \lambda\mu$. Taking this formula for granted, one may ask: what is the long-run average amount of time spent in the system per order? Denoting by W the long-run average sojourn time per order, the answer to the question is given by the famous formula

$$L = \lambda W. \tag{1.6.1}$$

This formula is known as *Little's formula* and is generally valid for queueing systems. To see the relation (1.6.1), assume that the system incurs for each order a cost according to a specific rule and use the general relation

> the long-run average cost incurred by the system per unit time
> = (the long-run average number of arrivals per unit time)
> $\times$ (the long-run average cost incurred per arrival). (1.6.2)

Choose the following cost structure. For each order the system incurs a cost at a rate of 1 while the order is in the system. In other words, the cost incurred for a given order is equal to the amount of time the order is in the system. Hence the long-run average cost incurred per arriving order equals the long-run average sojourn time per order. On the other hand, the system incurs a cost at a rate of j when j orders are present. Hence the long-run average cost incurred by the system per unit time gives the long-run average number of orders in the system. Applying relation (1.6.2) yields the desired result (1.6.1). It remains to prove that relation (1.6.2) is valid for the above cost structure. To do so, observe that the stochastic process $\{X(t), \ t \geq 0\}$ describing the number of orders present is regenerative. The regeneration epochs are the epochs at which an arriving order finds the system empty. Thus, by the renewal-reward theorem in Section 1.3,

$$\text{the long-run average cost per unit time} = \frac{E(\text{cost incurred in one cycle})}{E(\text{length of one cycle})}.$$

Similarly, using that the discrete-time process describing the sojourn times of the orders regenerates itself each time an arriving order finds the system empty, we have

$$\text{the long-run average cost per order} = \frac{E(\text{cost incurred in one cycle})}{E(\text{number of arrivals in one cycle})}.$$

The desired result (1.6.2) follows now from these two relations in conjunction with the relation

$$\text{the long-run average arrival rate} = \frac{E(\text{number of arrivals in one cycle})}{E(\text{length of one cycle})}.$$

The above derivation of (1.6.2) is not dependent on the specific problem considered, but uses only that the stochastic processes involved are regenerative and have a finite mean cycle length. Thus the principle (1.6.2) is generally applicable. By choosing appropriate cost structures, one may identify interesting relations. For example, assume that the system incurs a cost at rate 1 while an order is in production. Then the long-run average cost per unit time represents the long-run fraction of time the facility is in use. Since the cost incurred for a given order equals the production time of the order, the long-run average cost per order equals μ. This gives the interesting relation that the long-run fraction of time the facility is in use equals $\lambda\mu$ independently of the value of N.

1.7 POISSON ARRIVALS SEE TIME AVERAGES

In the analysis of queueing (and other) problems one needs sometimes the long-run fraction of time the system is some state and, other times, the long-run fraction of arrivals who find the system in some state. These averages can often be related to each other, but in general they are not equal to each other. However, there is one important exception in which the two averages are the same.

As a prelude, let us consider the following problem. At a processor jobs arrive according to a Poisson process with rate λ. An arriving job is only accepted for execution when the processor is idle. The processing times of the jobs are independent and identically distributed random variables with mean μ. What are the long-run fraction of time the processor is busy and the long-run fraction of jobs that are rejected? These two performance measures are easily obtained by using the renewal-reward theorem. Let us define the following random variables. For any $t \geq 0$, the indicator variable $I(t)$ is defined by

$$I(t) = \begin{cases} 1 & \text{if the processor is busy at time } t, \\ 0 & \text{otherwise.} \end{cases}$$

Also, for $n = 1, 2, \ldots$, the random variable I_n is defined by

$$I_n = \begin{cases} 1 & \text{if the processor is busy just prior to the } n\text{th arrival,} \\ 0 & \text{otherwise.} \end{cases}$$

The continuous-time process $\{I(t)\}$ and the discrete-time process $\{I_n\}$ are both regenerative. The epochs associated with arrivals who find the processor idle are regeneration epochs for each of the two processes. Thus a cycle starts each time an arriving job finds the processor idle. The long-run fraction of time the processor is busy is equal to the expected amount of time the processor is busy during one cycle divided by the expected length of one cycle. The numerator of this ratio is of course equal to μ. Since the Poisson arrival process is memoryless, it follows that the expected length of one cycle equals $\mu + 1/\lambda$. The waiting time from the service completion of a job until the next arrival of a job has the same exponential distribution with mean $1/\lambda$ as the time between two consecutive arrivals. Hence, with probability 1,

$$\text{the long-run fraction of time the processor is busy} = \frac{\mu}{\mu + 1/\lambda} = \frac{\lambda\mu}{1 + \lambda\mu}.$$

The long-run fraction of jobs that are rejected is equal to the expected number of jobs rejected during one cycle divided by the expected number of jobs arriving during one cycle. Thus, with probability 1,

$$\text{the long-run fraction of jobs that are rejected} = \frac{\lambda\mu}{1 + \lambda\mu}.$$

This shows that the long-run fraction of arrivals who find the processor busy is equal to the long-run fraction of time the processor is busy — a remarkable finding! This finding hinges upon the assumption that the arrival process is a Poisson process. To illustrate this, consider a deterministic arrival process. For example, suppose that a job arrives every one minute and that the processing time of a job is uniformly distributed between half a minute and one minute. Then the long-run fraction of time the processor is busy equals 3/4, whereas the long-run fraction of arrivals who find the processor busy is equal to zero.

The above example exhibits a property which is known as "Poisson arrivals see time averages". An intuitive explanation of this property is that Poisson events occur completely randomly in time. In any small time interval of the same length the occurrence of a Poisson event is equally likely.

Next we discuss the property of "Poisson arrivals see time averages" in a broader context. For ease of presentation we use the terminology of Poisson arrivals. However, the results below also apply to Poisson processes in other contexts.

For some specific problem let the continuous-time stochastic process $\{X(t),\ t \geq 0\}$ describe the evolution of the state of a system and let $\{N(t),\ t \geq 0\}$ be a renewal process describing arrivals to that system. As examples:

(a) $X(t)$ is the number of customers present at time t in a queueing system.

(b) $X(t)$ gives both the inventory level and the prevailing production rate at time t in a production/inventory problem with a variable production rate.

It is assumed that the arrival process $\{N(t),\ t \geq 0\}$ can be seen as an exogenous factor to the system and is not affected by the system itself. More precisely, the following assumption is made.

Lack of anticipation assumption   *For each $u \geq 0$ the future arrivals occurring after time u are independent of the history of the process $\{X(t)\}$ up to time u.*

It is not necessary to specify how the arrival process $\{N(t)\}$ precisely interacts with the state process $\{X(t)\}$. Denoting by τ_n the nth arrival epoch, let the random variable X_n be defined by $X(\tau_n^-)$. In other words,

$$X_n = \text{the state of the system just prior to the } n\text{th arrival epoch}.$$

Let B be any set of states for the $\{X(t)\}$ process. For each $t \geq 0$, define the indicator variable

$$I_B(t) = \begin{cases} 1 & \text{if } X(t) \in B, \\ 0 & \text{otherwise.} \end{cases}$$

Also, for each $n = 1, 2, \ldots$, define the random variable $I_n(B)$ by

$$I_n(B) = \begin{cases} 1 & \text{if } X_n \in B, \\ 0 & \text{otherwise.} \end{cases}$$

The technical assumption is made that the sample paths of the continuous-time process $\{I_B(t),\ t \geq 0\}$ are right-continuous and have left-hand limits. In practical situations this assumption is always satisfied.

Theorem 1.7.1 (Poisson arrivals see time averages)   *Suppose that the arrival process $\{N(t)\}$ is a Poisson process with rate λ. Then:*
  (a) For any $t > 0$,

   E[number of arrivals in $(0, t)$ finding the system in the set B]

$$= \lambda E \left[\int_0^t I_B(u)\, du \right].$$

(b) With probability 1,

$$\lim_{n \to \infty} \frac{1}{n} \sum_{k=1}^n I_k(B) = \lim_{t \to \infty} \frac{1}{t} \int_0^t I_B(u)\, du.$$

Proof   See Wolff (1982).

   In words, part (b) of Theorem 1.7.1 states that the long-run fraction of arrivals who find the system in the set B of states is equal to the long-run fraction of time the system is in the set B of states. This statement is usually abbreviated as PASTA (Poisson arrivals see time averages).
   Theorem 1.7.1 has a useful corollary when it is assumed that the continuous-time process $\{X(t),\ t \geq 0\}$ is a regenerative process whose cycle length has a finite and positive mean. Define the random variables T_B and N_B by

T_B = amount of time the process $\{X(t)\}$ is in the set B of states during
       one cycle,
N_B = number of arrivals during one cycle who find the process $\{X(t)\}$ in
       the set B of states.

The following corollary will be very useful in the algorithmic analysis of queueing systems in Chapter 4.

Corollary 1.7.2   *If the arrival process $\{N(t)\}$ is a Poisson process with rate λ, then*

$$E(N_B) = \lambda E(T_B).$$

Proof Denote by the random variables T and N the length of one cycle and the number of arrivals during one cycle. Then, by Theorem 1.3.2,

$$\lim_{t \to \infty} \frac{1}{t} \int_0^t I_B(u)\, du = \frac{E(T_B)}{E(T)} \quad \text{with probability 1}$$

and

$$\lim_{n \to \infty} \frac{1}{n} \sum_{k=1}^{n} I_k(B) = \frac{E(N_B)}{E(N)} \quad \text{with probability 1.}$$

Using part (b) of Theorem 1.7.1, it follows that $E(N_B)/E(N) = E(T_B)/E(T)$. Thus the corollary follows if we can verify that $E(N)/E(T) = \lambda$. To do so, note that the regeneration epochs for the process $\{X(t)\}$ are also regeneration epochs for the Poisson arrival process. Thus, by the renewal-reward theorem, the long-run average number of arrivals per unit time equals $E(N)/E(T)$ showing that $E(N)/E(T) = \lambda$.

To conclude this section, we demonstrate how the PASTA property can be used to establish relations between customer-average and time-average probabilities in queueing systems.

Example 1.7.1 The GI/M/1 queue

Suppose customers arrive at a single-server station according to a renewal process. Service is provided by a single server who can handle only one customer at a time, where the service time of each customer is exponentially distributed with mean $1/\beta$. The service times of the customers are independent of each other and are also independent of the arrival process. Denoting by λ the inverse of the mean interarrival time, it is assumed that $\lambda/\beta < 1$.

The continuous-time stochastic process $\{X(t),\ t \geq 0\}$ and the discrete-time stochastic process $\{X_n,\ n = 1, 2, \ldots\}$ are defined by

$$X(t) = \text{the number of customers in the system at time } t, t \geq 0,$$

and

$$X_n = \text{ the number of customers in the system just prior to the } n\text{th arrival}$$
$$\text{epoch, } n = 1, 2, \ldots .$$

The stochastic processes $\{X(t),\ t \geq 0\}$ and $\{X_n,\ n \geq 1\}$ are both regenerative. The regeneration epochs are the epochs at which an arriving customer finds the system empty. It is stated without proof that the assumption of $\lambda/\mu < 1$ implies that the processes have a finite mean cycle length. Thus we can define the time-average and customer-average probabilities p_j and π_j by

$$p_j = \text{the long-run fraction of time that } j \text{ customers are present}$$

and

π_j = the long-run fraction of customers who find upon arrival j other customers present.

These probabilities are related to each other by

$$\lambda \pi_{j-1} = \beta p_j, \quad j = 1, 2, \ldots . \tag{1.7.1}$$

The proof of this result is instructive and is based on three observations. Before giving the three steps, let us say that the continuous-time process $\{X(t)\}$ makes an upcrossing from state $j - 1$ to state j if a customer arrives and finds $j - 1$ other customers present. The process $\{X(t)\}$ makes a downcrossing from state j to state $j - 1$ if the service of a customer is completed and $j - 1$ other customers are left behind.

Observation 1 Since customers arrive singly and are served singly,

the long-run average number of upcrossings from $j - 1$ to j per unit time
 = the long-run average number of downcrossings from j to $j - 1$ per unit time.

This follows by noting that in any finite time interval the number of upcrossings from $j - 1$ to j and the number of downcrossings from j to $j - 1$ can differ at most by 1.

Observation 2

the long-run fraction of customers seeing upon arrival $j - 1$ other customers

$$= \frac{\text{the long-run average number of upcrossings from } j - 1 \text{ to } j \text{ per unit time}}{\text{the long-run average number of arrivals per unit time}}$$

for any $j \geq 1$. This intuitively obvious relation can be rigorously proved by using simple arguments from renewal-reward theory. The proof is very similar to that of relation (1.6.2). Thus, with probability 1,

the long-run average number of upcrossings from $j - 1$ to j per unit time

$= \lambda \pi_{j-1}$.

Observation 3 For exponential services,

the long-run average number of downcrossings from j to $j - 1$ per unit time

$= \beta p_j$

with probability 1 for each $j \geq 1$. The proof of this result relies heavily on the PASTA property. To make this clear, note that service completions occur according to a Poisson process with rate β as long as the server is busy. Equivalently, we can assume that an exogeneous Poisson process generates events at a rate of β, where a Poisson event results in a service completion only when there are j customers present. Thus, by part (a) of Theorem 1.7.1,

$$\beta E[I_j(t)] = E[D_j(t)] \quad \text{for } t > 0 \tag{1.7.2}$$

for any $j \geq 1$, where $I_j(t)$ is defined as the amount of time that j customers are present during $(0, t]$ and $D_j(t)$ is defined as the number of downcrossings from j to $j - 1$ in $(0, t]$. The random variable $I_j(t)$ can be interpreted as the cumulative reward earned up to time t when a reward at rate j is incurred whenever j customers are present. Thus, by the expected-value version of the renewal-reward theorem in Section 1.3,

$$\lim_{t \to \infty} \frac{E[I_j(t)]}{t} = p_j \tag{1.7.3}$$

for $j \geq 1$. Similarly, $D_j(t)$ can be interpreted as the cumulative reward earned up to the time t when a lump reward of 1 is incurred each time a downcrossing from j to $j - 1$ occurs. Thus, by the expected-value version of the renewal-reward theorem,

$$\lim_{t \to \infty} \frac{E[D_j(t)]}{t} = d_j \tag{1.7.4}$$

for all $j \geq 1$, where the constant d_j represents the long-run average number of downcrossings from j to $j - 1$ per unit time. From (1.7.2)–(1.7.4) we obtain the relation $d_j = \beta p_j$.

Together the observations 1, 2 and 3 yield the desired result (1.7.1). On purpose we have given in detail the derivation of the result (1.7.1). This derivation can be used in many other contexts in which a relation between the time-average and customer-average probabilities is required.

1.8   ASYMPTOTIC EXPANSION FOR RUIN AND WAITING-TIME PROBABILITIES

In many applied probability problems asymptotic expansions provide a simple alternative to computationally intractable solutions. A nice example is the ruin probability in risk theory. This probability arises in the following model. Claims arrive at an insurance company according to a Poisson process $\{N(t), \ t \geq 0\}$ with rate λ. The successive claim amounts $S_1, S_2, \ldots$ are positive, independent random variables having a common probability distribution function $B(x)$, and are also independent of the arrival process. In the absence of claims, the company's reserve increases at a constant rate of $\sigma > 0$ per unit time. It is assumed that $\sigma > \lambda E(S)$, i.e. the average premium received per unit time is larger than the average claim rate. Here the generic variable S denotes the individual claim amount. Denote by the compound Poisson variable

$$X(t) = \sum_{i=1}^{N(t)} S_i$$

the total amount claimed up to time t. If the company's initial reserve is $x > 0$ then the company's total reserve at time t is $x + \sigma t - X(t)$. We say that a ruin occurs at time t if $x + \sigma t - X(t) < 0$. Defining for each $x \geq 0$,

$$q(x) = P\{X(t) > x + \sigma t \quad \text{for some } t \geq 0\},$$

then $q(x)$ is the probability that a ruin will ever occur with initial capital x. Since a ruin can occur only at the claim epochs, we can equivalently write

$$q(x) = P\left\{ \sum_{j=1}^{k} S_j - \sigma \sum_{j=1}^{k} \tau_j > x \quad \text{for some } k \geq 1 \right\}, \tag{1.8.1}$$

where $\tau_1, \tau_2, \ldots$ denote the interoccurrence times of successive claims. We are interested in the asymptotic behaviour of $q(x)$ for large x.

The ruin probability $q(x)$ arises in a variety of contexts. As another example consider a production/inventory situation in which demands for a given product arrive according to a Poisson process. The successive demands are independent and identically distributed random variables. On the other hand, inventory replenishments of the product occur at a constant rate of $\sigma > 0$ per unit time. In this context, the ruin probability $q(x)$ represents the probability that a shortage will ever occur when the initial inventory is x.

The ruin probability as waiting-time probability

A less obvious context in which the ruin probability appears is the following standard queueing system with Poisson input and general service times. Customers arrive at a single-server station according to a Poisson process with rate λ. The service or work requirements $S_1, S_2, \ldots$ of the successive customers are independent of the arrival process. A customer who finds the server idle upon arrival immediately obtains service, otherwise the customer waits in line. The server works at a constant rate of $\sigma > 0$ per unit time whenever the system is not empty, i.e. the service requirement S of a customer has a processing time of S/σ. The customers are served in order of arrival. For ease it is assumed that the system is empty at epoch 0. Define for $n = 1, 2, \ldots$ the random variable D_n by

$D_n = $ the delay in queue of the nth customer (excluding service time).

Then the limiting distribution of D_n is given by

$$\lim_{n \to \infty} P\{D_n > x\} = q(\sigma x), \quad x \geq 0. \tag{1.8.2}$$

A proof of (1.8.2) proceeds as follows. Denoting by $\tau_1, \tau_2, \ldots$ the interarrival times for the successive customers, it is easily seen that,

$$D_{n+1} = \begin{cases} D_n + S_n/\sigma - \tau_{n+1} & \text{if } D_n + S_n/\sigma - \tau_{n+1} \geq 0, \\ 0 & \text{if } D_n + S_n/\sigma - \tau_{n+1} < 0. \end{cases}$$

Hence, letting $U_n = S_n/\sigma - \tau_{n+1}$ for $n \geq 1$, we have

$$D_{n+1} = \max(0, D_n + U_n).$$

Substituting this equation in itself, it follows that

$$D_{n+1} = \max\{0, U_n + \max(0, D_{n-1} + U_{n-1})\}$$
$$= \max(0, U_n, U_n + U_{n-1} + D_{n-1}), \quad n \geq 1.$$

By a repeated application of this equation and using that $D_1 = 0$, we find

$$\max\{0, U_n, U_n + U_{n-1} + D_{n-1}\}$$
$$= \max(0, U_n, U_n + U_{n-1}, \ldots, U_n + U_{n-1} + \cdots + U_1), \quad n \geq 1.$$

Since the random variables $U_1, U_2, \ldots$ are independent and identically distributed, we have that $(U_n, \ldots, U_1)$ has the same joint distribution as $(U_1, \ldots, U_n)$. Thus

$$D_{n+1} = \max\{0, U_1, U_1 + U_2, \ldots, U_1 + \cdots + U_n\}, \quad n \geq 1.$$

This implies that

$$P\{D_{n+1} > x\} = P\left\{ \sum_{j=1}^{k} U_j > x \text{ for some } 1 \leq k \leq n \right\}, \quad x \geq 0$$

and so

$$\lim_{n \to \infty} P\{D_n > x\} = P\left\{ \sum_{j=1}^{k} S_j - \sigma \sum_{j=1}^{k+1} \tau_{j+1} > x \text{ for some } k \geq 1 \right\}, \quad x \geq 0.$$

Together this relation and (1.8.1) prove the result (1.8.2).

A renewal equation for the ruin probability

We now turn to the determination of the ruin probability $q(x)$. For that purpose, we derive first an integro-differential equation for $q(x)$. For ease of presentation we assume that the probability distribution function $B(x)$ of the claim sizes has a probability density $b(x)$. Fix $x > 0$. To compute $q(x - \Delta x)$ with Δx small, we condition on what may happen in the first $\Delta x/\sigma$ time units. In the absence of claims the company's capital grows from $x - \Delta x$ to x. However, since the claims arrive according to a Poisson process with rate λ, a claim occurs in the first $\Delta x/\sigma$

time units with probability $\lambda \Delta x / \sigma + o(\Delta x)$, in which case the company's capital becomes $x - S$ if S is the size of that claim. In this case a ruin occurs only if $S > x$. Thus, we get for fixed $x > 0$,

$$q(x - \Delta x) = \left(1 - \frac{\lambda \Delta x}{\sigma}\right) q(x) + \frac{\lambda \Delta x}{\sigma} \int_x^\infty b(y)\, dy$$

$$+ \frac{\lambda \Delta x}{\sigma} \int_0^x q(x - y)b(y)\, dy + o(\Delta x).$$

Subtracting $q(x)$ from both sides of this equation, dividing by $h = -\Delta x$ and letting $h \to 0$, we obtain the integro-differential equation

$$q'(x) = -\frac{\lambda}{\sigma}\{1 - B(x)\} + \frac{\lambda}{\sigma}q(x) - \frac{\lambda}{\sigma} \int_0^x q(x - y)b(y)\, dy, \quad x > 0. \quad (1.8.3)$$

Equation (1.8.3) can be converted into an integral equation of the renewal-type. To do so, note that

$$\frac{d}{dx} \int_0^x q(x - y)\{1 - B(y)\}\, dy = q(0)\{1 - B(x)\} + \int_0^x q'(x - y)\{1 - B(y)\}\, dy$$

$$= q(0)\{1 - B(x)\} - q(x - y)\{1 - B(y)\}|_0^x - \int_0^x q(x - y)b(y)\, dy$$

$$= q(x) - \int_0^x q(x - y)b(y)\, dy.$$

Thus, by integrating both sides of (1.8.3) over x, we find

$$q(x) = q(0) - \frac{\lambda}{\sigma} \int_0^x \{1 - B(y)\}\, dy + \frac{\lambda}{\sigma} \int_0^x q(x - y)\{1 - B(y)\}\, dy, \quad x \geq 0.$$

Using Laplace transforms, it is shown in Appendix C that $q(0) = \rho$, where

$$\rho = \lambda E(S)/\sigma.$$

Hence the integro-differential equation (1.8.3) is equivalent to

$$q(x) = a(x) + \int_0^x q(x - y)h(y)\, dy, \quad x \geq 0, \quad (1.8.4)$$

where the functions $a(x)$ and $h(x)$ are given by

$$a(x) = \rho - \frac{\lambda}{\sigma} \int_0^x \{1 - B(y)\}\, dy \quad \text{and} \quad h(x) = \frac{\lambda}{\sigma}\{1 - B(x)\}, \quad x \geq 0.$$

This equation has the form of a standard renewal equation except that the function $h(x)$, $x \geq 0$, is not a proper probability density. It is true that the function h is non-negative, but

$$\int_0^\infty h(x)\, dx = \frac{\lambda}{\sigma} \int_0^\infty \{1 - B(x)\}\, dx = \frac{\lambda E(S)}{\sigma} < 1.$$

Thus h is the density of a distribution whose total mass is less than 1 with a defect of $1 - \rho$. Equation (1.8.4) is called a *defective renewal equation*. This equation can be numerically solved by the discretization method discussed in Section 1.1.3.

Asymptotic expansion for the ruin probability

A very useful asymptotic expansion of $q(x)$ can be given when it is assumed that the probability density of the claim size (service time) does not have an extremely long tail. To be more precise, the following assumption is made.

Assumption 1.8.1 *There are positive numbers a and b such that the complementary distribution function $1 - B(y) \leq ae^{-by}$ for all y sufficiently large.*

This assumption excludes probability distributions with extremely long tails like the lognormal distribution. The assumption implies that the number s_0 defined by

$$s_0 = \sup \left\{ s \left| \int_0^\infty e^{sy} \{1 - B(y)\} \, dy < \infty \right. \right\}$$

exists and is positive (possibly $s_0 = \infty$). In addition to Assumption 1.8.1 we make the technical assumption

$$\lim_{s \to s_0} \frac{\lambda}{\sigma} \int_0^\infty e^{sy} \{1 - B(y)\} \, dy > 1.$$

Then it is readily verified that the equation

$$\frac{\lambda}{\sigma} \int_0^\infty e^{\delta y} \{1 - B(y)\} dy = 1. \tag{1.8.5}$$

has a unique solution on the interval $(0, s_0)$. Next we convert the defective renewal equation (1.8.4) into a standard renewal equation. This enables us to apply the Key Renewal Theorem to obtain the asymptotic behaviour of $q(x)$. Let $h^*(x)$ be defined by

$$h^*(x) = \frac{\lambda}{\sigma} e^{\delta x} \{1 - B(x)\}, \quad x \geq 0.$$

Then $h^*(x)$, $x \geq 0$ is a probability density with finite mean. Multiplying both sides of equation (1.7.4) by $e^{\delta x}$ and defining the functions

$$q^*(x) = e^{\delta x} q(x) \text{ and } a^*(x) = e^{\delta x} a(x), \quad x \geq 0,$$

we find that the defective renewal function (1.8.4) is equivalent to

$$q^*(x) = a^*(x) + \int_0^x q^*(x - y) h^*(y) \, dy, \quad x \geq 0. \tag{1.8.6}$$

This is a standard renewal equation to which we can apply the Key Renewal Theorem. The function $a^*(x)$ is directly Riemann integrable as can be shown by verifying that $|a^*(x)| \leq ce^{-(a-\delta)x}$ as $x \to \infty$ for finite constants $c > 0$ and $a > \delta$. Using equation (1.8.5) and relation (A.9) in Appendix A, we find

$$\int_0^\infty a^*(x)\, dx = \int_0^\infty e^{\delta x} \left[\frac{\lambda}{\sigma} \int_x^\infty \{1 - B(y)\}dy \right] dx$$

$$= \frac{\lambda}{\sigma} \int_0^\infty \{1 - B(y)\} \left[\int_0^y e^{\delta x}\, dx \right] dy = \frac{1-\rho}{\delta}.$$

Applying the Key Renewal Theorem in Section 1.1.2 to the renewal equation (1.8.6), we find

$$\lim_{x\to\infty} q^*(x) = \gamma,$$

where the constant γ is given by

$$\gamma = (1-\rho) \left[\frac{\lambda\delta}{\sigma} \int_0^\infty y e^{\delta y}\{1 - B(y)\}\, dy \right]^{-1}.$$

This yields the asymptotic expansion

$$q(x) \approx \gamma e^{-\delta x} \quad \text{for } x \text{ large enough.} \tag{1.8.7}$$

This is an extremely important result. The asymptotic expansion is very useful for practical purposes in view of the remarkable finding that already for relatively small values of x the asymptotic estimate predicts quite well the exact value of $q(x)$ when the load factor ρ is not very small. To illustrate this, we give in Table 1.8.1 the

Table 1.8.1 Exact and asymptotic values for $q(x)$

		$c_s^2 = 0$		$c_s^2 = 0.5$		$c_s^2 = 1.5$	
	x	$q(x)$	$q_{asy}(x)$	$q(x)$	$q_{asy}(x)$	$q(x)$	$q_{asy}(x)$
$\rho = 0.2$	0.5	0.115 86	0.077 55	0.124 62	0.144 78	0.136 67	0.097 37
	1	0.022 88	0.030 07	0.071 46	0.077 12	0.096 69	0.076 30
	2	0.001 96	0.002 10	0.021 44	0.021 88	0.052 34	0.046 85
	3	0.000 15	0.000 15	0.006 17	0.006 21	0.030 25	0.028 77
	5	—	—	0.000 50	0.000 50	0.010 95	0.010 85
$\rho = 0.5$	0.5	0.357 99	0.306 73	0.372 85	0.386 08	0.393 90	0.340 55
	1	0.175 64	0.188 17	0.266 17	0.269 47	0.316 29	0.286 32
	2	0.053 04	0.053 56	0.131 06	0.131 26	0.211 86	0.202 39
	5	0.001 24	0.001 24	0.015 17	0.015 17	0.071 79	0.071 49
	10	—	—	0.000 42	0.000 42	0.012 62	0.012 62
$\rho = 0.8$	0.5	0.701 64	0.671 19	0.711 97	0.717 09	0.727 05	0.702 04
	1	0.554 89	0.563 12	0.624 30	0.625 49	0.665 22	0.650 40
	2	0.365 48	0.366 01	0.475 82	0.475 89	0.563 45	0.558 25
	5	0.100 50	0.100 50	0.209 59	0.209 59	0.353 22	0.352 99
	10	0.011 66	0.011 66	0.053 43	0.053 43	0.164 44	0.164 44

numerical values of $q(x)$ and the asymptotic estimate $q_{asy}(x)$ for several examples. The squared coefficient of variation c_S^2 of the claim size (service requirement) is varied as $c_S^2 = 0$ (deterministic distribution), $c_S^2 = 0.5$ (Erlang-2 distribution) and $c_S^2 = 2$ (hyperexponential distribution with balanced means). The load factor ρ is varied as $\rho = 0.2$, 0.5 and 0.8. It turns out that the closer ρ is to 1, the earlier the asymptotic expansion applies.

EXERCISES

1.1 An item deteriorates gradually in time by aging. The failure rate of an item having age x is $\alpha(x) = \alpha x$ for some constant $\alpha > 0$, i.e. an item with age x will fail in the next Δt time units with probability $\alpha(x)\Delta t$ for small Δt. The item is replaced upon failure or upon reaching the age T, whichever occurs first. Determine an asymptotic expansion for the expected number of replacements up to time t.

1.2 A machine is subject to breakdowns. Each breakdown requires an exponentially distributed repair time with mean $1/\mu$. The running time of the machine until the next breakdown is exponentially distributed with mean $1/\lambda$. The repair times and the running times are assumed to be independent of each other. The machine is in working condition at time 0. Determine an asymptotic expansion for the expected number of breakdowns up to time t.

1.3 (a) Verify for the Erlang-r distribution with mean r/λ that the associated renewal function $M(t)$ can be computed from the rapidly converging series

$$M(t) = \sum_{n=1}^{\infty} \left\{ 1 - \sum_{j=0}^{nr-1} e^{-\lambda t} \frac{(\lambda t)^j}{j!} \right\}, \quad t \geq 0.$$

(*Hint*: an Erlang-r distributed random variable can be represented as the sum of r independently exponentially distributed random variables.)

(b) Verify for the probability density $f(t) = p\lambda e^{-\lambda t} + (1 - p)\lambda^2 t e^{-\lambda t}$ that the associated renewal function is given by

$$M(t) = \frac{t}{E(X_1)} + \left\{ \frac{E(X_1^2)}{2E^2(X_1)} - 1 \right\} (1 - e^{-\lambda(2-p)t}), \quad t \geq 0.$$

1.4 Consider a renewal process in which the interoccurrence times between renewals have an Erlang-r distribution. Verify that the limiting distribution of the excess life has a probability density of the form $\sum_{j=1}^{r} p_j f_j(x)$ where $p_j = 1/r$ for all j and $f_j(x)$ is the density of an Erlang-j distribution. Can you give an heuristic explanation of this result?

1.5 For a renewal process let $M_2(t) = E[N^2(t)]$ be the second moment of the number of renewals up to time t. Verify that $M_2(t)$ satisfies the renewal equation

$$M_2(t) = 2M(t) - F(t) + \int_0^t M_2(t - x) f(x) \, dx, \quad t \geq 0,$$

where $f(x)$ is the density of the interoccurrence times of the renewals. Next verify that

$$\lim_{t \to \infty} E[N^2(t)] - \left\{ \frac{t^2}{\mu_1^2} + \left(\frac{2\mu_2}{\mu_1^3} - \frac{3}{\mu_1} \right) t \right\} = \frac{3\mu_2^2}{2\mu_1^4} - \frac{2\mu_3}{3\mu_1^3} - \frac{3\mu_2}{2\mu_1^2} + 1,$$

where μ_k denotes the kth moment of the density $f(x)$. Also, prove that

$$\lim_{t \to \infty} \int_0^t E[N^2(y)] \, dy - \left[\frac{t^3}{3\mu_1^3} + \left(\frac{\mu_2}{\mu_1^3} - \frac{3}{2\mu_1} \right) t^2 + \left(\frac{3\mu_2^2}{2\mu_1^4} - \frac{2\mu_3}{3\mu_1^3} - \frac{3\mu_2}{2\mu_1^2} + 1 \right) t \right]$$

$$= \frac{\mu_4}{6\mu_1^3} - \frac{\mu_2\mu_3}{\mu_1^4} + \frac{\mu_2^3}{\mu_1^5} + \frac{\mu_3}{2\mu_1^2} - \frac{3\mu_2^2}{4\mu_1^3}.$$

1.6 Consider a renewal process generated by the interoccurrence times $X_1, X_2, \ldots$ that have mean μ_1 and second moment μ_2. Let L_t be the length of the interoccurence time covering epoch t. Derive a renewal equation for $E(L_t)$. Verify the following results:

(a) $E(L_t) = 2\mu_1 - \mu_1 e^{-t/\mu_1}$ for all t when the X_i's are exponentially distributed.

(b) $\lim_{t \to \infty} E(L_t) = \mu_2/\mu_1$ when the X_i's are continuously distributed.

Also derive a renewal equation for $P\{L_t > x\}$. Prove that the limiting distribution of L_t has the density $xf(x)/\mu_1$ when the X_i's have a probability density $f(x)$. Can you give an heuristic explanation of why $E(L_t) \geq \mu_1$?

1.7 Let $\{N(t), \ t \geq 0\}$ be a Poisson process with interarrival times $X_1, X_2, \ldots$. Prove for any $t, h > 0$ that

$$P\{N(t + s) - N(t) \leq k, N(t) = n\} = P\{N(s) \leq k\} P\{N(t) = n\} \quad \text{for } n, k = 0, 1, \ldots .$$

In other words, the process has stationary and independent increments. (*Hint*: evaluate the probability $P\{X_1 + \cdots + X_n < t < X_1 + \cdots + X_{n+1}, X_1 + \cdots + X_{n+k+1} > t + s\}$).

1.8 At a shuttle station passengers arrive according to a Poisson process with rate λ. A shuttle departs as soon as seven passengers have arrived. There is an ample number of shuttles at the station. What is the probability that an arbitrary passenger finds upon arrival j other passengers waiting for departure? Next calculate the probability distribution of the time an arbitrary passenger has to wait until departure.

1.9 Consider a production line with two stations in series. The items produced at station 1 are passed to station 2 where they are assembled. There is a buffer space at station 2. Each station can handle only one item at a time. The production time of an item at station 1 is exponentially distributed with mean $1/\mu_1$, while the assembly time of an item at station 2 is exponentially distributed with mean $1/\mu_2$. The production times and the assembly times are assumed to be mutually independent. Suppose that presently an item is in production at station 1, while the buffer at station 2 holds $K \geq 1$ items including the one in assembly. What is the probability that the buffer becomes empty before the current production at station 1 is completed?

1.10 Orders for a certain product can be placed during a random time τ having an exponential distribution with mean $1/\mu$. Customers place orders for the product according to a Poisson process with rate λ. Each order is for j units of the product with probability $\varphi_j, j = 1, 2, \ldots$. Verify that the mean and variance of the total demand D in the random time τ are given by

$$E(D) = \frac{\lambda}{\mu} E(O_1) \quad \text{and} \quad \text{var}(D) = \frac{\lambda}{\mu} E(O_1^2) + \frac{\lambda^2}{\mu^2} E^2(O_1)$$

where O_1 denotes the individual order size. Denote by $\{a_k, k = 0, 1, \ldots\}$ the probability distribution of the total demand D. Use conditioning on the first demand epoch to verify that the a_k's can be recursively computed from $a_0 = \mu/(\lambda + \mu)$ and

$$a_k = \frac{\lambda}{\lambda + \mu} \sum_{j=1}^{k} a_{k-j} \varphi_j \qquad \text{for } k = 1, 2, \ldots .$$

1.11 Passengers arrive at a bus stop according to a Poisson process with rate λ. Buses depart from the stop according to a renewal process with interdeparture time A. Using renewal-reward processes, prove that the long-run average waiting time per passenger equals $E(A^2)/2E(A)$. Can you give an heuristic explanation of why this answer is the same as the average residual life in a renewal process?

1.12 (a) Customers arrive in batches at a service facility. The batch sizes are independent random variables having a common discrete probability distribution $\{\beta_k, k = 1, 2, \ldots\}$. Use renewal-reward theory to verify that the long-run fraction of customers belonging to a batch of size k is given by $k\beta_k/\beta$ for $k = 1, 2, \ldots$, where β denotes the mean batch size.

(b) A shuttle departs every 15 minutes from a station. The shuttle has ample capacity and takes upon departure all passengers present at the station. The passengers arrive at the shuttle station according to a compound Poisson process. What is the long-run fraction of passengers who are in the shuttle with j fellow-passengers?

1.13 At a production facility orders arrive according to a renewal process with a mean interarrival time $1/\lambda$. A production is started only when N orders have accumulated. The production time is negligible. A fixed cost of $K > 0$ is incurred for each production setup and holding costs are incurred at the rate of hj when j orders are waiting to be processed. Verify that the value of N for which the long-run average cost per unit time is minimal is one of the two integers nearest to $\sqrt{2K\lambda/h}$.

1.14 A tour boat that makes trips through the canals of Amsterdam leaves every T minutes from the Central Station. Potential customers for a trip arrive at the Central Station according to a Poisson process with rate λ. A potential customer waits until the next departure with probability $e^{-\mu t}$ if upon arrival of the customer the time until the next departure is t minutes. A fixed cost of $K > 0$ is incurred for each trip of the tour boat and a reward of $r > 0$ is earned for each customer joining a trip. Show that the value of T maximizing the long-run average reward per unit time is the unique solution to the equation $e^{-\mu T}(r\lambda T + r\lambda/\mu) = r\lambda/\mu - K$ provided $r\lambda/\mu > K$.

1.15 In an inventory system for a single product the demand process for that product is a Poisson process with rate λ. Each demand which cannot be satisfied directly from inventory on hand is lost. Opportunities to replenish the inventory on hand occur according to a Poisson process with rate μ. For technical reasons a replenishment can only be made when the inventory equals zero. Each replenishment order consists of Q units. The lead time of a replenishment order is negligible. There is a fixed cost of $K > 0$ per replenishment order, a holding cost of h per unit inventory per unit time and a lost-sales cost of π per unit demand lost. Determine the long-run average cost per unit time. Also, determine the fraction of demand lost.

1.16 Potential building projects of types $1, \ldots, N$ are offered to a constructor according to independent Poisson processes with respective rates $\lambda_1, \ldots \lambda_N$. Each offer should be either accepted or rejected upon its occurrence. The constructor is only able to handle one project at a time. An accepted project of type j occupies the constructor for a random time τ_j and yields a random pay-off ξ_j. The successive offers are assumed to be independent of each other. The constructor accepts only jobs of type $j \in A$ whenever he is idle, where A is a given subset of $\{1, \ldots, N\}$. Verify that the average payoff per unit time and the fraction of time the constructor is idle are given by

$$\frac{\displaystyle\sum_{j \in A} \lambda_j E(\xi_j)}{1 + \displaystyle\sum_{j \in A} \lambda_j E(\tau_j)} \quad \text{and} \quad \frac{1}{1 + \displaystyle\sum_{j \in A} \lambda_j E(\tau_j)} \quad \text{respectively.}$$

1.17 At the beginning of each day a batch of containers arrives at a stackyard having ample capacity to store any number of containers. The batch size has a discrete probability distribution $\{q_k, \ k \geq 1\}$. Each container stays an exponentially distributed time at the stackyard, where the holding times of the various containers are independent of each other and have the same mean $1/\mu$. Prove that the long-run average number of containers present at the end of the day equals $e^{-\mu} E(Q)/(1 - e^{-\mu})$ with probability 1, where $E(Q)$ denotes the batch size. (*Hint*: define the random variable A_{kn} as the number of containers that belong to the batch arriving at the beginning of the kth day and are still present at the end of the nth day.)

1.18 Consider the inventory control of an expensive item that is infrequently demanded. At most one unit of this item is kept in stock. Each time a demand is satisfied from stock on hand, a replenishment order for one unit of the item is placed. The replenishment lead time is a positive constant L. The demand process can be described by a renewal process, where the times between the unit demands are Erlang-r distributed. Each demand occurring while the system is out of stock is lost. What is the long-run fraction of time the system is out of stock? What is the long-run fraction of demand lost? (*Hint*: use the results in Exercise 1.3 to compute the expected number of lost demands in one cycle.)

1.19 A production machine gradually deteriorates in time. The machine has N possible working conditions $1, \ldots, N$ which describe increasing degrees of deterioration. Here working condition 1 represents a new system and working condition N represents a failed system. The system remains in each working condition i, with $1 \leq i < N$, during an exponentially distributed time with mean $1/\mu$. The changes to the working condition cannot be observed except for a failure which is detected immediately. The machine is replaced by a new one upon failure or upon having worked during a time T, whichever occurs first. Each planned replacement involves a fixed cost of $J_1 > 0$, whereas a replacement because of a failure involves a fixed cost of $J_2 > 0$. The replacement time is negligible in both cases. Also, the system incurs an operating cost of $a_i > 0$ for each time unit the system is operating in working condition i. Use Lemma 1.2.4 to verify that the long-run average cost per unit time is given by

$$\left\{ T \sum_{k=0}^{N-2} p_k + \frac{N-1}{\mu} \left(1 - \sum_{k=0}^{N-1} p_k \right) \right\}^{-1}$$

$$\times \left\{ J_2 + (J_2 - J_1) \sum_{k=0}^{N-2} p_k + \sum_{k=0}^{N-2} p_k \sum_{i=1}^{k+1} a_i \frac{T}{k+1} + \sum_{i=1}^{N-1} a_i \sum_{k=N-1}^{\infty} p_k \frac{T}{k} \right\},$$

where $p_k = e^{-\mu T} (\mu T)^k / k!, \ k \geq 0$. (This problem is motivated by one in Luss (1976).)

1.20 We consider again the production system from Example 1.3.4 except that the system is now controlled in a different way when it becomes idle. Each time the production facility becomes empty of orders, the facility is used during a period of fixed length T for some other work in order to utilize the idle time. After this vacation period the facility is reactivated for servicing the orders only when at least one order is present; otherwise the facility is used again for some other work during a vacation period of length T. This utilization of idle time is continued until at least one order is present after the end of a vacation period. This control policy is called the T-policy. The cost structure is the same as in Example 1.3.4. Verify that the long-run average cost per unit time is minimal for $T = [2K\{1 - \lambda E(\tau_1)\}/(h\lambda)]^{1/2}$.

1.21 Suppose that at a communication channel messages of the types 1 and 2 arrive according to independent Poisson processes with respective rates λ_1 and λ_2. Message of type 1, finding the channel occupied upon arrival, are lost, whereas messages of type 2 are temporarily stored in a buffer and wait until the channel becomes available. The channel can transmit only one message at a time. The transmission time of a message of type i has a general probability distribution with mean τ_i and the transmission times are independent from each other. It is assumed that $\lambda_2\tau_2 < 1$. Prove that the long-run fraction of time the channel is busy equals $(\rho_1 + \rho_2)/(1 + \rho_1)$, where $\rho_i = \lambda_i\tau_i$ for $i = 1, 2$. (*Hint*: use results from Example 1.3.4 to obtain the expected amount of time elapsed between two arrivals finding the channel free.)

1.22 Every time unit an item arrives at a work station. An item finding the work station busy upon arrival is lost. The processing times of the accepted items are independent random variables having a common probability distribution function $B(x)$ with mean μ. Verify that the long-run fraction of items lost and the long-run fraction of time the work station is busy are given by $1 - 1/L$ and μ/L, where $L = \Sigma_{k=0}^{\infty}(k + 1)[B(k + 1) - B(k)]$.

1.23 Consider the continuous-review (s, Q) inventory model with Poisson demand, where any excess demand is backlogged. Assume that the replenishment lead time is a constant L. Denote by p_j and r_j the long-run fraction of time the inventory position is j and the long-run fraction of time the net stock is j.
   (a) Prove that $p_j = 1/Q$ for $j = s + 1, \ldots, s + Q$.
   (b) Denoting by λ the average demand rate, verify that

$$r_j = \frac{1}{Q} \sum_{k=\max(j,s+1)}^{s+Q} p_k e^{-\lambda L} \frac{(\lambda L)^{k-j}}{(k - j)!}, \quad j \le s + Q.$$

(*Hint*: use that the net stock at time t equals the inventory position at time $t - L$ minus the total demand between $t - L$ and t.)
   (c) Give an expression for the long-run fraction of demand not satisfied directly from stock on hand.

1.24 Consider an age-replacement model in which preventive replacements are only possible at special times. Opportunities for preventive replacements occur according to a Poisson process with rate λ. The item is replaced by a new one upon failure or upon a preventive replacement opportunity occurring when the age of the item is T or more, whichever occurs first. The lifetime of the item has a probability density $f(x)$. The cost of replacing the item upon failure is c_0 and the cost of a preventive replacement is c_1 with $0 < c_1 < c_0$. Determine the long-run average cost per unit time. (This problem is motivated by Dekker and Smeitink (1994).)

1.25 A group of N identical machines is maintained by a single repairman. The machines operate independently of each other and each machine has a constant failure rate μ. Repair is done only if the number of failed machines has reached some critical level. Then all failed machines are repaired simultaneously. Any repair takes a negligible time and a repaired machine is again as good as new. The cost of the simultaneous repair of j machines is $K + cj$, where $K, c > 0$. Also there is an idle-time cost of $\alpha > 0$ per unit time for each failed machine. Determine the long-run average cost per unit time when the critical level for simultaneous repair is R with $1 \leq R \leq N$.

1.26 Consider a two-unit reliability model with one operating unit and one unit in warm standby. The operating unit has a constant failure rate of λ_0, while the unit in warm standby has a constant failure rate of λ_1. Upon failure of the operating unit, the unit in warm standby is put into operation if available. The repair time of a failed unit has a general probability distribution function $G(x)$ with density $g(x)$ and mean μ_R. The system is down when both units have failed. For the case of a single repair facility, prove that the long-run fraction of time the system is down is given by

$$\frac{\mu_R - \displaystyle\int_0^\infty \{1 - G(x)\} e^{-\lambda_0 x}\, dx}{\mu_R + (\lambda_0 + \lambda_1)^{-1} \displaystyle\int_0^\infty e^{-\lambda_0 x} g(x)\, dx}.$$

(This exercise is based on Gaver (1963).)

1.27 An alternating renewal process is a two-state process alternating between an on-state and an off-state, where both the on-times and the off-times are independent and identically distributed random variables. Assume that the on-times and off-times are mutually independent. Let μ_{on} and μ_{off} denote the respective means of an on-time and an off-time. Denote by $G_{\text{on}}(x, t)$ the joint probability that the system is on at time t and that the residual on-time at time t is no more than x. Assuming that the distribution functions of the on-time and off-time are not both arithmetic, prove that

$$\lim_{t \to \infty} G_{\text{on}}(x, t) = \frac{\mu_{\text{on}}}{\mu_{\text{on}} + \mu_{\text{off}}} \times \frac{1}{\mu_{\text{on}}} \int_0^x [1 - F_{\text{on}}(y)]\, dy, \quad x \geq 0,$$

where $F_{\text{on}}(x)$ denotes the probability distribution function of the on-time.

1.28 Consider the alternating renewal process from Exercise 1.27. Assuming that an on-time starts at epoch 0, denote by the random variable $U(t)$ the cumulative time the system is in the on-state during $[0, t]$.

(a) Use Theorem 1.3.5 to verify that $U(t)$ is asymptotically normally distributed with mean $\mu_{\text{on}} t / (\mu_{\text{on}} + \mu_{\text{off}})$ and variance $(\mu_{\text{on}}^2 \sigma_{\text{off}}^2 + \mu_{\text{off}}^2 \sigma_{\text{on}}^2) t / (\mu_{\text{on}} + \mu_{\text{off}})^3$, where σ_{on}^2 and σ_{off}^2 denote the variances of the on-time and off-time.

(b) Assuming that the on-time distribution and the off-time distribution are not both arithmetic, prove that

$$\lim_{t \to \infty} \left[E(U(t)) - \frac{\mu_{\text{on}}}{\mu_{\text{on}} + \mu_{\text{off}}} t \right] = \frac{\mu_{\text{on}} \sigma_{\text{off}}^2 - \mu_{\text{off}} \sigma_{\text{on}}^2}{2(\mu_{\text{on}} + \mu_{\text{off}})^2} + \frac{\mu_{\text{on}} \mu_{\text{off}}}{2(\mu_{\text{on}} + \mu_{\text{off}})}.$$

1.29 Consider an unreliable production unit whose output is temporarily stored in a finite buffer with capacity K. The buffer serves for the demand process as protection against random interruptions in the production process. For the output there is a constant demand at rate v. When operating, the production unit produces at a constant rate $P > v$ if the buffer is not full and produces at the demand rate v otherwise. Demand occurring while the unit is down and the buffer is empty is lost. The operating time of the unit is exponentially distributed with mean $1/\lambda$. If a failure occurs, the unit enters repair for an exponentially distributed time with mean $1/\mu$. Determine the long-run fraction of demand lost and determine the average inventory level in the buffer. (*Hint*: Define the state of the system as $(1, x)$ respectively $(0, x)$ when the inventory in the buffer is x and the unit is operating respectively down. The process regenerates itself each time the system enters state $(0, 0)$. Use differential equations to get the desired performance measures. (This exercise is motivated by Wijngaard (1979).)

BIBLIOGRAPHIC NOTES

The very readable monograph of Cox (1962) contributed much to the popularization of renewal theory. A good account of renewal theory can also be found in the texts of Heyman and Sobel (1982), Karlin and Taylor (1975) and Ross (1983). A basic paper on renewal theory and regenerative processes is that of Smith (1958), a paper which recognized the usefulness of renewal-reward processes in the analysis of applied probability problems. The book of Ross (1970) was influential in promoting the application of renewal-reward processes. Renewal-reward processes have many applications in inventory and queueing. The illustrative Example 1.3.4 is taken from the paper of Yadin and Naor (1963) which initiated the study of control rules for queueing systems. A good account of the application of renewal theory in maintenance and reliability is given in the books of Barlow and Proschan (1975) and Beichelt and Franken (1983). The reliability Examples 1.3.3 and 1.4.3 are based on the papers of Vered and Yechiali (1979) and Van der Heyden (1987). The results for the alternating renewal process in Example 1.4.3 are proved in greater generality in Takács (1957). Section 1.5 on inventory control is partly based on Tijms and Groenevelt (1984). An excellent textbook on inventory theory is Silver and Peterson (1985). The first rigorous proof of $L = \lambda W$ was given by Little (1961) under rather strong conditions. The simple and transparent proof in Section 1.6 is based on Jewell (1967). Under very weak conditions a sample-path proof of $L = \lambda W$ was given by Stidham (1974). The important result Poisson arrivals see time averages was taken for granted by earlier practitioners. A rigorous proof was given in the paper of Wolff (1982). A nice account of relations between time-average and customer-average probabilities is given in Stidham and El Taha (1989). The technique used in Section 1.8 to derive asymptotic estimates for ruin and waiting-time probabilities comes from Feller (1971). This powerful technique is applied in De Kok, Tijms and Van der Duyn Schouten (1984) to obtain an asymptotic estimate for the overflow probability in a finite-capacity dam model with compound Poisson input and a controllable continuous release rate.

REFERENCES

Abramowitz, M., and Stegun, I. (1965). *Handbook of Mathematical Functions*, Dover, New York.

Adelson, R.M. (1966). 'Compound Poisson distributions', *Operat. Res. Quart.*, **17**, 73-75.

Barlow, R.E. and Proschan, F. (1975). *Statistical Theory of Reliability and Life Testing*, Holt, Rinehart and Winston, New York.

Beichelt, F. and Franken, P. (1983). *Zuverlässigkeit und Instandhaltung* (in German), VEB Verlag Technik, Berlin.

Cox, D.R. (1962). *Renewal Theory*, Methuen, London.

Dekker, R. and Smeitink, E. (1994). 'Preventive maintenance at opportunities of restricted duration', *Naval Res. Logist.* (to appear).

De Kok, A.G., Tijms, H.C. and Van der Duyn Schouten, F.A. (1984). 'Approximations for the single-product production-inventory problem with compound Poisson demand and service-level constraints', *Adv. Appl. Prob.*, **16**, 378-401.

Feller, W. (1971). *An Introduction to Probability Theory and Its Applications*, Vol. II, 2nd edn., Wiley, New York.

Gaver, D.P. (1963). 'Time to failure and availability of paralleled systems with repair', *IEEE Trans. Reliab.*, **R-12**, 30-38.

Heyman, D.P., and Sobel, M.J. (1982). *Stochastic Models in Operations Research*, Vol. I: *Stochastic Processes and Operating Characteristics*, McGraw-Hill, New York.

Jewell, W.S. (1967). 'A simple proof of $L = \lambda W$', *Operat. Res.*, **15**, 1109-1116.

Karlin S. and Taylor, H.M. (1975). *A First Course in Stochastic Processes*, 2nd edn., Academic Press, New York.

Keilson, J. (1979). *Markov Chain Models — Rarity and Exponentiality*, Springer-Verlag, Berlin.

Little, J.D.C. (1961). 'A proof for the queueing formula $L = \lambda W$', *Operat. Res.*, **9**, 383-387.

Lorden, G. (1970). 'On excess over the boundary', *Ann. Math. Statist.*, **41**, 520-527.

Luss, H. (1976). 'Maintenance policies when deterioration can be observed by inspections', *Operat. Res.*, **24**, 359-366.

Miller, D.R. (1972). 'Existence of limits in regenerative processes', *Ann. Math. Statist.*, **43**, 1275-1282.

Press, W.H., Flannery, B.P. Teukolsky, S.A. and Vetterling, W.T. (1986). *Numerical Recipes*, Cambridge University Press, Cambridge.

Ross, S.M. (1970). *Applied Probability Models with Optimization Applications*, Holden-Day, San Francisco.

Ross, S.M. (1983). *Stochastic Processes*, Wiley, New York.

Schneider, H. (1979). *Servicegrade in Lagerhaltungsmodellen* (in German), G Marchal and H-J Matzenbacher, Wissenschaftsverlag, Berlin.

Silver, E.A., and Peterson, R. (1985). *Decision Systems for Inventory Management and Production Planning*, 2nd edn., Wiley, New York.

Smeitink, E., and Dekker, R. (1990). 'A simple approximation to the renewal function', *IEEE Trans. Reliab.*, **R-39**, 71-75.

Smith, W.L. (1958). 'Renewal theory and its ramifications', *J. Roy. Statist. Soc. B*, **20**, 243-302.

Solovyez, A.D. (1971). 'Asymptotic behaviour of the time of first occurrence of a rare event in a regenerating process', *Engineering Cybernetics*, **9**, 1038-1048.

Stidham, S., Jr., (1984). 'A last word on $L = \lambda W$', *Operat. Res.*, **22**, 417-421.

Stidham, S., Jr., and El Taha, M. (1989). 'Sample-path analysis of processes with embedded point processes', *Queueing Systems*, **5**, 131-166.

Takács, L.J. (1957). 'On certain sojourn time problems in the theory of stochastic processes', *Acta Mathematica Academiae Scientiarum Hungaricae*, **8**, 169-191.

Tijms, H.C., and Groenevelt, H. (1984). 'Simple approximations for the reorder point in periodic review and continuous review (s, S) inventory systems', *European J. Operat. Res.*, **17**, 175-190.

Van der Heyden, M.C. (1987). 'Interval availability distribution for a 1-out-of-2 reliability system', *Prob. Engineering Inform. Sci.*, **2**, 211–224.

Vered, G., and Yechiali, U. (1979). 'Optimal structures and maintenance policies for PABX power systems', *Operat. Res.*, **27**, 37–47.

Wijngaard, J. (1979). 'The effect of interstage buffer storage on the output of two unreliable production units in series with different production rates', *AIIE Trans.*, **11**, 42–47.

Wolff, R.W. (1982). 'Poisson arrivals see time averages', *Operat. Res.*, **30**, 223–231.

Wolff, R.W. (1989). *Stochastic Modeling and the Theory of Queues*, Prentice-Hall, Englewood Cliffs, NJ.

Xie, M. (1989). 'On the solution of renewal-type integral equations', *Commun. Statist., B* **18**, 281–293.

Yadin, M., and Naor, P. (1963). 'Queueing systems with removable service station', *Operat. Res. Quart.*, **14**, 393–405.

CHAPTER 2

Markov Chains: Theory and Applications

2.0 INTRODUCTION

The notion of what is nowadays called a Markov chain was devised by the Russian mathematician A. A. Markov when, at the beginning of the twentieth century, he investigated the alternation of vowels and consonants in Poeshkin's poem *Onegin*. He developed a probability model in which the outcomes of successive trials are allowed to be dependent on each other such that each trial depends only on its immediate predecessor. This model, being the simplest generalization of the probability model of independent trials, appeared to give an excellent description of the alternation of vowels and consonants and enabled Markov to calculate a very accurate estimate of the frequency at which consonants occur in the above-mentioned poem of Poeshkin.

The Markov model is no exception to the rule that simple models are often the most useful models for analysing practical problems. A Markov process allows us to model the uncertainty in many real-world systems that evolve dynamically in time. The basic concepts of a Markov process are those of a *state* of a system and a state *transition*. In specific applications the modelling 'art' is to find an adequate state description such that the associated stochastic process has indeed the Markovian property that the knowledge of the present state is sufficient to predict the future stochastic behaviour of the process. The theory of Markov processes has applications to a wide variety of fields including biology, economics and physics among others, but in this chapter we primarily deal with applications to operations research and computer science. As in the previous chapter, the emphasis in this chapter is again on basic results and modelling aspects. We discuss a wide variety of practical problems that can be solved by Markov chain analysis. In each problem much attention is paid to an appropriate identification of the state and the state transition. The student cannot be urged enough to try the problems at the end of this chapter in order to acquire skills to model new situations. The modelling is often more difficult than the mathematics!

In Section 2.1 we introduce the discrete-time Markov chain model. In this model state transitions can occur only at given, discrete points in time. The main interest is in the long-run behaviour of the Markov chain. In Section 2.3 we establish the ergodic theorems that are needed for computing long-run averages in specific applications. The proofs of these theorems require some basic results for state classification that are discussed in Section 2.2. In Section 2.3 we establish not only the existence of an equilibrium distribution but discuss also how to compute the equilibrium probabilities. Applications of the discrete-time Markov chain model are presented in Section 2.4 and include applications to insurance, inventory and telecommunication.

Sections 2.5–2.9 deal with the continuous-time analogue of the discrete-time Markov chain. In the continuous-time Markov chain model the holding times in the various states are exponentially distributed rather than deterministic, but the state transitions are again governed by a Markov chain. Equivalently, a continuous-time Markov chain can be represented by infinitesimal transition rates (generalizing the Poisson process). This representation allows us to work with the flowrate equation technique. This technique is easy to visualize by a graphical representation and is widely used by practitioners. The continuous-time Markov chain model is introduced in Section 2.5. In Section 2.6 we focus on the calculation of the equilibrium probabilities by using the flow-equating technique. In Section 2.7 we model and analyse several practical problems. In some of these problems it will be seen that the measures of system performance are (fairly) insensitive to the distributional form of the underlying random phenomena and require only their average values. Section 2.8 deals with the uniformization technique for analysing the transient behaviour of the process (including the calculation of the first-passage time distribution). The powerful phase method for formulating practical problems in the framework of continuous-time Markov chains is discussed in Section 2.9. An application to the single-server queue with Poisson arrivals and general service times is given.

2.1 DISCRETE-TIME MARKOV CHAINS

A discrete-time Markov chain is a stochastic process which is the simplest generalization of a sequence of independent random variables. A Markov chain is a random sequence in which the dependency of the successive events goes back only one unit in time. In other words, the future probabilistic behaviour of the process depends only on the present state of the process and is not influenced by its past history. This is called the *Markovian* property. Despite the very simple structure of Markov chains, the Markov model is extremely useful in a wide variety of practical probability problems.

Formally, let $\{X_n, \ n = 0, 1, \ldots\}$ be a sequence of random variables with a *discrete* state space I. We interpret the random variable X_n as the state of some dynamic system at time n. The set of possible values of the process is denoted by I and is assumed to be finite or countably infinite.

Definition 2.1.1 *The stochastic process* $\{X_n, \ n = 0, 1, \ldots\}$ *with state space I is called a discrete-time Markov chain if, for each $n = 0, 1, \ldots,$*

$$P\{X_{n+1} = i_{n+1}|X_0 = i_0, \ldots, X_n = i_n\} = P\{X_{n+1} = i_{n+1}|X_n = i_n\} \qquad (2.1.1)$$

for all possible values of $i_0, \ldots, i_{n+1} \in I$.

In the following we consider only Markov chains with time-homogeneous transition probabilities; that is, we assume that

$$P\{X_{n+1} = j|X_n = i\} = p_{ij}, \quad i, j \in I,$$

independently of the time parameter n. The probabilities p_{ij} are called the *one-step transition probabilities* and satisfy

$$p_{ij} \geq 0, \quad i, j \in I, \quad \text{and} \quad \sum_{j \in I} p_{ij} = 1, \quad i \in I.$$

The Markov chain $\{X_n, n = 0, 1, \ldots\}$ is completely determined by the probability distribution of the initial state X_0 and the one-step transition probabilities p_{ij}.

We now demonstrate how the one-step transition probabilities can be used to determine the probability that the process goes from state i to state j in the coming n transitions. The n-step transition probabilities are defined by

$$p_{ij}^{(n)} = P\{X_n = j|X_0 = i\}, \quad i, j \in I$$

for any $n = 1, 2, \ldots$. Note that $p_{ij}^{(1)} = p_{ij}$. It is convenient to define

$$p_{ij}^{(0)} = \begin{cases} 1 & \text{if } j = i, \\ 0 & \text{if } j \neq i. \end{cases}$$

Theorem 2.1.1 (Chapman–Kolmogoroff equations)   *For all n, $m = 0, 1, \ldots$,*

$$p_{ij}^{(n+m)} = \sum_{k \in I} p_{ik}^{(n)} p_{kj}^{(m)}, \quad i, j \in I. \qquad (2.1.2)$$

Proof   A formal proof is as follows. By conditioning on the state of the Markov chain at time $t = n$, we find

$$P\{X_{n+m} = j|X_0 = i\} = \sum_{k \in I} P\{X_{n+m} = j|X_0 = i, X_n = k\}P\{X_n = k|X_0 = i\}$$

$$= \sum_{k \in I} P\{X_{n+m} = j|X_n = k\}P\{X_n = k|X_0 = i\}$$

$$= \sum_{k \in I} P\{X_m = j|X_0 = k\}P\{X_n = k|X_0 = i\},$$

which verifies (2.1.2). Note that the second equality uses the Markovian property and the last equality uses the assumption of time-homogenity.

In words, the theorem states that the probability of going from i to j in $n + m$ steps is obtained by summing the probabilities of the mutually exclusive events of going first from state i to some state $k \in I$ in n steps and then going from state k to state j in m steps. This explanation is helpful to memorize the equation (2.1.2). In particular, we have for any $n = 1, 2, \ldots$,

$$p_{ij}^{(n+1)} = \sum_{k \in I} p_{ik}^{(n)} p_{kj}, \quad i, j \in I. \tag{2.1.3}$$

Thus the n-step transition probabilities $p_{ij}^{(n)}$ can be recursively computed from the one-step transition probabilities p_{ij}.

A problem of considerable importance is to find the asymptotic behaviour of $p_{ij}^{(n)}$ as $n \to \infty$. The long-run behaviour of the Markov chain will be analysed in Section 2.3 after we have discussed some fundamental concepts in Section 2.2. To conclude this section, we give two examples of a Markov chain.

Example 2.1.1 A periodic-review (s, S) inventory system

Consider a single-product inventory system in which the demands for the product in the successive weeks $n = 0, 1, \ldots$ are independent and identically distributed random variables. The probability of a demand for j units during a week is given by ϕ_j, $j = 0, 1, \ldots$, where the first moment of the weekly demand is finite and positive. It is assumed that any demand in excess of current inventory is lost. The inventory position is reviewed only at the beginning of each week and at each review there is an opportunity for placing a replenishment order. It is supposed that the delivery of each replenishment order is instantaneous. The inventory level is controlled by an (s, S) rule with $0 < s \leq S$. Under this rule the inventory position is ordered up to the level S if at a review the inventory position is below the reorder point s, and no ordering is done otherwise. For $n = 0, 1, \ldots$, define

$X_n =$ the inventory position at the beginning of the nth week just prior to review.

The process $\{X_n, n = 0, 1, \ldots\}$ is a discrete-time Markov chain with state space $I = \{0, 1, \ldots, S\}$. The Markovian property (2.1.1) is an immediate consequence of the independence of the weekly demands and the form of the control rule used. Specifically, denoting by the random variable D_n the demand to occur during the nth week, we have

$$X_{n+1} = \begin{cases} \max(X_n - D_n, 0) & \text{if } X_n \geq s, \\ \max(S - D_n, 0) & \text{if } X_n < s. \end{cases}$$

The one-step transition probabilities of the Markov chain $\{X_n\}$ are easily found. Distinguish between the cases $i \geq s$ and $i < s$. For $i \geq s$,

$$p_{ij} = \begin{cases} P\{\text{the demand in the coming week is } i - j\}, & 1 \leq j \leq i, \\ P\{\text{the demand in the coming week is } i \text{ or more}\}, & j = 0. \end{cases}$$

Hence, for $i \geq s$,

$$p_{ij} = \begin{cases} \phi_{i-j}, & 1 \leq j \leq i, \\ \sum_{k=i}^{\infty} \phi_k, & j = 0. \end{cases}$$

Similarly, we find for $0 \leq i < s$ that

$$p_{ij} = \begin{cases} \phi_{S-j}, & 1 \leq j \leq S, \\ \sum_{k=S}^{\infty} \phi_k, & j = 0. \end{cases}$$

The other p_{ij}'s are equal to zero.

The following example illustrates the powerful technique of *embedded Markov chains*. Many stochastic processes can be analysed by using properly chosen embedded stochastic processes that are discrete-time Markov chains.

Example 2.1.2 A single-server queue with exponential services

Consider a single-server station at which customers arrive according to a renewal process with interarrival times $A_1, A_2, \ldots$. A customer who finds upon arrival that the server is idle enters service immediately; otherwise the customer waits in line. The service times of the successive customers are independent random variables having a common exponential distribution with mean $1/\mu$, and are also independent of the arrival process. A customer leaves the system upon service completion. This queueing system is usually abbreviated as the *GI/M/*1 queue. For any $t \geq 0$, define the random variable $X(t)$ by

$$X(t) = \text{the number of customers present at time } t.$$

The continuous-time stochastic process $\{X(t), \ t \geq 0\}$ does not possess the Markovian property that the future behaviour of the process depends only on its present state. Clearly, to predict the future behaviour of the process, the knowledge of the number of customers present does not suffice in general but the knowledge of the time elapsed since the last arrival is required too. Note that, by the memoryless property of the exponential distribution, the elapsed service time of the service in progress (if any) is not relevant. In fact, by the lack of memory of the exponential services, we can find an embedded Markov chain for the continuous-time process $\{X(t)\}$. Consider the process embedded at the epochs when customers arrive. At these epochs the time elapsed since the last arrival is known and equals zero. Define for $n = 0, 1, \ldots$,

$$X_n = \text{the number of customers present just prior to the } n\text{th arrival epoch}$$

with $X_0 = 0$ by convention. The embedded stochastic process $\{X_n, \ n = 0, 1, \ldots\}$ is a discrete-time Markov chain. This Markov chain has the countably infinite state space $I = \{0, 1, \ldots\}$. In Example 1.7.1 in Section 1.7 it has been shown how the probability distributions of the continuous-time process $\{X(t)\}$ and the discrete-time process $\{X_n\}$ can be related to each other.

Denoting by C_{n+1} the number of customers served during the interarrival time A_{n+1} between the nth and $(n + 1)$th customer, we have that $X_{n+1} = X_n + 1 - C_{n+1}$. Clearly, the distribution of C_{n+1} depends on X_n. This distribution follows by observing that as long as the server is busy the number of service completions is described by a Poisson process with rate μ, since the service times are independent and have a common exponential distribution with mean $1/\mu$. To give the one-step transition probabilities of the Markov chain $\{X_n\}$, we distinguish between the cases $j \neq 0$ and $j = 0$. We assume for ease that the probability distribution function of the interarrival times has a probability density $g(t)$. By conditioning on the interarrival time, we find for any $i \geq 0$ that

$$p_{ij} = P\{X_{n+1} = j | X_n = i\}$$

$$= \int_0^\infty P\{i + 1 - j \text{ service completions occur during } A_{n+1} | A_{n+1} = t\}g(t)\,dt$$

$$= \int_0^\infty e^{-\mu t} \frac{(\mu t)^{i+1-j}}{(i + 1 - j)!} g(t)\,dt, \quad 1 \leq j \leq i + 1.$$

The probability p_{i0} is easiest to compute from $p_{i0} = 1 - \Sigma_{j=1}^{i+1} p_{ij}$, $i \geq 0$.

2.2 STATE CLASSIFICATION

This section introduces some concepts that will be needed in Section 2.3 to analyse the long-run behaviour of the Markov chain. For the classification of the states of the Markov chain, we need the first-passage time probabilities. For any $n = 1, 2, \ldots$, let $f_{ij}^{(n)}$ be defined by

$$f_{ij}^{(n)} = P\{X_n = j,\ X_k \neq j \text{ for } 1 \leq k \leq n - 1 | X_0 = i\}, \quad i, j \in I,$$

with the convention that $f_{ij}^{(0)} = 0$ for all $i, j \in I$. In other words, $f_{ij}^{(n)}$ is the probability that, starting from state i, the first transition into state j occurs at time $t = n$. Next define the probabilities f_{ij} by

$$f_{ij} = \sum_{n=1}^{\infty} f_{ij}^{(n)}, \quad i, j \in I.$$

Then f_{ij} denotes the probability that, starting from state i, the process ever makes a transition into state j. We now come to a very important definition.

Definition 2.2.1 *A state i is said to be recurrent if $f_{ii} = 1$. A nonrecurrent state is said to be transient.*

The following lemma gives another characterization of recurrence.

Lemma 2.2.1 *(a) For any state i, let*

$$Q_{ii} = P\{\text{the process makes infinitely often a transition into state } i | X_0 = i\}.$$

Then

$$Q_{ii} = \begin{cases} 1 & \text{if } i \text{ recurrent,} \\ 0 & \text{if } i \text{ transient.} \end{cases}$$

(b) A state i is transient if and only if $\sum_{n=1}^{\infty} p_{ii}^{(n)} < \infty$.

Proof To prove part (a), define

$$Q_{ii}^{(k)} = P\{\text{the process makes at least } k \text{ transitions into state } i | X_0 = i\}$$

for $k = 1, 2, \ldots$ and $i \in I$. Using that $\lim_{n \to \infty} P(A_n) = P(\lim_{n \to \infty} A_n)$ when $\{A_n\}$ is a sequence of decreasing events, we have

$$\lim_{k \to \infty} Q_{ii}^{(k)} = Q_{ii}.$$

By conditioning on the time of the first return to state i, we have

$$Q_{ii}^{(k)} = \sum_{n=1}^{\infty} f_{ii}^{(n)} Q_{ii}^{(k-1)}, \quad k = 1, 2, \ldots .$$

This yields $Q_{ii}^{(k)} = f_{ii} Q_{ii}^{(k-1)}$ and so $Q_{ii}^{(k)} = f_{ii}^k$ for $k = 1, 2, \ldots$. By letting $k \to \infty$, we get the desired result. To prove part (b), fix $i \in I$. Define for $n = 1, 2, \ldots$ the indicator variable I_n by

$$I_n = \begin{cases} 1 & \text{if } X_n = i, \\ 0 & \text{otherwise.} \end{cases}$$

Then $E(\sum_{n=1}^{\infty} I_n | X_0 = i) = \sum_{n=1}^{\infty} E(I_n | X_0 = i) = \sum_{n=1}^{\infty} p_{ii}^{(n)}$. In other words, the expected number of returns to state i over infinitely many transitions is equal to $\sum_{n=1}^{\infty} p_{ii}^{(n)}$. On the other hand, using that $E(N) = \sum_{k=1}^{\infty} P[N \geq k]$ for any non-negative, integer-valued random variable N, we have

$$E(\text{number of returns to state } i | X_0 = i) = \sum_{k=1}^{\infty} Q_{ii}^{(k)}.$$

Thus we find $\sum_{n=1}^{\infty} p_{ii}^{(n)} = \sum_{k=1}^{\infty} Q_{ii}^{(k)} = \sum_{k=1}^{\infty} f_{ii}^k$. Using Definition 2.2.1 we next obtain part (b).

Definition 2.2.2 *For a recurrent state j the mean recurrence time μ_{jj} is defined by*

$$\mu_{jj} = \sum_{n=1}^{\infty} n f_{jj}^{(n)}.$$

In other words, μ_{jj} denotes the expected number of transitions needed to return in state j when the process starts in state j.

Theorem 2.2.2   *For any $j \in I$,*

$$\lim_{n \to \infty} \frac{1}{n} \sum_{k=1}^{n} p_{jj}^{(k)} = \begin{cases} 1/\mu_{jj} & \text{if } j \text{ is recurrent,} \\ 0 & \text{if } j \text{ is transient.} \end{cases}$$

Also, for any $i, j \in I$,

$$\lim_{n \to \infty} \frac{1}{n} \sum_{k=1}^{n} p_{ij}^{(k)} = f_{ij} \lim_{n \to \infty} \frac{1}{n} \sum_{k=1}^{n} p_{jj}^{(k)}.$$

Proof   If j is transient, then $\lim_{n \to \infty} p_{jj}^{(n)} = 0$ by part (b) of Lemma 2.2.1 and so the Cesaro limit (see Appendix A) is also equal to 0. Next fix a recurrent state j. The times between successive visits to state j are independent and identically distributed random variables. Thus we can say that a renewal occurs each time the Markov chain makes a transition into state j. Denote by $N_j(n)$ the number of renewals up to time $t = n$ when $X_0 = j$. Then, by Lemma 1.1.6 in Section 1.1,

$$\lim_{n \to \infty} \frac{N_j(n)}{n} = \frac{1}{\mu_{jj}} \quad \text{with probability 1.}$$

This result holds for both $\mu_{jj} < \infty$ and $\mu_{jj} = \infty$. Since $N_j(n)/n$ is bounded by 1, we have by the bounded convergence theorem that

$$\lim_{n \to \infty} \frac{E[N_j(n)]}{n} = \frac{1}{\mu_{jj}}.$$

The first assertion now follows by noting that $E[N_j(n)] = \Sigma_{k=1}^{n} p_{jj}^{(k)}$. To prove the other assertion, we use the relation

$$p_{ij}^{(n)} = \sum_{k=1}^{n} f_{ij}^{(k)} p_{jj}^{(n-k)}, \quad i, j \in I \text{ and } n \geq 1. \tag{2.2.1}$$

This relation follows by conditioning on the first epoch at which the Markov chain makes a transition into state j. Averaging this equation over $n = 1, \ldots, m$ and defining $p_{jj}^{(k)} = 0$ for $k < 0$, we find

$$\frac{1}{m} \sum_{n=1}^{m} p_{ij}^{(n)} = \sum_{k=1}^{\infty} f_{ij}^{(k)} \frac{1}{m} \sum_{n=1}^{m} p_{jj}^{(n-k)}, \quad m \geq 1,$$

for any $i, j \in I$. Letting $m \to \infty$ and using the convergence Theorem A.2 in Appendix A, the desired result easily follows.

In general we cannot replace the Cesaro limit by the ordinary limit in Theorem 2.2.2. To see this, consider the Markov chain with $I = \{1, 2\}$ and $p_{12} = p_{21} = 1$. Then, for any i, j, the sequence $\{p_{ij}^{(n)}, n \geq 1\}$ alternates between the values 0 and 1. Periodicity is the reason that the ordinary limit does not exist in this example. If the ordinary limit exists, it is equal to the Cesaro limit.

Definition 2.2.3   *A nonempty set C of states is said to be closed if*

$$p_{ij} = 0 \quad \text{for } i \in C \text{ and } j \notin C.$$

A set C of states is said to be irreducible if the set C is closed and contains no smaller closed subset.

Definition 2.2.4   *State j is said to be accessible from state i if $p_{ij}^{(n)} > 0$ for some $n \geq 0$. Two states i and j, each accessible to the other, are said to communicate.*

It is convenient to write $i \to j$ if state j is accessible from state i and to write $i \Leftrightarrow j$ if the two states i and j communicate. Since $p_{ii}^{(0)} = 1$, we have always $i \Leftrightarrow i$. Note that $i \to j$ for $i \neq j$ only if $p_{ij}^{(n)} > 0$ for some positive value of n.

Lemma 2.2.3   *Let C be a closed set of states. The set C is irreducible if and only if all states in C communicate with each other.*

Proof   For each $i \in C$, define the set $S(i)$ by

$$S(i) = \{j \mid i \to j\}.$$

The set $S(i)$ is not empty since $i \to i$ (by $p_{ii}^{(0)} = 1$). Since the set C is closed, we have $S(i) \subseteq C$. First suppose that C is irreducible. The 'only if' part of the lemma then follows by showing that $S(i) = C$ for all i. To do so, it suffices to show that $S(i)$ is closed. Assume now to the contrary that $S(i)$ is not closed. Then there is a state $r \in S(i)$ and a state $s \notin S(i)$ with $p_{rs} > 0$. Since $r \in S(i)$ we have $p_{ir}^{(n)} > 0$ for some $n \geq 0$ and so $p_{is}^{(n+1)} \geq p_{ir}^{(n)} p_{rs} > 0$ (use relation (2.1.3)). The inequality $p_{is}^{(n+1)} > 0$ contradicts the fact that $s \notin S(i)$. This completes the proof of the 'only if' part of the Lemma. To prove the other part, assume to the contrary that C is not irreducible. Then there is a closed set $S \subseteq C$ with $S \neq C$. Choose $i \in S$ and let the set $S(i)$ be as above. Since S is closed, we have $S(i) \subseteq S$. Hence $S(i) \neq C$ which contradicts the assumption that all states in C communicate.

We are now able to prove the following interesting theorem.

Theorem 2.2.4   *(a) Let C be an irreducible set of states. Then either all states in C are recurrent or all states in C are transient.*

(b) Let C be an irreducible set consisting of recurrent states. Then $f_{ij} = 1$ for all i, $j \in C$. Moreover, either $\mu_{jj} < \infty$ for all $j \in C$ or $\mu_{jj} = \infty$ for all $j \in C$.

Proof   (a) Choose $i, j \in C$ with $j \neq i$. By Lemma 2.2.3, we have $i \Leftrightarrow j$. Hence there are integers $v \geq 1$ and $w \geq 1$ such that $p_{ij}^{(v)} > 0$ and $p_{ji}^{(w)} > 0$. Next, observe that for any $n \geq 0$,

$$p_{ii}^{(n+v+w)} \geq p_{ij}^{(v)} p_{jj}^{(n)} p_{ji}^{(w)} \quad \text{and} \quad p_{jj}^{(n+v+w)} \geq p_{ji}^{(w)} p_{ii}^{(n)} p_{ij}^{(v)}. \tag{2.2.2}$$

These inequalities imply that $\sum_{n=1}^{\infty} p_{jj}^{(n)} < \infty$ if and only if $\sum_{n=1}^{\infty} p_{ii}^{(n)} < \infty$. Part (a) next follows from part (b) of Lemma 2.2.1.

   (b) Since the states of C are recurrent, $f_{ii} = 1$ for all $i \in C$. Choose now $i, j \in C$ with $j \neq i$. By Lemma 2.2.3, $j \to i$. Hence there is an integer $m \geq 1$ with $p_{ji}^{(m)} > 0$. Let r be the smallest integer $m \geq 1$ for which $p_{ji}^{(m)} > 0$. Then

$$1 - f_{jj} = P\{X_n \neq j \text{ for all } n \geq 1 | X_0 = j\} \geq p_{ji}^{(r)}(1 - f_{ij}).$$

Since $f_{jj} = 1$, we get from this inequality that $f_{ij} = 1$. The inequalities in (2.2.2) imply that the sequence $\{p_{jj}^{(k)}, \ k \geq 1\}$ has a positive Cesaro limit if and only if the sequence $\{p_{ii}^{(k)}, \ k \geq 1\}$ has a positive Cesaro limit. Applying Theorem 2.2.2 shows now that $\mu_{jj} < \infty$ if and only if $\mu_{ii} < \infty$.

Definition 2.2.5   *Let i be a recurrent state. The period of state i is said to be d if d is the greatest common divisor of the indices $n \geq 1$ for which $p_{ii}^{(n)} > 0$. A state i with period $d = 1$ is said to be aperiodic.*

Periodicity is not an important issue in operations research applications of Markov chains, since the analysis usually concerns long-run averages. However, the concept of aperiodicity will be needed in the next section to establish a basic limit theorem.

Lemma 2.2.5   *(a) Let C be an irreducible set consisting of recurrent states. Then all states in C have the same period.*

   (b) If state i is aperiodic, then there is an integer n_0 such that $p_{ii}^{(n)} > 0$ for all $n \geq n_0$.

Proof   (a) Denote by $d(k)$ the period of state $k \in C$. Choose $i, j \in C$ with $j \neq i$. By Lemma 2.2.3, we have $i \Leftrightarrow j$. Hence there are integers, $v, \ w \geq 1$ such that $p_{ij}^{(v)} > 0$ and $p_{ji}^{(w)} > 0$. Let n be any positive integer with $p_{jj}^{(n)} > 0$. Then the first inequality in (2.2.2) implies that $p_{ii}^{(n+v+w)} > 0$ and so $n + v + w$ is divisible by $d(i)$. Further, $p_{ii}^{(v+w)} \geq p_{ij}^{(v)} p_{ji}^{(w)} > 0$ and so $v + w$ is divisible by $d(i)$. Thus we find that n is divisible by $d(i)$ whenever $p_{jj}^{(n)} > 0$. This implies that $d(i) \leq d(j)$. For reasons of symmetry, $d(j) \leq d(i)$. Hence $d(i) = d(j)$ which verifies part (a).

   (b) Let $A = \{n \geq 1 | p_{ii}^{(n)} > 0\}$. The index set A is closed in the sense that $n + m \in A$ when $n \in A$ and $m \in A$. This follows from $p_{ii}^{(n+m)} \geq p_{ii}^{(n)} p_{ii}^{(m)}$. Since state i is aperiodic, there are integers $a \in A$ and $b \in A$ whose greatest common divisor is equal to 1. An elementary result in number theory states that there exist integers r and s such that g.c.d. $(a, b) = ar + bs$. The integers r and s are not necessarily non-negative. Let p and q be any positive integers such that p, q are larger than $a \times \max(|r|, |s|)$. Take $m = pa + qb$. Since $m + a = (p + 1)a + qb$, part (b) of the lemma follows by proving that $m + k \in A$ for $k = 0, \ldots, a-1$. Noting that $ar + bs = 1$, it follows that $m + k = pa + qb + k(ar + bs) = (p + kr)a + (q + ks)b$. The integers $p + kr$ and $q + ks$ are positive. Hence, by the closedness of A, the integers $(p + kr)a$ and $(q + ks)b$ belong to A and so the integer $m + k \in A$ for any $k = 0, \ldots, a - 1$.

Finite state space

There are a number of basic results that hold for finite-state Markov chains but not for Markov chains with infinitely many states. In an infinite-state Markov chain it may happen that there is no recurrent state. The next lemma shows that this cannot happen in a finite-state Markov chain

Lemma 2.2.6 *Each finite closed set of states has at least one recurrent state.*

Proof Let C be a closed set of states. Then, for any $i \in C$,

$$\sum_{j \in C} p_{ij}^{(n)} = 1, \quad n = 1, 2, \ldots . \tag{2.2.3}$$

It follows from part (b) of Lemma 2.2.1 that $\lim_{n \to \infty} p_{jj}^{(n)} = 0$ if state j is transient. Next, using relation (2.2.1), we find that $\lim_{n \to \infty} p_{ij}^{(n)} = 0$ for all $i \in I$ when state j is transient. Assume now that all states $j \in C$ are transient. Let $n \to \infty$ in (2.2.3). By the finiteness of C it is allowed to interchange the order of limit and summation. Hence we obtain the contradiction '$0 = 1$' when all states in C are transient. This ends the proof.

The following counterexample shows that Lemma 2.2.6 is not necessarily true when the state space I is infinite. Suppose $I = \{1, 2, \ldots\}$ and $p_{i,i+1} = 1$ for all $i \in I$. Then I is a closed set of states but all states in I are transient.

Next we show that the set of recurrent states can be split into disjoint irreducible classes. Let R be the set of recurrent states. By Lemma 2.2.6 the set R is not empty for a finite-state Markov chain. To prove that the set R can be uniquely split in disjoint irreducible subsets $R_1, \ldots, R_f$, we first make the following two observations.

(a) If state i communicates with state j and state i communicates with state k, then the states j and k communicate.

(b) If state i is recurrent and state k accessible from state i, then state i is accessible from state k (if not, there would be a positive probability of never returning in i).

Define now for each $i \in R$ the set $C(i)$ as the set of all states j that communicate with state i. Note that $C(i)$ is not empty since i communicates with itself by definition. Since state $i \in R$ is recurrent, it follows from (b) that $C(i)$ is closed. From (a) it follows that any two states $j, k \in C(i)$ communicate. Hence, by Lemma 2.2.3, the set $C(i)$ is irreducible for each $i \in R$. Also, using (a) and (b), it is readily verified that $C(i) = C(j)$ if states i and j communicate and $C(i) \cap C(j)$ is empty otherwise. This completes the verification that the set R can be uniquely split into disjoint irreducible subsets $R_1, \ldots, R_f$. Each of those subsets is called a recurrent subclass.

Illustration

Consider a Markov chain with five states. The matrix $P = (p_{ij})$ of one-step transition probabilities is given by

$$P = \begin{bmatrix} 0.2 & 0.8 & 0 & 0 & 0 \\ 0.7 & 0.3 & 0 & 0 & 0 \\ 0.1 & 0 & 0.2 & 0.3 & 0.4 \\ 0 & 0.4 & 0.3 & 0 & 0.3 \\ 0 & 0 & 0 & 0 & 1 \end{bmatrix}.$$

For such small examples, a state diagram is useful for doing the state classification. The state diagram uses a Boolean representation of the p_{ij} 's. An arrow is drawn from state i to state j only if $p_{ij} > 0$. The state diagram is given in Figure 2.2.1. By inspection it is seen that the set of transient states is $T = \{3, 4\}$ and the set of recurrent states is $R = \{1, 2, 5\}$. The set R of recurrent states can be split into two disjoint recurrent classes $R_1 = \{1, 2\}$ and $R_2 = \{5\}$. The state 5 is called absorbing. An interesting question is in which recurrent subclass is the process ultimately absorbed when it starts in a transient state. Define for a given irreducible set C of states, the absorption probabilities

f_{iC} = the probability that the process ever makes a transition into the set C starting from state i

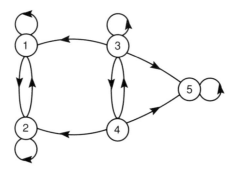

Figure 2.2.1   The state diagram for a Markov chain

for $i \in T$. Then the probabilities f_{iC} can be computed as the unique solution to the system of linear equations

$$f_{iC} = \sum_{j \in C} p_{ij} + \sum_{j \in T} p_{ij} f_{jC}, \quad i \in T.$$

For example, taking $C = R_1$, we find $f_{3C} = 0.3099$ and $f_{4C} = 0.4930$. The mean times until absorption in the set C are infinite in this example. The mean absorption times are only well defined when the Markov chain has a single recurrent class. Then they can be computed from a system of linear equations similar to the system of equations for the absorption probabilities.

A systematic procedure for identifying the transient states and the recurrent subclasses is given in Fox and Landi (1968).

In most applications the Markov chain has no two disjoint closed sets (usually there is a state that is accessible from any other state). The next theorem summarizes a number of useful results for finite-state Markov chains having no two disjoint closed sets.

Theorem 2.2.7   *Let $\{X_n\}$ be a finite-state Markov chain. Suppose that the Markov chain has no two disjoint closed sets. Denote by R the set of recurrent states. Then*
   (a) $f_{ij} = 1$ for all $i \in I$ and $j \in R$.
   (b) $\mu_{ij} < \infty$ for all $i \in I$ and $j \in R$, where the mean first-passage times μ_{ij} are defined by $\mu_{ij} = \Sigma_{n=1}^{\infty} n f_{ij}^{(n)}$.
   (c) If the recurrent states are aperiodic, then there is an integer $v \geq 1$ such that $p_{ij}^{(v)} > 0$ for all $i \in I$ and $j \in R$.

Proof   Since the Markov chain has no two disjoint closed sets, the closed set R of recurrent states cannot be split into two or more disjoint recurrent subclasses. Thus all states in R communicate with each other. This implies that for any $i, j \in R$ there is an integer $n \geq 1$ such that $p_{ij}^{(n)} > 0$. Next we prove that for any $i \in I$ and $j \in R$ there is an integer $n \geq 1$ such that $p_{ij}^{(n)} > 0$. To verify this, assume to the contrary that there is a transient state $i \in I$ such that no state $j \in R$ is accessible from i. Then there is a closed set that contains i and is disjoint from R. This contradicts the assumption that the Markov chain has no two disjoint closed sets. Hence for any transient state $i \in I$ there is a state $j \in R$ that is accessible from i. Thus any state $j \in R$ is accessible from any $i \in I$.

To verify parts (a) and (b), define under the condition $X_0 = i$ the random variable N_{ij} by

$$N_{ij} = \min\{n \geq 1 | X_n = j\}.$$

Fix now $j \in R$. For each $i \in I$, let r_i be the smallest positive integer n for which $p_{ij}^{(n)} > 0$. Define

$$r = \max_{i \in I} r_i \quad \text{and} \quad \rho = \min_{i \in I} p_{ij}^{(r_i)}.$$

Since I is finite, we have $r < \infty$ and $\rho > 0$. Next observe that

$$P\{N_{ij} > r\} \le P\{N_{ij} > r_i\} = 1 - p_{ij}^{(r_i)} \le 1 - \rho, \quad i \in I.$$

Thus, for any $i \in I$,

$$P\{N_{ij} > kr\} \le (1 - \rho)^k, \quad k = 0, 1, \dots .$$

Note that $1 - f_{ij} = \lim_{n\to\infty} P\{N_{ij} > n\}$ and $\mu_{ij} = \Sigma_{n=0}^{\infty} P\{N_{ij} > n\}$. Since the probability $P\{N_{ij} > n\}$ is decreasing in n, it now follows that $f_{ij} = 1$ and $\mu_{ij} \le r/\rho < \infty$.

It remains to prove (c). Fix $i \in I$ and $j \in R$. As shown above, there is an integer $v \ge 1$ such that $p_{ij}^{(v)} > 0$. By part (b) of Lemma 2.2.5 there is an integer $n_0 \ge 1$ such that $p_{jj}^{(n)} > 0$ for all $n \ge n_0$. Hence, by $p_{ij}^{(v+n)} \ge p_{ij}^{(v)} p_{jj}^{(n)}$, it follows that $p_{ij}^{(n)} > 0$ for all $n \ge v + n_0$. Using the finiteness of I, part (c) of the theorem now follows.

The theorem is in general not valid when the state space I is infinite. To illustrate this, take $I = \{0, \pm 1, \pm 2, \dots\}$ and $p_{i,i+1} = p_{i,i-1} = \frac{1}{2}$ for all $i \in I$. This Markov chain corresponds to the symmetric random walk. It can be shown for the symmetric random walk that $f_{ij} = 1$ for all $i, j \in I$ but $\mu_{ij} = \infty$ for all $i, j \in I$. Such an infinite-state Markov chain is called *null-recurrent*. Null-recurrent Markov chains occur very seldomly in applications.

2.3   LONG-RUN ANALYSIS OF DISCRETE-TIME MARKOV CHAINS

As pointed out before, the theoretical analysis of Markov chains is much more subtle for the case of infinitely many states than for the case of finitely many states. A finite-state Markov chain is always a regenerative process with a finite mean cycle length. This is not true for infinite-state Markov chains. Recall the example with $I = \{1, 2, \dots\}$ and $p_{i,i+1} = 1$ for all $i \in I$ and recall the example of the symmetric random walk with $I = \{0, \pm 1, \pm 2, \dots\}$ and $p_{i,i+1} = p_{i,i-1} = \frac{1}{2}$ for all i. In the first example the Markov chain is not regenerative, while in the other example the Markov chain is regenerative but has an infinite mean cycle length. In practical applications these pathological situations occur very rarely. Typically there is a state that can be reached from any other state in a finite expected time. In the practical analysis of Markov chains the distinction between the case of finitely many states and the case of infinitely many states is usually not important.

This section lays the groundwork for the applications to be dealt with in the next sections. A number of ergodic theorems will be established. These ergodic theorems provide the basis for the computation of long-run averages in applications. The analysis is first given for the case of a finite state space. The results are next generalized for infinite-state Markov chains.

An outline of the analysis for the finite-state Markov chain in Section 2.3.1 is as follows. First, we establish the convergence of the n-step transition probabilities for an aperiodic Markov chain. Secondly, it is verified that this limiting distribution constitutes an equilibrium distribution for the Markov chain. The equilibrium probabilities play a prominent role in the subsequent analysis, since they also represent the long-run average occupancies of the various states. Thirdly, it is shown that the aperiodicity assumption is superfluous for the existence of an equilibrium distribution. Finally, we establish an ergodic theorem expressing the long-run average reward per unit time in terms of the equilibrium probabilities. The reader may skip the proofs but should read Section 2.3.1. Next the reader may go directly to the applications Section 2.4. Sections 2.3.2 and 2.3.3 dealing with the infinite-state Markov chain may be omitted at first reading.

2.3.1 Finite-state Markov chains

We restrict this analysis to Markov chains having no two disjoint closed sets. This restriction ensures that the influence of the initial state of the process ultimately fades out. This phenomenon does not occur when the Markov chain has two disjoint closed sets. Then the initial state will typically be reflected in the long-run behaviour of the process for the simple reason that the process stays forever in a closed set once it starts there. In most applications, however, there is some state that is accessible from any other state.

A finite-state Markov chain with no two disjoint closed sets has a unique irreducible set of states and this set consists of recurrent states. The Markov chain is said to be *aperiodic* when the period of the recurrent states is equal to 1. Although the concept of aperiodicity is not relevant for the analysis of long-run averages, it is needed to establish the convergence of the n-step transition probabilities.

In the next theorem it will not only be proved that the n-step transition probabilities $p_{ij}^{(n)}$ converge as $n \to \infty$, but the convergence rate will be established as well. It will be seen that the convergence is geometrically fast. The proof of the theorem goes back to Markov (1906). This proof is not just of historical interest but the ideas used in the proof are still very much alive.

Theorem 2.3.1   *Let $\{X_n\}$ be a finite-state Markov chain with no two disjoint closed sets. Suppose that the Markov chain is aperiodic. Then there exists a probability distribution $\{\pi_j, \ j \in I\}$ and numbers $\alpha > 0$ and $0 < \beta < 1$ such that, for all $i, j \in I$,*

$$|p_{ij}^{(n)} - \pi_j| \le \alpha\beta^n, \quad n = 1, 2, \ldots .$$

In particular,

$$\lim_{n \to \infty} p_{ij}^{(n)} = \pi_j \quad \text{for all } i, j \in I.$$

Proof   Since the Markov chain is aperiodic, we have by part (c) of Theorem 2.2.7 that there exist an integer $v \ge 1$, a number $\rho > 0$ and a state s such that

$$p_{is}^{(v)} \geq \rho \quad \text{for all } i \in I.$$

For any $j \in I$, define the sequences $\{M_j^{(n)}\}$ and $\{m_j^{(n)}\}$ by

$$M_j^{(n)} = \max_{i \in I} p_{ij}^{(n)} \quad \text{and} \quad m_j^{(n)} = \min_{i \in I} p_{ij}^{(n)}.$$

Applying relation (2.1.2), we find

$$M_j^{(n+1)} = \max_{i \in I} \sum_{k \in I} p_{ik} p_{kj}^{(n)} \leq \max_{i \in I} \sum_{k \in I} p_{ik} M_j^{(n)} = M_j^{(n)} \max_{i \in I} \sum_{k \in I} p_{ik},$$

and so, for any $j \in I$,

$$M_j^{(n+1)} \leq M_j^{(n)}, \quad n = 0, 1, \dots \, .$$

Similarly we find for any $j \in I$ that

$$m_j^{(n+1)} \geq m_j^{(n)}, \quad n = 0, 1, \dots \, .$$

Since the sequences $\{M_j^{(n)}\}$ and $\{m_j^{(n)}\}$ are bounded and monotone, they have finite limits. Next we establish the inequality

$$0 \leq M_j^{(n)} - m_j^{(n)} \leq (1 - \rho) \left[M_j^{(n-v)} - m_j^{(n-v)} \right], \quad n \geq v \qquad (2.3.1)$$

for any $j \in I$. Suppose for the moment we have proved this inequality. A repeated application of the inequality shows that

$$0 \leq M_j^{(n)} - m_j^{(n)} \leq (1 - \rho)^{[n/v]} (M_j^{(0)} - m_j^{(0)}), \quad n = 0, 1, \dots, \qquad (2.3.2)$$

where $[x]$ denotes the largest integer contained in x. Here we used the fact that $M_j^{(n)} - m_j^{(n)}$ is decreasing in n. By (2.3.2), we have that the limits of the monotone sequences $\{M_j^{(n)}\}$ and $\{m_j^{(n)}\}$ coincide. Denote the common limit by π_j. Hence

$$\lim_{n \to \infty} M_j^{(n)} = \lim_{n \to \infty} m_j^{(n)} = \pi_j.$$

Using the inequalities $m_j^{(n)} \leq p_{ij}^{(n)} \leq M_j^{(n)}$ and $m_j^{(n)} \leq \pi_j \leq M_j^{(n)}$, we find

$$\left| p_{ij}^{(n)} - \pi_j \right| \leq M_j^{(n)} - m_j^{(n)}, \quad n = 0, 1, \dots \qquad (2.3.3)$$

for any $i, j \in I$. Together the inequalities (2.3.2) and (2.3.3) yield the assertion of the theorem except that we have still to verify that $\{\pi_j\}$ represents a probability distribution. Obviously, the π_j's are non-negative. From $\Sigma_{j \in I} p_{ij}^{(n)} = 1$ for all n and $p_{ij}^{(n)} \to \pi_j$ as $n \to \infty$ we obtain that the π_j's sum up to 1.

It remains to verify (2.3.1). To do so, fix $j \in I$ and $n \geq v$. Let x and y be the states for which $M_j^{(n)} = p_{xj}^{(n)}$ and $m_j^{(n)} = p_{yj}^{(n)}$. Then

$$0 \leq M_j^{(n)} - m_j^{(n)} = p_{xj}^{(n)} - p_{yj}^{(n)} = \sum_{k \in I} p_{xk}^{(v)} p_{kj}^{(n-v)} - \sum_{k \in I} p_{yk}^{(v)} p_{kj}^{(n-v)}$$

$$= \sum_{k \in I} \{p_{xk}^{(v)} - p_{yk}^{(v)}\} p_{kj}^{(n-v)}$$

$$= \sum_{k \in I} \{p_{xk}^{(v)} - p_{yk}^{(v)}\}^+ p_{kj}^{(n-v)} - \sum_{k \in I} \{p_{xk}^{(v)} - p_{yk}^{(v)}\}^- p_{kj}^{(n-v)},$$

where $a^+ = \max(a, 0)$ and $a^- = -\min(a, 0)$. Hence, using that a^+, $a^- \geq 0$,

$$0 \leq M_j^{(n)} - m_j^{(n)} \leq \sum_{k \in I} \{p_{xk}^{(v)} - p_{yk}^{(v)}\}^+ M_j^{(n-v)} - \sum_{k \in I} \{p_{xk}^{(v)} - p_{yk}^{(v)}\}^- m_j^{(n-v)}$$

$$= \sum_{k \in I} \{p_{xk}^{(v)} - p_{yk}^{(v)}\}^+ \left[M_j^{(n-v)} - m_j^{(n-v)} \right],$$

where the last equality uses that $\Sigma_k a_k^+ = \Sigma_k a_k^-$ if $\Sigma_k a_k = 0$. Using the relation $(a - b)^+ = a - \min(a, b)$, we next find

$$0 \leq M_j^{(n)} - m_j^{(n)} \leq \left[1 - \sum_{k \in I} \min(p_{xk}^{(v)}, p_{yk}^{(v)}) \right] \left[M_j^{(n-v)} - m_j^{(n-v)} \right].$$

Since $p_{is}^{(v)} \geq \rho$ for all i, we find

$$1 - \sum_{k \in I} \min(p_{xk}^{(v)}, p_{yk}^{(v)}) \leq 1 - \rho,$$

which implies the inequality (2.3.1). This completes the proof.

We now come to a very important definition. This definition applies to both finite-state Markov chains and infinite-state Markov chains.

Definition 2.3.1 *A probability distribution $\{\pi_j, j \in I\}$ is said to be an equilibrium distribution for the Markov chain $\{X_n\}$ if*

$$\pi_j = \sum_{k \in I} \pi_k p_{kj}, \quad j \in I. \tag{2.3.4}$$

An explanation of the term equilibrium distribution is as follows. Suppose that the initial state of the process $\{X_n\}$ is chosen according to

$$P\{X_0 = j\} = \pi_j, \quad j \in I.$$

Then, for each $n = 1, 2, \ldots,$

$$P\{X_n = j\} = \pi_j, \quad j \in I.$$

In other words, starting the process according to the equilibrium distribution leads to a process that operates in an equilibrium mode. The proof is simple. The result follows by induction on n from

$$P\{X_n = j\} = \sum_{k \in I} P\{X_{n-1} = k\} p_{kj}.$$

An important question is: Does the Markov chain have an equilibrium distribution and, if it has, is this equilibrium distribution unique? To answer this question, we first prove:

Lemma 2.3.2 *Let $\{X_n\}$ be a finite-state Markov chain with no two disjoint closed sets. Suppose that the Markov chain is aperiodic. Then the probability distribution $(\pi_j, \ j \in I)$ from Theorem 2.3.1 is an equilibrium distribution for the Markov chain.*

Proof Letting $n \to \infty$ in both sides of equation (2.1.3), the desired result follows. The interchange of the order of limit and summation is justified by the finiteness of the state space.

The aperiodicity assumption is superfluous to establish the existence of an equilibrium distribution. We now introduce a simple but useful trick to get rid of the aperiodicity condition in Lemma 2.3.2. For a fixed number τ with $0 < \tau < 1$, define the probabilities $\overline{p}_{ij}$ by

$$\overline{p}_{ij} = \begin{cases} \tau p_{ij} & \text{if } j \neq i, \\ 1 - \tau + \tau p_{ii} & \text{if } j = i, \end{cases} \tag{2.3.5}$$

for all $i, j \in I$. The numbers $\overline{p}_{ij}$ are indeed probabilities since they are all non-negative and $\Sigma_j \overline{p}_{ij} = 1$ for all i. The probabilities $\overline{p}_{ij}$ can be interpreted as the one-step transition probabilities of a Markov chain that arises when each transition of the original Markov chain happens unchanged with probability τ and is cancelled in favour of a self-transition with probability $1 - \tau$. The effect of this modification is that each transition of the original Markov chain is delayed on average with the same time factor $1/\tau$. It is immediately seen that the modified Markov chain has the same transient states and the same recurrent classes as the original Markov chain. However, the modified Markov chain is guaranteed to be aperiodic, since $\overline{p}_{ii} > 0$ for all i. Using the transformation (2.3.5), we can now extend Lemma 2.3.2. This leads to the following main result.

Theorem 2.3.3 *Let $\{X_n\}$ be a finite-state Markov chain with no two disjoint closed sets. Then:*

(a) The Markov chain $\{X_n\}$ has a unique equilibrium distribution $\{\pi_j\}$. For all $j \in I$,

$$\lim_{n \to \infty} \frac{1}{n} \sum_{k=1}^{n} p_{ij}^{(k)} = \pi_j, \text{ independently of } i. \tag{2.3.6}$$

(b) Let $\{x_j, \; j \in I\}$ be any finite solution to $x_j = \Sigma_k x_k p_{kj}, \; j \in I$, then, for some constant c, $x_j = c\pi_j$ for all $j \in I$.

Proof   Let $\{\overline{X}_n\}$ be the Markov chain associated with the one-step transition probabilities $\overline{p}_{ij}$ defined in (2.3.5). The Markov chain $\{\overline{X}_n\}$ has no two disjoint closed sets and is aperiodic. Hence, by Lemma 2.3.2, an equilibrium distribution exists for the Markov chain $\{\overline{X}_n\}$. We now verify that the Markov chains $\{\overline{X}_n\}$ and $\{X_n\}$ have identical solutions to the equilibrium equations. To see this, note that

$$x_j = \sum_{k \in I} x_k \overline{p}_{kj} \Leftrightarrow x_j = (1 - \tau + \tau p_{jj})x_j + \tau \sum_{k \neq j} x_k p_{kj}$$

$$\Leftrightarrow \tau x_j = \tau \sum_{k \in I} x_k p_{kj} \Leftrightarrow x_j = \sum_{k \in I} x_k p_{kj}.$$

This proves that the Markov chain $\{X_n\}$ has also an equilibrium distribution $\{\pi_j, \; j \in I\}$.

To verify the other assertions of the theorem, let $\{x_j, \; j \in I\}$ be any finite solution to $x_j = \Sigma_{k \in I} x_k p_{kj}, \; j \in I$. Substituting this equation repeatedly into itself, interchanging the order of summation and using the relation (2.1.3), it is readily verified that

$$x_j = \sum_{k \in I} x_k p_{kj}^{(n)}, \quad j \in I \text{ and } n = 1, 2, \ldots .$$

Averaging this equation over $n = 1, \ldots, m$, we find

$$x_j = \sum_{k \in I} x_k \frac{1}{m} \sum_{n=1}^{m} p_{kj}^{(n)}, \quad j \in I \text{ and } m = 1, 2 \ldots . \tag{2.3.7}$$

Next we let $m \to \infty$ in this equation. We can interchange the order of limit and summation by the finiteness of the state space. Using the second relation in Theorem 2.2.2 and part (a) of Theorem 2.2.7, it is easily verified that

$$\lim_{m \to \infty} \frac{1}{m} \sum_{n=1}^{m} p_{kj}^{(n)} = \lim_{m \to \infty} \frac{1}{m} \sum_{n=1}^{m} p_{jj}^{(n)} \quad \text{for all } j, k \in I.$$

Thus we find

$$x_j = \left(\sum_{k \in I} x_k \right) \lim_{m \to \infty} \frac{1}{m} \sum_{n=1}^{m} p_{jj}^{(n)}, \quad j \in I.$$

Taking the particular solution $x_j = \pi_j, \; j \in I$ and using that $\Sigma_{j \in I} \pi_j = 1$, it follows that $\pi_j = \lim_{m \to \infty} (1/m) \Sigma_{n=1}^{m} p_{jj}^{(n)}, \; j \in I$, showing the uniqueness of the equilibrium distribution. Moreover, the above relation verifies part (b).

The above proof of the important Theorem 2.3.3 is based on the ergodic Theorem 2.3.1 and the transformation (2.3.5). An alternative proof of the theorem is given in Section 2.3.2 in the more general context of an infinite-state Markov chain.

Many practical applications involve a Markov chain on which a reward structure is imposed. Suppose that a reward $f(j)$ is earned each time the Markov chain makes a transition into state j. The next ergodic theorem shows how to compute the long-run average reward per unit time in terms of the equilibrium probabilities π_j.

Theorem 2.3.4 *Let $\{X_n\}$ be a finite-state Markov chain with no two disjoint closed sets. Suppose that $f(j)$, $j \in I$ is a given finite function defined on the state space I. Then, for each initial state $X_0 = i$,*

$$\lim_{n\to\infty} \frac{1}{n} \sum_{k=1}^{n} f(X_k) = \sum_{j\in I} \pi_j f(j) \quad \text{with probability 1.}$$

Proof We first show that $\lim_{n\to\infty}(1/n)\Sigma_{k=1}^{n} f(X_k)$ exists and is equal to a constant with probability 1. Next we establish the constant. Choose a recurrent state s. Then the Markov chain $\{X_n\}$ regenerates itself each time a transition is made into state s. Denoting by a cycle the time between two successive visits to state s, it follows from part (b) of Theorem 2.2.7 that the expected number of transition epochs in one cycle is finite. Thus, by the renewal-reward theorem in Section 1.3, we have for $X_0 = s$ that

$$\lim_{n\to\infty} \frac{1}{n} \sum_{k=1}^{n} f(X_k) = c \quad \text{with probability 1}$$

where the constant c is given by

$$c = \frac{E(\text{reward earned in one cycle})}{E(\text{number of transitions in one cycle})}.$$

Next we prove that $\lim_{n\to\infty}(1/n)\Sigma_{k=1}^{n} f(X_k)$ is also equal to the constant c with probability 1 for any initial state $X_0 = i$. To prove this, consider a realization $\omega = (i_0, i_1, \ldots)$ of the Markov chain $\{X_n\}$ with $i_k = X_k(\omega)$ denoting the realized state at the kth transition. By part (a) of Theorem 2.2.7, there is a finite integer $t = t(\omega)$ such that $i_t = s$. Then

$$\frac{1}{n} \sum_{k=1}^{n} f(X_k(\omega)) = \frac{1}{n} \sum_{k=1}^{t(\omega)} f(X_k(\omega)) + \frac{1}{n} \sum_{k=t(\omega)+1}^{n} f(X_k(\omega)).$$

Next, by letting $n \to \infty$, we find

$$\lim_{n\to\infty} \frac{1}{n} \sum_{k=1}^{n} f(X_k(\omega)) = c \tag{2.3.8}$$

for almost all ω. This shows that $\lim_{n\to\infty}(1/n)\Sigma_{k=1}^{n} f(X_k)$ is equal to the constant c with probability 1 for each $X_0 = i$. To find the constant c, we first use the bounded convergence theorem to conclude that

$$\lim_{n\to\infty} E\left[\frac{1}{n}\sum_{k=1}^{n} f(X_k)\right] = c$$

for each $X_0 = i$. Note that $\{(1/n)\Sigma_{k=1}^n f(X_k)\}$ is a bounded sequence of random variables by the finiteness of the state space. Thus we find for each $X_0 = i$ that

$$c = \lim_{n\to\infty} \frac{1}{n} \sum_{k=1}^{n} E[f(X_k)] = \lim_{n\to\infty} \frac{1}{n} \sum_{k=1}^{n} \sum_{j\in I} f(j) p_{ij}^{(k)}$$

$$= \lim_{n\to\infty} \frac{1}{n} \sum_{j\in I} f(j) \sum_{k=1}^{n} p_{ij}^{(k)} = \sum_{j\in I} f(j) \lim_{n\to\infty} \frac{1}{n} \sum_{k=1}^{n} p_{ij}^{(k)},$$

where the various operations on summation and limit are justified by the finiteness of the state space. Using the relation (2.3.6), the desired result next follows.

In words, the theorem states that the long-run average reward per unit time is equal to $\Sigma_{j\in I} f(j)\pi_j$ with probability 1 when a reward $f(j)$ is earned each time the process makes a transition into state j. The time between two transitions is taken as time unit.

Remark 2.3.1 Modification of the ergodic Theorem 2.3.4

In Theorem 2.3.4 the function f refers to an immediate reward $f(j)$ that is earned each time the Markov chain makes a transition into state j. However, in applications it happens often that rewards are earned gradually during the time between two transitions of the Markov chain. Define for these situations $f(j)$ as the expected value of the reward earned until the next state transition when the current state is j. Then it remains true that the long-run average reward per unit time equals $\Sigma_j f(j)\pi_j$ with probability 1. This can be directly seen from the proof of Theorem 2.3.4. The long-run average reward per unit time equals E(reward earned during one cycle) divided by E(length of one cycle). The expression for E(reward earned during one cycle) is not affected whether $f(j)$ represents an immediate reward or an expected reward.

Interpretation of the equilibrium probabilities

Applying Theorem 2.3.4 with the particular reward function

$$f(j) = \begin{cases} 1 & \text{if } j = k, \\ 0 & \text{if } j \neq k, \end{cases}$$

for some fixed $k \in I$, it follows that

the long-run fraction of time the process is in state $k = \pi_k$

with probability 1. In other words, the long-run fraction of transitions into state k is equal to π_k with probability 1. This interpretation is helpful in memorizing the equilibrium equations

$$\pi_k = \sum_{j \in I} \pi_j p_{jk}, \quad k \in I.$$

To explain this, note that in any finite time interval the number of transitions into state k can differ by at most 1 from the number of transitions out of state k. Hence the long-run fraction of transitions into state k must be equal to the long-run fraction of transitions out of state k. The left side of the equilibrium equation can be interpreted as the long-run fraction of transitions out of state k. The right side of the equilibrium equation can be interpreted as the long-run fraction of transitions into state k, since

the long-run fraction of transitions into state k

$$= \sum_{j \in I} \text{(the long-run fraction of transitions from state } j \text{ to state } k)$$

$$= \sum_{j \in I} p_{jk} \times \text{(the long-run fraction of transitions out of state } j).$$

Computation of the equilibrium probabilities

The equilibrium (or balance) probabilities π_j are the unique solution to the system of linear equations

$$\pi_j = \sum_{k \in I} \pi_k p_{kj}, \quad j \in I, \tag{2.3.9}$$

$$\sum_{j \in I} \pi_j = 1. \tag{2.3.10}$$

Equations (2.3.9) are called the equilibrium (or balance) equations and equation (2.3.10) is called the normalizing equation.

A Gaussian elimination method such as the Gauss–Jordan method is the recommended method to solve the equations when the number of states is in the order of 1–200. This direct method has the advantage that fast and reliable codes are widely available. In applying the Gaussian elimination method one of the equilibrium equations is omitted in order to obtain a square system of linear equations. It is left to the reader to verify that any of the equilibrium equations is implied by the other equilibrium equations in conjunction with the normalizing equation.

For larger systems of linear equations, the Gaussian elimination method is too time-consuming and requires too much computer memory. Then an iterative method is recommended. A convenient and effective method is Gauss–Seidel or successive overrelaxation. These iterative methods are described in Appendix D. They are easy to program and usually do not require memory for declaring the coefficient matrix when the structure of the specific application is exploited.

Sometimes it is possible to compute the π_j's by a recursion scheme. To explain this, we first state the following lemma.

Lemma 2.3.5   *Let the finite-state Markov chain $\{X_n\}$ have no two disjoint closed sets. Then for any nonempty subset A of states with $A \neq I$,*

$$\sum_{j \in A} \pi_j \sum_{k \notin A} p_{jk} = \sum_{k \notin A} \pi_k \sum_{j \in A} p_{kj}. \tag{2.3.11}$$

Proof   A formal proof is as follows. Using (2.3.9), we have

$$\sum_{j \in A} \pi_j \sum_{k \notin A} p_{jk} = \sum_{k \notin A} \sum_{j \in A} \pi_j p_{jk} = \sum_{k \notin A} \left[\sum_{j \in I} \pi_j p_{jk} - \sum_{j \notin A} \pi_j p_{jk} \right]$$

$$= \sum_{k \notin A} \left[\pi_k - \sum_{j \notin A} \pi_j p_{jk} \right] = \sum_{k \notin A} \pi_k - \sum_{j \notin A} \pi_j \sum_{k \notin A} p_{jk}$$

$$= \sum_{k \notin A} \pi_k - \sum_{j \notin A} \pi_j \left(1 - \sum_{k \in A} p_{jk} \right) = \sum_{j \notin A} \pi_j \sum_{k \in A} p_{jk}.$$

An intuitive explanation of (2.3.11) can be given. This explanation helps to memorize the balance equation. By physical considerations,

the long-run fraction of transitions from inside A to outside A
= the long-run fraction of transitions from outside A to inside A.

This balance principle enables us to write down directly the equation (2.3.11). Lemma 2.3.5 has an important corollary.

Corollary 2.3.6   *Let the finite-state Markov chain $\{X_n\}$ have no two disjoint closed sets. Suppose that the state space $I = \{0, 1, \ldots, N\}$. If the one-step transition probabilities have the property $p_{ij} = 0$ for all $j \leq i - 2$ then the π_j's satisfy the recursion relation*

$$p_{i,i-1}\pi_i = \sum_{k=0}^{i-1} \pi_k \sum_{j=i}^{N} p_{kj}, \quad i = 1, \ldots, N. \tag{2.3.12}$$

This corollary follows by applying Lemma 2.3.5 with $A = \{i, \ldots, N\}$. To compute the π_i's from (2.3.12), we use the result that the equilibrium equations (2.3.9) determine the π_i's uniquely up to a multiplicative constant. The multiplicative constant is found by using the normalizing equation. Thus we proceed as follows. Initialize $\overline{\pi}_0 := 1$. Then compute $\overline{\pi}_1, \ldots, \overline{\pi}_N$ recursively from (2.3.12). Finally, normalize the $\overline{\pi}_i$'s according to

$$\pi_i := \overline{\pi}_i \bigg/ \sum_{k=0}^{N} \overline{\pi}_k, \quad i = 0, 1, \ldots, N. \tag{2.3.13}$$

In the recursion (2.3.12) we assume that $p_{i,i-1} > 0$ for all $i \geq 1$. The recursion scheme (2.3.12) involves no subtractions. Basically, it is numerically stable.

However, since the initialization is $\overline{\pi}_0 = 1$, very large numbers $\overline{\pi}_i$ may build up. Therefore, to avoid overflow, it may be wise to do a renormalization at intermediate steps of the recursion. Also, it may occur that $p_{i,i-1}$ becomes extremely close to 0. Then it may be helpful to take the logarithms of both sides of (2.3.12) and to calculate $\overline{\pi}_i$ indirectly via $\ln(\overline{\pi}_i)$.

The recursion scheme (2.3.12) is computationally less demanding than solving the linear equations (2.3.9)–(2.3.10) by a general code. The computational complexity of the recursion scheme is $O(N^2)$, while the computational complexity of the Gaussian elimination method is $O(N^3)$. The linear equations (2.3.9)–(2.3.10) are of the Hessenberg type when the p_{ij}'s satisfy the condition of Corollary 2.3.6. It should be noted that an efficient code for solving linear equations of the Hessenberg type is competitive with the recursion scheme (2.3.12) provided that N is not too large.

2.3.2 Infinite-state Markov chains

In the previous subsection the analysis for the finite-state Markov model rested upon the ergodic Theorem 2.3.1. This theorem is in general not valid for an infinite-state Markov chain. It is only valid under a strong recurrence condition.

Theorem 2.3.7 *Let the Markov chain $\{X_n\}$ have no two disjoint closed sets. Suppose that the Markov chain is aperiodic. Then a necessary and sufficient condition for the result of Theorem 2.3.1 to hold is the existence of a finite set K of states, a number $\rho > 0$ and an integer $\nu \geq 1$ such that*

$$\sum_{j \in K} p_{ij}^{(\nu)} \geq \rho \quad for\ all\ i \in I.$$

Proof   See Federgruen and Tijms (1978).

The recurrence condition in Theorem 2.3.7 is trivially satisfied when the state space I is finite, but it is too strong for many applications of infinite-state Markov chains. The existence of a unique equilibrium distribution can be established under much weaker recurrence conditions. We now introduce the following recurrence condition.

Assumption 2.3.1   *The Markov chain $\{X_n\}$ has some state r such that $f_{ir} = 1$ for all $i \in I$ and $\mu_{rr} < \infty$.*

In other words, the Markov chain has a regeneration state r that is ultimately reached from each initial state with probability 1, where the mean recurrence time from state r to itself is finite. The assumption is typically satisfied in practical applications. Note that the assumption implies that the Markov chain has no two disjoint closed sets. Moreover, state r is recurrent.

Lemma 2.3.8   *Suppose that the Markov chain $\{X_n\}$ satisfies Assumption 2.3.1. Then, for any recurrent state j, $f_{ij} = 1$ for all $i \in I$ and $\mu_{jj} < \infty$.*

Proof   The set R of recurrent states is not empty. Since the Markov chain has no two disjoint closed sets, the set R is irreducible. The lemma now follows from Theorem 2.2.4.

We now come to a generalization of Theorem 2.3.3.

Theorem 2.3.9   *Suppose that the Markov chain $\{X_n\}$ satisfies Assumption 2.3.1. Then*
   (a) The Markov chain $\{X_n\}$ has a unique equilibrium distribution $\{\pi_j,\ j \in I\}$. For all $i, j \in I$,

$$\lim_{n \to \infty} \frac{1}{n} \sum_{k=1}^{n} p_{ij}^{(k)} = \pi_j. \tag{2.3.14}$$

   (b) Let $\{x_j\}$ be any solution to $x_j = \Sigma_{k \in I} x_k p_{kj}$, $j \in I$ with $\Sigma_j |x_j| < \infty$. Then, for some constant c, $x_j = c\pi_j$ for all $j \in I$.

Proof   Define $\pi_j = \lim_{n \to \infty} (1/n) \Sigma_{k=1}^n p_{jj}^{(k)}$, $j \in I$. The result (2.3.14) next follows from Theorem 2.2.2, Lemma 2.3.8, and the fact that $\pi_j = 0$ if j is transient. Next we average both sides of (2.1.3) over $n = 1, \ldots, m$ and let $m \to \infty$. We cannot interchange the order of limit and summation, but applying Fatou's lemma in Appendix A gives $\pi_j \geq \Sigma \pi_k p_{kj}$ for all j. The equality sign must hold for each j, otherwise summing both sides of the inequality over $j \in I$ would lead to the contradiction $\Sigma \pi_j > \Sigma \pi_k$. To prove that the π_j's sum up to 1, note that $\pi_j = 1/\mu_{jj}$ for any recurrent state j by Theorem 2.2.2. Since $\mu_{rr} < \infty$, we have $\pi_r > 0$ and so $\Sigma_i \pi_i > 0$. This shows that $x_j^* = \pi_j / \Sigma_i \pi_i$, $j \in I$, is a solution to $x_j = \Sigma_k x_k p_{kj}$, $j \in I$. Let x_j be any solution to this equation such that $\Sigma |x_j| < \infty$. In the same way as in the proof of part (b) of Theorem 2.3.3, we derive relation (2.3.7). The derivation involves an interchange of the order of summation that is justified by Theorem A.1 in Appendix A. Letting $m \to \infty$ in (2.3.7) and using the convergence Theorem A.2 in Appendix A, it now follows that $x_j = \pi_j \Sigma_k x_k$, $j \in I$. This result proves not only part (b), but shows as well that $\Sigma_j \pi_j = 1$ by taking the particular solution $x_j = x_j^*$ and using that $\pi_r > 0$.

Lemma 2.3.10   *Suppose that the Markov chain $\{X_n\}$ satisfies Assumption 2.3.1. Then the equilibrium probabilities π_j have the properties:*
   (a) For each $X_0 = i$,

   the long-run fraction of time the process is in state $k = \pi_k$

with probability 1 for any $k \in I$.
   (b) For any recurrent state s, $\pi_s = 1/\mu_{ss}$ and $\pi_s > 0$.
   (c) For any recurrent state s and any state $j \in I$,

$$E(\textit{number of visits to state } j \textit{ between two returns to state } s) = \frac{\pi_j}{\pi_s}.$$

Proof The assertion (a) follows by mimicking the proof of Theorem 2.3.4 with $f(j) = 1$ for $j = k$ and $f(j) = 0$ for $j \neq k$. Use that the function f is bounded. Also use that $\pi_k = 0$ for k transient. Part (b) follows by using Theorem 2.2.2 and Lemma 2.3.8. To prove (c), fix a recurrent state s. The Markov chain regenerates itself each time it makes a transition into state s. Thus a cycle starts each time the process makes a transition into state s. Then the expected cycle length is given by μ_{ss}. By Lemma 2.3.8, $\mu_{ss} < \infty$. Impose now the following reward structure on the Markov chain. For a given state j, a reward of 1 is earned each time the Markov chain visits state j. Then the long-run average reward per unit time represents the long-run fraction of time the process is in state j and is thus equal to π_j. On the other hand, by the renewal-reward theorem, the long-run average reward per unit time equals E(number of visits to state j in one cycle) divided by E(number of transitions in one cycle). The two alternative expressions for the long-run average reward per unit time show that E(number of visits to state j in one cycle) equals $\pi_j \mu_{ss}$ and is thus given by π_j/π_s.

Next we generalize the ergodic Theorem 2.3.4. To do so, let us assume that a reward $f(j)$ is earned each time the Markov chain makes a transition into state j. In addition to Assumption 2.3.1 involving the regeneration state r, we make the following assumption.

Assumption 2.3.2 *(a)* $\Sigma_{j\in I}|f(j)|\pi_j < \infty$.
(b) For each nonrecurrent initial state $X_0 = i$ the total expected reward earned until the first visit to the regeneration state r is finite.

Theorem 2.3.11 *Suppose that the Markov chain $\{X_n\}$ satisfies the Assumptions 2.3.1 and 2.3.2. Then, for each $X_0 = i$,*

$$\lim_{n\to\infty} \frac{1}{n} \sum_{k=1}^{n} f(X_k) = \sum_{j\in I} f(j)\pi_j \quad \text{with probability 1.}$$

Proof First consider the case of a recurrent initial state $X_0 = s$. Let us say that a cycle starts each time the Markov chain makes a transition into state s. By Lemma 2.3.8 the mean cycle length μ_{ss} is finite. By the renewal-reward theorem,

$$\lim_{n\to\infty} \frac{1}{n} \sum_{k=1}^{n} f(X_k) = \frac{E(\text{reward earned during one cycle})}{E(\text{length of one cycle})}$$

with probability 1. Using part (c) of Lemma 2.3.10, we have

$$E(\text{reward earned during one cycle}) = \sum_{j\in I} f(j)\frac{\pi_j}{\pi_s}.$$

Also, by Lemma 2.3.10, $\pi_s = 1/\mu_{ss}$ and so E(length of one cycle) $= 1/\pi_s$. Together the above relations imply the desired result. For a nonrecurrent initial state $X_0 = i$ the proof of the result follows by mimicking the proof of relation (2.3.8).

2.3.3 A numerical approach for the infinite-state balance equations

Throughout this section it is supposed that Assumption 2.3.1 is satisfied. Then the equilibrium probabilities $\{\pi_j, \ j \in I\}$ are the unique solution to the system of linear equations

$$\pi_j = \sum_{k \in I} \pi_k p_{kj}, \quad j \in I, \tag{2.3.15}$$

$$\sum_{j \in I} \pi_j = 1. \tag{2.3.16}$$

The equilibrium equations determine the π_j's uniquely up to a multiplicative constant. The multiplicative constant follows by using the normalizing equation. In the last paragraph of Section 2.3.1 we discussed already how to compute the π_j's when the state space I is finite.

Consider now the case of an infinite state space I. What one usually does to solve numerically the infinite set of linear equations is to approximate the infinite-state Markov model by a truncated model with finitely many states so that the probability mass of the deleted states is very small. Indeed, for a finite-state truncation with a sufficiently large number of states, the difference between the two models will be negligible from a computational point of view. However, such a truncation often leads to a finite but very large system of linear equations whose numerical solution will be quite time-consuming, although an arsenal of good methods is available to solve the equilibrium equations of a finite Markov chain. Moreover, it is somewhat disconcerting that we need a brute-force approximation to solve numerically the infinite-state model. Usually we introduce infinite-state models to obtain mathematical simplification, and now in its numerical analysis using a brute-force truncation we are proceeding in the reverse direction. Fortunately, many applications allow for a much simpler and more satisfactory approach to solve the infinite set of linear equations. The basic idea of the approach is to reduce the infinite system of linear equations to a finite one by exploiting the geometric tail behaviour of the equilibrium probabilities when such a behaviour exists. The geometric tail approach results in a finite system of linear equations whose size is usually much smaller than the size of the finite system obtained from a brute-force truncation.

Suppose now that the state space I is given by

$$I = \{0, 1, \ldots\}.$$

Further, suppose that the π_j's exhibit the geometric tail behaviour

$$\pi_j \approx \gamma \eta^j \quad \text{for } j \text{ large enough} \tag{2.3.17}$$

for some constants $\gamma > 0$ and $0 < \eta < 1$. Sufficient conditions for this tail behaviour are stated in Appendix C. These conditions involve the generating function of the state probabilities $\{\pi_j, \ j \geq 0\}$. It is also shown in Appendix C that the decay factor η is the root of a nonlinear equation in a single variable. The computation of the constant η appears to be quite simple in specific applications. This is usually not the case for the computation of the constant γ. Fortunately, we do not need the constant γ in our approach. The asymptotic expansion is only used by

$$\lim_{j \to \infty} \frac{\pi_j}{\pi_{j-1}} = \eta.$$

In other words, for a sufficiently large integer M,

$$\pi_j \approx \pi_M \eta^{j-M} \quad \text{for } j \geq M.$$

Replacing π_j by $\pi_M \eta^{j-M}$ for $j \geq M$ in the equations (2.3.15) and (2.3.16) leads to the following finite set of linear equations:

$$\pi_j = \sum_{k=0}^{M} a_{jk} \pi_k, \quad j = 0, 1, \ldots, M - 1,$$

$$\sum_{j=0}^{M-1} \pi_j + \frac{\pi_M}{1 - \eta} = 1,$$

where for any $j = 0, 1, \ldots, M - 1$ the coefficients a_{jk} are given by

$$a_{jk} = \begin{cases} p_{kj}, & k = 0, 1, \ldots, M - 1, \\ \sum_{i=M}^{\infty} \eta^{i-M} p_{ij} & k = M. \end{cases}$$

How large an M should be chosen has to be determined experimentally and depends, of course, on the required accuracy in the equilibrium probabilities. However, empirical investigations show that remarkably small values of M are already good enough for practical purposes. We found in all practical examples that the system of linear equations is nonsingular, irrespective of the value chosen for M. An appropriate value of M is often in the range of 1–200 when a reasonable accuracy (say, a seven-digit accuracy) is required for the equilibrium probabilities. A Gaussian elimination method is a convenient method for solving linear equations of this size. Fast and reliable codes for Gaussian elimination are widely available. The geometric tail approach combines effectivity with simplicity. Several applications of this approach will be encountered in Chapter 4.

2.4 APPLICATIONS OF DISCRETE-TIME MARKOV CHAINS

In this section we give some applications of discrete-time Markov chains to a variety of applied probability problems.

Example 2.1.1 (continued) The periodic-review (s, S) inventory system

Operating characteristics of interest for the lost-sales (s, S) inventory system include the long-run frequency of ordering and the long-run fraction of weeks in which a shortage occurs. These performance measures can be calculated using Markov chain theory. In Section 2.1 we showed that the discrete-time process $\{X_n\}$ is a Markov chain when X_n is defined as the stock on hand just prior to review at the beginning of the nth week. This Markov chain has the finite state space $I = \{0, 1, \ldots, S\}$ and has no two disjoint closed sets. Hence the Markov chain has a unique equilibrium distribution $\{\pi_j, \ 0 \le j \le S\}$.

To find the long-run frequency of ordering, it is convenient to assume that a cost of 1 is incurred each time a replenishment order is placed. In other words, $f(j) = 1$ for $j < s$ and $f(j) = 0$ for $j \ge s$. Then the average cost per unit time represents the fraction of weeks in which a replenishment is done. Hence an application of the ergodic Theorem 2.3.4 yields that

the long-run fraction of weeks in which a replenishment is done

$$= \sum_{j=0}^{s-1} \pi_j \text{ with probability 1.}$$

To find the long-run fraction of weeks in which a shortage occurs, assume that a cost of 1 is incurred for each week in which a shortage occurs. Then the average cost per unit time represents the fraction of weeks in which a shortage occurs. We have

$$E(\text{shortage costs in the } n\text{th week}|X_n = j) = \begin{cases} \displaystyle\sum_{k=S+1}^{\infty} \phi_k & \text{if } 0 \le j < s, \\ \displaystyle\sum_{k=j+1}^{\infty} \phi_k & \text{if } s \le j \le S. \end{cases}$$

Here we have the situation that the function f from the ergodic Theorem 2.3.4 represents the expected value of the cost incurred between two state transitions. It now follows from Theorem 2.3.4 (cf. also Remark 2.3.1) that

the long-run fraction of weeks in which a shortage occurs

$$= \sum_{j=0}^{s-1} \pi_j \left(\sum_{k=S+1}^{\infty} \phi_k \right) + \sum_{j=s}^{S} \pi_j \left(\sum_{k=j+1}^{\infty} \phi_k \right) \quad \text{with probability 1.}$$

Note that in computations $\sum_{k=j+1}^{\infty} \phi_k$ should be replaced by $1 - \sum_{k=0}^{j} \phi_k$. The reader is asked to verify that

the long-run average amount of demand lost per week

$$= \sum_{j=0}^{s-1} \pi_j \left\{ \sum_{k=S+1}^{\infty} (k-S)\phi_k \right\} + \sum_{j=s}^{S} \pi_j \left\{ \sum_{k=j+1}^{\infty} (k-j)\phi_k \right\} \qquad (2.4.1)$$

with probability 1. To verify this relation, assume that a cost of 1 is incurred for each unit of demand that is lost. An expression for the long-run fraction of demand lost next follows from (2.4.1) and the relation

the long-run fraction of demand lost

$$= \frac{\text{(the long-run average amount of demand lost per week)}}{\text{(the long-run average demand per week)}}.$$

Note that the long-run average demand per week equals $\mu = \Sigma_j j\phi_j$.

To conclude this example, we specify the equilibrium equations. In Section 2.1 we determined already the one-step transition probabilities p_{ij}. Thus we obtain the equilibrium equations

$$\pi_j = \sum_{k=0}^{s-1} \pi_k \phi_{S-j} + \sum_{k=j}^{S} \pi_k \phi_{k-j}, \qquad s \le j \le S,$$

$$\pi_j = \sum_{k=0}^{s-1} \pi_k \phi_{S-j} + \sum_{k=s}^{S} \pi_k \phi_{k-j}, \qquad 1 \le j < s,$$

$$\pi_j = \sum_{k=0}^{s-1} \pi_k \sum_{d=S}^{\infty} \phi_d + \sum_{k=s}^{S} \pi_k \sum_{d=k}^{\infty} \phi_d, \qquad j = 0.$$

The equilibrium equations can be recursively solved, although the p_{ij}'s do not have the property stated in Corollary 2.3.6. The recursive computation is possible due to the special structure of the Markov model. The Markov chain has the property $p_{kj} = p_{Sj}$ for $k < s$. Note from the first equation that $\pi_S = \phi_0 \Sigma_{k=0}^{s-1} \pi_k / (1 - \phi_0)$. For ease let us assume that $\phi_0 > 0$ (if $\phi_0 = 0$, then $\pi_S = 0$). Starting with $\overline{\pi}_S := 1$ and replacing in the equations $\Sigma_{k=0}^{s-1} \pi_k$ by $(1 - \phi_0)\pi_S / \phi_0$, we successively compute $\overline{\pi}_{S-1}, \ldots, \overline{\pi}_0$. Next the π_j's follow by normalizing the $\overline{\pi}_j$'s. This example demonstrates that one should always look for special structure that can be exploited to facilitate the computations.

Example 2.4.1 A vehicle insurance problem

A transport firm has effected an insurance contract for a fleet of vehicles. The premium payment is due at the beginning of each year. There are four possible premium classes with a premium payment of P_i in class i where $P_{i+1} < P_i$, $i = 1, 2, 3$. The size of the premium depends on the previous premium and the claim history during the past year. In case no damage is claimed in the past year and the previous premium is P_i, the next premium payment is P_{i+1} (with $P_5 = P_4$, by convention); otherwise the highest premium P_1 is due. Since the insurance contract is for a whole fleet of vehicles, the transport firm has obtained the option to decide only at the end of the year whether the accumulated damage during that year should be claimed or not. In case a claim is made, the insurance company compensates the accumulated damage minus an own risk which amounts to r_i for premium class i.

The total damages in the successive years are independent random variables having a common probability distribution function $G(s)$ with density $g(s)$.

We are interested in the claim limits which minimize the yearly cost for the transport firm. Under a claim rule $(\alpha_1, \ldots, \alpha_4)$ the transport firm claims only damages larger than α_i when the current premium class is i. To find the optimal claim limits, we first derive for a given claim rule $(\alpha_1, \ldots, \alpha_4)$ the average cost per year and next minimize this cost function with respect to the parameters $\alpha_1, \ldots, \alpha_4$. Consider a given claim rule $(\alpha_1, \ldots, \alpha_4)$ where $\alpha_i > r_i$, $i = 1, \ldots, 4$. The average cost per year can be obtained by considering the Markov chain which describes the evolution of the premium class for the transport firm. Let

$X_n =$ the premium class for the transport firm at the beginning of the nth year.

Then the stochastic process $\{X_n\}$ is a Markov chain with four possible states $i = 1, \ldots, 4$. The one-step transition probabilities p_{ij} are easily found. A one-step transition from state i to state 1 occurs only if at the end of the present year a damage is claimed; otherwise a transition from state i to state $i + 1$ occurs (with state 5 $\equiv$ state 4). Since for premium class i only cumulative damages larger than α_i are claimed, it follows that

$$p_{i1} = 1 - G(\alpha_i), \quad i = 1, \ldots, 4,$$

$$p_{i,i+1} = G(\alpha_i), \quad i = 1, 2, 3 \text{ and } p_{44} = G(\alpha_4).$$

The other one-step transition probabilities p_{ij} are equal to zero. The Markov chain has no two disjoint closed sets. Hence the equilibrium probabilities π_j, $1 \leq j \leq 4$ are the unique solution to the linear equations

$$\pi_4 = G(\alpha_3)\pi_3 + G(\alpha_4)\pi_4$$

$$\pi_3 = G(\alpha_2)\pi_2,$$

$$\pi_2 = G(\alpha_1)\pi_1,$$

$$\pi_1 = \{1 - G(\alpha_1)\}\pi_1 + \{1 - G(\alpha_2)\}\pi_2 + \{1 - G(\alpha_3)\}\pi_3 + \{1 - G(\alpha_4)\}\pi_4,$$

together with the normalizing equation $\pi_1 + \pi_2 + \pi_3 + \pi_4 = 1$. These linear equations can be solved recursively. Starting with $\overline{\pi}_4 := 1$, we recursively compute $\overline{\pi}_3$, $\overline{\pi}_2$ and $\overline{\pi}_1$ from the first three equations and then we obtain the π_j's by using the normalizing equation. Next we determine the expected costs incurred during a year in which premium P_j is paid. Denoting these costs by $f(j)$, we have by the ergodic Theorem 2.3.4 that the long-run average cost per year is

$$g(\alpha_1, \ldots, \alpha_4) = \sum_{j=1}^{4} f(j)\pi_j$$

with probability 1. The one-year cost $f(j)$ consists of the premium P_j and any damages not compensated that year by the insurance company. By conditioning on

Table 2.4.1 The optimal claim limits and the minimal costs

	Gamma			Lognormal		
	$c_D^2 = 1$	$c_D^2 = 4$	$c_D^2 = 25$	$c_D^2 = 1$	$c_D^2 = 4$	$c_D^2 = 25$
α_1^*	5908	6008	6280	6015	6065	6174
α_2^*	7800	7908	8236	7931	7983	8112
α_3^*	8595	8702	9007	8717	8769	8890
α_4^*	8345	8452	8757	8467	8519	8640
g^*	9058	7698	6030	9174	8318	7357

the cumulative damage in a year, it follows that

$$f(j) = P_j + \int_0^{\alpha_j} sg(s)\,ds + r_j[1 - G(\alpha_j)].$$

The optimal claim limits follow by minimizing the function $g(\alpha_1, \ldots, \alpha_4)$ with respect to the parameters $\alpha_1, \ldots, \alpha_4$. Efficient numerical procedures are widely available to minimize a function of several variables. In Table 2.4.1 we give for a number of examples the optimal claim limits $\alpha_1^*, \ldots, \alpha_4^*$ together with the minimal average cost g^*. In all examples we take

$$P_1 = 10\,000, \quad P_2 = 7500, \quad P_3 = 6000, \quad P_4 = 5000,$$
$$r_1 = 1500, \quad r_2 = 1000, \quad r_3 = 750, \quad r_4 = 500.$$

The average damage size is 5000 in each example, but the squared coefficient of variation of the damage size D is varied as $c_D^2 = 1$, 4 and 25. To see the effect of the shape of the probability density of the damage size on the claim limits, we take the gamma distribution and the lognormal distribution both having the same first two moments. In particular, the minimal average cost becomes increasingly sensitive to the distributional form of the damage size D when c_D^2 gets larger.

Example 2.4.2 *Buffer design for a communication system*

Consider a communication system with a finite-capacity buffer and multiple transmission channels. Messages arrive at the buffer according to a Poisson process with rate λ. The messages are first stored in the buffer which has capacity for only N messages (exclusive of the messages in transmission). Each message which finds, upon arrival, that the buffer is full is lost and does not influence the system. At fixed clock times $t = 0, 1, \ldots$ messages are taken out from the buffer and are synchronously transmitted. Each transmission channel can transmit only one message at a time. The transmission time is constant and equals one time unit for each message. There are c transmission channels and so at most c messages can synchronously be transmitted. The transmission of a message can only start at a clock time so that a message which finds, upon arrival, that a transmission line is idle has to wait until a subsequent clock time. We are interested in the following question. How should we choose the buffer size so that the long-run fraction of messages lost does not exceed a prespecified value? In practice one typically wishes to design the buffer size in such a way that the loss probability is very small (say, of the order of 10^{-9}).

To find the overflow probability for a given buffer size N with $N \geq c$, we define, for each $n = 0, 1, \ldots$,

X_n = the number of messages in the buffer at clock time n just prior to transmission.

Using the lack of memory of the Poisson arrival process of the messages, it follows that the process $\{X_n\}$ is a discrete-time Markov chain with state space $I = \{0, 1, \ldots, N\}$. Denoting by A_n the number of messages that will arrive at the buffer during the period between the clock times n and $n + 1$, we have

$$X_{n+1} = \begin{cases} \min(A_n, N) & \text{if } X_n < c, \\ \min(X_n - c + A_n, N) & \text{if } X_n \geq c. \end{cases}$$

Since the messages arrive according to a Poisson process with rate λ, the probability of k arrivals during the transmission time of one time unit is given by

$$a_k = e^{-\lambda} \frac{\lambda^k}{k!}, \quad k = 0, 1, \ldots .$$

Denote by p_{ij} the one-step transition probabilities of the Markov chain $\{X_n\}$. Then, for $0 \leq i < c$,

$$p_{ij} = \begin{cases} a_j & 0 \leq j < N, \\ \displaystyle\sum_{k=N}^{\infty} a_k, & j = N, \end{cases}$$

while, for $c \leq i \leq N$,

$$p_{ij} = \begin{cases} a_{j-(i-c)} & i - c \leq j < N, \\ \displaystyle\sum_{k=N-(i-c)}^{\infty} a_k, & j = N, \end{cases}$$

The Markov chain has no two disjoint closed sets. Hence the equilibrium probabilities π_j, $j = 0, 1, \ldots, N$, are the unique solution of the linear equations

$$\pi_j = a_j \sum_{k=0}^{c} \pi_k + \sum_{k=c+1}^{\min(c+j,N)} \pi_{k+c} a_{j-k}, \quad 0 \leq j \leq N - 1,$$

$$\pi_j = \sum_{k=0}^{c-1} \pi_k \left(\sum_{j=N}^{\infty} a_j \right) + \sum_{k=0}^{N-c} \pi_{k+c} \left(\sum_{j=N-k}^{\infty} a_j \right), \quad j = N,$$

$$\sum_{j=0}^{N} \pi_j = 1.$$

When solving this system of linear equations, the infinite sums $\Sigma_{j=k}^{\infty} a_j$ are of course replaced by $1 - \Sigma_{j=0}^{k-1} a_j$. The Poisson probabilities a_j are computed beforehand by a recursion scheme; cf. Section 1.2.2. Depending on the magnitude of N, the system of linear equations is solved by a Gaussian elimination method or by an iterative method.

To find the long-run fraction of messages lost, we need the relation

the long-run fraction of messages lost

$= \frac{1}{\lambda}$ (the long-run average number of messages lost per unit time),

where λ is the average number of messages arriving per unit time. The long-run average number of messages lost per unit time can be calculated in several ways. A direct derivation proceeds as follows. Using the shorthand notation $j_c = \max(j - c, 0)$, we have

E(number of messages lost between the clock times n and $n + 1 | X_n = j$)

$$= \sum_{k=N-j_c}^{\infty} [k - (N - j_c)]a_k = \lambda - N + j_c + \sum_{k=0}^{N-j_c} (N - j_c - k)a_k.$$

Thus, by applying the ergodic Theorem 2.3.4,

the long-run average number of messages lost per unit time

$$= \sum_{j=0}^{N} \pi_j \left\{ \lambda - N + j_c + \sum_{k=0}^{N-j_c} (N - j_c - k)a_k \right\} \quad \text{with probability 1.}$$

A more subtle derivation of an expression for the long-run average number of messages lost per unit time is as follows. First, we make the obvious observation

the average number of messages arriving per unit time
= (the average number of messages transmitted per unit time)
+ (the average number of messages lost per unit time).

Further, since each channel can transmit only one message at a time,

the average number of messages transmitted per unit time
= the average number of busy transmission channels.

An application of the ergodic Theorem 2.3.4 gives

the long-run average number of busy channels

$$= \sum_{j=0}^{c-1} j\pi_j + c \sum_{j=c}^{N} \pi_j \quad \text{with probability 1.}$$

Thus we find the alternative expression

the long-run average number of messages lost per unit time

$$= \lambda - \sum_{j=0}^{c-1} j\pi_j - c \sum_{j=c}^{N} \pi_j \text{ with probability 1.}$$

The two different expressions for the long-run average number of messages lost per unit time lead to the identity

$$\sum_{j=0}^{N} \pi_j \left\{ \lambda - N + j_c + \sum_{k=0}^{N-j_c} (N - j_c - k)a_k \right\} = \lambda - \sum_{j=0}^{c-1} j\pi_j - c \sum_{j=c}^{N} \pi_j.$$

It is very useful to have such an identity as an extra check whether the π_j's have been correctly calculated. Summarizing, we find

the long-run fraction of messages lost

$$= \frac{1}{\lambda} \left[\lambda - \sum_{j=0}^{c-1} j\pi_j - c \sum_{j=c}^{N} \pi_j \right] \text{ with probability 1.}$$

In Table 2.4.2 we give for several values of c and $\rho = \lambda/c$ the minimal buffer size $N(\alpha)$ that is required to achieve a loss probability less than α, where α is varied as $\alpha = 10^{-6}$, 10^{-8} and 10^{-10}. Note that the minimal buffer size sharply increases as ρ gets close to 1. Also the results in the table indicate that $N(\alpha)$ increases logarithmically in α. This behaviour is typical for finite-buffer systems.

To conclude this example, we discuss the calculation of the long-run average number of messages in the buffer. An exact expression for this long-run average can be obtained by applying the ergodic Theorem 2.3.4 for the following cost structure. A waiting cost at rate k is incurred whenever there are k messages in the buffer. Using Lemma 1.2.4 in Section 1.2 it is not very difficult to derive an expression for the expected waiting cost incurred between the clock times n and $n + 1$ given that $X_n = j$. The reader is asked to work out the details in Exercise 2.9. Here we give an interesting approximation for the long-run average number of messages in the buffer. This long-run average is approximated by $\frac{1}{2}(L_1 + L_2)$, where L_1 and

Table 2.4.2 The minimal buffer size

$c \backslash \rho$	$\alpha = 10^{-6}$				$\alpha = 10^{-8}$				$\alpha = 10^{-10}$			
	0.5	0.8	0.9	0.95	0.5	0.8	0.9	0.95	0.5	0.8	0.9	0.95
1	11	29	56	107	15	40	78	152	18	50	101	197
2	12	30	57	108	15	40	79	153	19	51	101	198
5	14	32	59	110	18	42	81	155	21	53	103	200
10	18	36	63	114	22	46	85	159	25	57	107	204
25	30	49	76	127	35	59	98	172	39	70	120	217

L_2 denote the long-run average number of messages in the buffer at the beginning respectively at the end of a time slot. Obviously, $L_1 = \Sigma_{j=c}^{N}(j - c)\pi_j$. If the loss probability is very small, it is reasonable to approximate L_2 by $L_1 + \lambda$ with λ denoting the average number of messages arriving in one time slot. This suggests the approximation

$$\text{the long-run average number of messages in the buffer} \approx \sum_{j=c}^{N}(j - c)\pi_j + \frac{\lambda}{2}.$$

This is an excellent approximation provided the loss probability is very small.

Example 2.1.2 (continued) A single-server queue with exponential services

For the single-server queue with exponential services, we defined in Section 2.1 an embedded Markov chain $\{X_n\}$ by taking X_n as the number of customers present just prior to the nth arrival epoch. The Markov chain $\{X_n\}$ has the infinite state space $I = \{0, 1, \ldots\}$. We now assume that $\lambda/\beta < 1$, where $1/\beta$ is the mean service time and the average arrival rate λ is defined as the inverse of the mean interarrival time. Under the condition $\lambda/\beta < 1$ it can be shown that the Markov chain satisfies Assumption 2.3.1 when state 0 is taken for the regeneration state r. The proof is not given here.

By Theorem 2.3.9, the Markov chain $\{X_n\}$ has a unique equilibrium distribution $\{\pi_j, j = 0, 1, \ldots\}$. The probabilities π_j are the unique solution to the linear equations (2.3.15)–(2.3.16). In particular, using the specification of the p_{ij}'s given in Section 2.1,

$$\pi_j = \sum_{k=j-1}^{\infty} \pi_k \int_0^{\infty} e^{-\beta t} \frac{(\beta t)^{k+1-j}}{(k+1-j)!} g(t)\, dt, \quad j \geq 1. \tag{2.4.2}$$

The Markov chain $\{X_n\}$ provides an example of an infinite-state Markov chain for which the equilibrium probabilities π_j have the property (2.3.17). In fact, we have the stronger result that $\pi_j = \gamma \tau^j$ for all $j \geq 0$ for constants $\gamma > 0$ and $0 < \tau < 1$. This result can be proved in two ways. A direct way is to verify the result by substitution in the equilibrium equations. A more elegant proof is by probabilistic reasoning. Since Assumption 2.3.1 is satisfied and any two states in I communicate, each state $j \in I$ is recurrent by Theorem 2.2.4. Hence, by Lemma 2.3.10,

$$\frac{\pi_{j+1}}{\pi_j} = E(\text{number of visits to state } j + 1 \text{ between two returns to state } j)$$

for each $j = 0, 1, \ldots$. Some reflections show that the right side of this equation is independent of j by the memoryless property of the exponential services. Thus, for some constant η, $\pi_{j+1}/\pi_j = \eta$ for all $j \geq 0$. This implies that $\pi_j = \pi_0 \eta^j$ for $j \geq 0$. The constant η is easily found. By substitution of $\pi_j = \pi_0 \tau^j$ for $j \geq 0$ into

(2.4.2), we find after some algebra that

$$\eta = \int_0^\infty e^{-\beta t(1-\eta)} g(t)\, dt.$$

It is readily verified that this equation has a unique solution η with $0 < \eta < 1$. The constant η can be numerically determined by bisection. Using the normalizing equation $\Sigma_{j=0}^\infty \pi_j = 1$, it follows that $\pi_0 = 1 - \eta$. Hence we find that the equilibrium distribution of the embedded Markov chain $\{X_n\}$ is given by

$$\pi_j = (1 - \eta)\eta^j, \quad j = 0, 1, \ldots.$$

The solution of the next example uses the powerful phase method. This method is discussed in more detail in Section 2.9.

Example 2.4.3 A manufacturing problem

Each time unit a job arrives at a conveyor with a single work station. The work station can process only one job at a time and has a finite buffer to store temporarily other jobs. The buffer has only capacity to store N items. A job finding upon arrival the buffer full is not accepted. The processing times of the accepted jobs are independent random variables having a common Erlang-r distribution with mean r/μ. What are the long-run fractions of jobs rejected and the long-run fraction of time the work station is busy?

The key to the solution of the problem is the fact that the Erlang-r distributed processing time of a job can be seen as the sum of r independent phases, where the phases must be sequentially processed and the durations of the phases are exponentially distributed with common mean $1/\mu$. In other words, each job in the system can be represented by its number of uncompleted phases. Thus we define the process $\{X_n, \ n \geq 1\}$ by

$X_n =$ the number of uncompleted phases in the system just prior to the arrival of the nth job.

The process $\{X_n\}$ is a discrete-time Markov chain by the memoryless property of the exponential distribution. The state space of the process is $I = \{0, 1, \ldots (N + 1)r\}$.

To specify the one-step transition probabilities, note that a job is rejected and leaves the state of the system unchanged when it finds upon arrival the system in a state i with $Nr < i \leq (N + 1)r$; otherwise the job is accepted and increases the number of uncompleted phases in the system by r. For any i with $Nr < i \leq (N + 1)r$ and $j \neq 0$, it follows that p_{ij} equals the probability that $j - i$ phases are completed during the time until the next arrival when an arrival has just occurred and leaves the system in state i. Since the processing times of the phases are exponentially distributed, the completions of phases occur according to a Poisson process with rate μ as long as the work station is busy. Thus, for any

$Nr < i \le (N+1)r$,

$$p_{ij} = e^{-\mu} \frac{\mu^{i-j}}{(i-j)!} \quad \text{for } 1 \le j \le i.$$

For any $0 \le i \le Nr$, we have

$$p_{ij} = e^{-\mu} \frac{\mu^{i+r-j}}{(i+r-j)!} \quad \text{for } 1 \le j \le i+r.$$

The probability p_{i0} follows from $p_{i0} = 1 - \Sigma_{j \ge 1} p_{ij}$, $i \ge 0$. The Markov chain $\{X_n\}$ has no two disjoint closed sets. Its equilibrium distribution $\{\pi_j\}$ is the unique solution to the system of linear equations (2.3.9) and (2.3.10) in which the expressions for the p_{ij}'s have been substituted. Once the probabilities π_j have been computed, we obtain

$$\text{the long-run fraction of jobs rejected} = \sum_{i=Nr+1}^{(N+1)r} \pi_i.$$

The long-run fraction of time the work station is busy can be derived by using the ergodic Theorem 2.3.4. A more elegant derivation is based on Little's formula. The general cost principle (1.6.2) in Section 1.6 is applied. It is assumed that the system incurs a cost at rate 1 whenever an (accepted) job is being processed. Then the long-run average cost per unit time gives the long-run fraction of time the work station is busy. For the chosen cost structure the long-run average cost per accepted job is the average processing time r/μ. Further, the long-run average arrival rate of accepted jobs is $\Sigma_{i=0}^{Nr} \pi_i$. Thus, by applying the general principle (1.6.2) in Section 1.6, we find

$$\text{the long-run fraction of time the work station is busy} = \frac{r}{\mu} \sum_{i=0}^{Nr} \pi_i.$$

2.5   CONTINUOUS-TIME MARKOV CHAINS

So far we have considered Markov processes in which the changes of the state only occurred at fixed times $t = 0, 1, \ldots$. However, in numerous practical situations changes of the state may occur at each point of time. One of the most appropriate models for analysing such situations is the continuous-time Markov chain model. In this model the times between successive transitions are exponentially distributed, while the succession of states is described by a discrete-time Markov chain. A wide variety of applied probability problems can be modelled as a continuous-time Markov chain by a proper state description.

In analogy with the definition of a discrete-time Markov chain, a continuous-time Markov chain is defined as follows.

Definition 2.5.1   *A continuous-time stochastic process $\{X(t),\ t \geq 0\}$ with discrete state space I is said to be a continuous-time Markov chain if*

$$P\{X(t_n) = i_n | X(t_0) = i_0, \ldots, X(t_{n-1}) = i_{n-1}\}$$

$$= P\{X(t_n) = i_n | X(t_{n-1}) = i_{n-1}\}$$

for all $0 \leq t_0 < \cdots < t_{n-1} < t_n$ and $i_0, \ldots, i_{n-1}, i_n \in I$.

Just as in the discrete-time case the Markov property expresses that the conditional distribution of a future state given the present state and past states depends only on the present state and is independent of the past. In the following we consider time-homogeneous Markov chains for which the transition probability $P\{X(t + u) = j | X(u) = i\}$ is independent of $u > 0$. We write

$$p_{ij}(t) = P\{X(t + u) = j | X(u) = i\}.$$

The theory of continuous-time Markov chains is much more intricate than the theory of discrete-time Markov chains. There are very difficult technical problems and some of them are not even solved at present time. Fortunately, the staggering technical problems do not occur in almost all practical applications. In our treatment of continuous-time Markov chains we proceed pragmatically. We impose a regularity condition that is not too strong from a practical point of view but avoids all technical problems.

As an introduction to the modelling by a continuous-time Markov chain, let us construct the following jump process. A stochastic system with a discrete state space I jumps from state to state according to the following rules:

Rule (a)   If the system jumps to state i, it stays in state i an exponentially distributed time with mean $1/\nu_i$ independently of how the system reached state i and how long it took to get there.

Rule (b)   If the system leaves state i, it jumps to state j with probability $p_{ij} (j \neq i)$ independently of the duration of the stay in state i, where $\Sigma_{j \neq i} p_{ij} = 1$ for all $i \in I$.

The following assumption is now made.

Assumption 2.5.1   *In any finite time interval the number of jumps is finite with probability 1.*

Define now the continuous-time stochastic process $\{X(t),\ t \geq 0\}$ by

$$X(t) = \text{the state of the system at time } t, t \geq 0,$$

where the process is taken to be right-continuous. Then the process $\{X(t)\}$ can be shown to be a continuous-time Markov chain. It will be intuitively clear that the process has the Markov property in view of the assumption of the exponentially distributed sojourn times in the states. Note the convention that the transition from a state is always to a different state. This convention ensures that the sojourn time in a state is unambiguously defined. The convention is primarily for convenience.

Assumption 2.5.1 is essential to exclude pathological cases. For example, suppose that $I = \{1, 2, \ldots\}$, $p_{i,i+1} = 1$ and $v_i = i^2$ for all $i \in I$. Then the process will ultimately face an explosion of jumps.

In modelling continuous-time Markov processes one typically uses the so-called infinitesimal transition rates. To introduce these rates, let us consider the process $\{X(t)\}$ as constructed above. In view of Assumption 2.5.1 and the memoryless property of the exponentially distributed sojourn times, the probability of two or more state transitions within a time Δt is negligibly small compared to Δt as $\Delta t \to 0$. Using the failure rate representation of the exponential distribution, we have for any $i \in I$ that

$$P\{X(t + \Delta t) = j | X(t) = i\} = \begin{cases} v_i \Delta t \times p_{ij} + o(\Delta t) & \text{for } j \neq i, \\ 1 - v_i \Delta t + o(\Delta t) & \text{for } j = i, \end{cases}$$

as $\Delta t \to 0$. Denote by $q_{ij} (j \neq i)$ the quantity

$$q_{ij} = v_i p_{ij}, \quad i, j \in I, \ j \neq i.$$

The numbers q_{ij} are called the infinitesimal transition rates of the continuous-time Markov chain $\{X(t)\}$. Although $q_{ij} \Delta t$ can be interpreted as a probability of moving from state i to state j in the next Δt time units, the q_{ij}'s themselves are not probabilities but transition rates. Note that the q_{ij}'s uniquely determine the v_i's and the p_{ij}'s by $v_i = \Sigma_{j \neq i} q_{ij}$ and $p_{ij} = q_{ij}/v_i$.

In applications one usually proceeds in the reverse direction of the way followed above. The infinitesimal transition rates q_{ij} are determined directly in applications. They are typically the result of the interaction of two or more elementary processes. Contrary to the discrete-time case in which the one-step transition probabilities determine unambiguously a discrete-time Markov chain, it is not generally true that the infinitesimal transition rates determine a unique continuous-time Markov chain. Here we run into subtleties that are well beyond the scope of this book. The reader should be aware that fundamental difficulties can arise when the state space is infinite, but on the other hand these difficulties are absent in almost all practical applications. To avoid the technical problems, we make the following assumption for the given data q_{ij}.

Assumption 2.5.2   *The rates $v_i = \Sigma_{j \neq i} q_{ij}$, are bounded in $i \in I$. Moreover, $v_i > 0$ for all $i \in I$.*

The boundedness assumption is trivially satisfied when I is finite and holds in most applications with an infinite state space.

Using very deep mathematics it has been shown that under Assumption 2.5.2 there is a unique continuous-time Markov chain $\{X(t), \ t \geq 0\}$ having the transition rates $q_{ij} (j \neq i)$. This continuous-time Markov chain is precisely the Markov jump process constructed according to the rules (a) and (b) given above, where the leaving rate of any state i is taken as v_i and the jump probabilities p_{ij} are taken as

$$p_{ij} = \begin{cases} \frac{q_{ij}}{v_i}, & j \neq i, \\ 0 & j = i. \end{cases} \qquad (2.5.1)$$

Then, for any $t > 0$,

$$P\{X(t + \Delta t) = j | X(t) = i\} = \begin{cases} q_{ij}\Delta t + o(\Delta t), & j \neq i, \\ 1 - v_i\Delta t + o(\Delta t), & j = i. \end{cases} \qquad (2.5.2)$$

It is noted that Assumption 2.5.2 implies that the constructed continuous-time Markov chain $\{X(t)\}$ automatically satisfies Assumption 2.5.1.

The probabilities p_{ij} defined by (2.5.1) are said to be the one-step transition probabilities of the embedded Markov chain $\{X_n\}$ in the continuous-time process $\{X(t)\}$. The discrete-time Markov chain $\{X_n\}$ is embedded at the jump epochs of the continuous-time process.

In solving specific problems it suffices to specify the infinitesimal transition rates q_{ij}. We now give two examples. In these examples the q_{ij}'s are determined as the result of the interaction of several elementary processes of the Poisson type. The q_{ij}'s are found by using the interpretation that $q_{ij}\Delta t$ for Δt small represents the probability of making a transition to state j in the next time Δt when the current state is i.

Example 2.5.1 The M/M/1 queue

Consider a single-server station at which customers arrive according to a Poisson process with rate λ. The service times of the customers are independent random variables having a common exponential distribution with mean $1/\mu$. The arrival process and the service times are independent of each other. An arriving customer who finds the server busy joins the queue. As soon as a service is completed a next customer enters service provided the queue is not empty. This model is the simplest model in queueing theory and is usually abbreviated as the *M/M/1* queue.

An interesting stochastic process is the process describing the number of customers in the system. For any $t \geq 0$, let

$$X(t) = \text{the number of customers present at time } t.$$

The process $\{X(t), t \geq 0\}$ is a continuous-time Markov chain with infinite state space $I = \{0, 1, \ldots\}$. To show that $\{X(t)\}$ is indeed a Markov process, let us verify that the specification (2.5.2) applies. By the failure rate representations for the Poisson process and the exponential distribution, the probability of one arrival within a small time interval of length Δt is $\lambda\Delta t + o(\Delta t)$ and the probability that a service in progress is completed within the next Δt time units is $\mu\Delta t + o(\Delta t)$. The probability that two or more events happen within a time Δt is of the order of magnitude $(\Delta t)^2$ and therefore negligibly small compared with Δt. The process

$\{X(t)\}$ jumps from i to $i+1$ if an arrival occurs and jumps from i to $i-1$ if a service completion occurs. It now follows that

$$P\{X(t+\Delta t) = i+1 | X(t) = i\} = \lambda \Delta t (1 - \mu \Delta t) + o(\Delta t) = \lambda \Delta t + o(\Delta t), \quad i \geq 1,$$

and

$$P\{X(t+\Delta t) = i-1 | X(t) = i\} = \mu \Delta t (1 - \lambda \Delta t) + o(\Delta t) = \mu \Delta t + o(\Delta t), \quad i \geq 1.$$

Also, $P\{X(t+\Delta t) = 1 | X(t) = 0\} = \lambda \Delta t + o(\Delta t)$. Hence we find the infinitesimal transition rates

$$q_{ij} = \begin{cases} \lambda, & j = i+1 \text{ and } i \geq 0, \\ \mu, & j = i-1 \text{ and } i \geq 1, \\ 0, & \text{otherwise.} \end{cases}$$

Example 2.5.2 Response areas for emergency units

Consider a city with two emergency units that cooperate in responding to some type of alarms arising in that city. The alarms arrive at a central dispatcher who sends exactly one unit to each alarm. The two units in the city only respond to the alarms within that city. For the given dispatch strategy the city is divided into two districts. The emergency unit i is the first-due unit for response area i. This means that an alarm arriving when both units are available is served by unit 1 if the alarm is in district 1 and is served by unit 2 otherwise, while an alarm arriving when only one of the units is available is served by the available unit. In case both units are not available, the alarm is settled by some unit from outside the city.

Suppose that for the given design of the two response areas the alarms from the areas 1 and 2 arise according to independent Poisson processes with respective rates λ_1 and λ_2. The service times of the alarms are independent random variables, where the time needed to serve an alarm from district j by unit i has an exponential distribution with mean $1/\mu_{ij}$. The service times include travel times.

Let for $i = 1, 2$,

$$X_i(t) = \begin{cases} 0 & \text{if unit } i \text{ is free at time } t, \\ 1 & \text{if unit } i \text{ is servicing an alarm from district 1 at time } t, \\ 2 & \text{if unit } i \text{ is servicing an alarm from district 2 at time } t. \end{cases}$$

The stochastic process $\{X(t) = (X_1(t), X_2(t)), t \geq 0\}$ has a finite state space I with nine states. To show that the process $\{X(t), t \geq 0\}$ is a continuous-time Markov chain, we verify the specification (2.5.2). The transition rates q_{ij} are again the result of the interaction of the Poisson process generating the alarms and the service time process with exponentially distributed services. For example, suppose that the process $\{X(t)\}$ is in state $(0, 2)$ at the current time t. Then in the time interval $(t, t + \Delta t]$ the process jumps to state $(1, 2)$ only if an alarm arises from district 1 in the next Δt time units and the unit 2 does not complete its service in district 2 within the next Δt time units. Hence

$$P\{X(t + \Delta t) = (1, 2) | X(t) = (0, 2)\} = \lambda_1 \Delta t (1 - \mu_{22} \Delta t) + o(\Delta t),$$

showing that $q_{(0,2)(1,2)} = \lambda_1$. The other transition rates follow in a similar way. Let us give one other example. The process jumps from state (2, 1) to state (0, 1) only if unit 1 completes its service in district 2 while the other unit continues its service in district 2. Thus $q_{(2,1)(0,1)} = \mu_{12}$. It is left to the reader to give the other transition rates.

2.6 LONG-RUN ANALYSIS OF CONTINUOUS-TIME MARKOV CHAINS

This section discusses the continuous-time analogues of the ergodic theorems given for the discrete-time Markov model. The derivation of the continuous-time analogues is based on the results for the discrete-time case.

Let $\{X(t), \ t \geq 0\}$ be a continuous-time Markov chain with state space I and transition rates $q_{ij} (j \neq i)$ and $v_i = \Sigma_{j \neq i} q_{ij}, \ i \in I$. Throughout this section it is assumed that the v_i's are positive and bounded. Similarly as in the discrete-time case, a recurrence condition is imposed on the model.

Assumption 2.6.1   *The process $\{X(t), \ t \geq 0\}$ has a regeneration state r such that*

$$P\{\tau_r < \infty | X(0) = i\} = 1 \quad \text{for all } i \in I \quad \text{and} \quad E(\tau_r | X(0) = r) < \infty$$

where τ_r is the first epoch at which the process $\{X(t)\}$ makes a transition into state r.

In other words, state r will ultimately be reached with probability 1 from any other state and the mean recurrence time from state r to itself is finite. This assumption is usually satisfied in applications. In our analysis, we need the embedded Markov chain $\{X_n\}$, where X_n is defined by

X_n = the state of the process $\{X(t)\}$ just after the nth state transition

with the convention that $X_0 = X(0)$. The one-step transition probabilities p_{ij} of the discrete-time Markov chain $\{X_n\}$ are given by (2.5.1). It is readily verified that Assumption 2.6.1 implies that the embedded Markov chain $\{X_n\}$ satisfies the corresponding Assumption 2.3.1.

Definition 2.6.1   *A probability distribution $\{p_j, \ j \in I\}$ is said to be an equilibrium distribution for the continuous-time Markov chain $\{X(t)\}$ if*

$$v_j p_j = \sum_{k \neq j} p_k q_{kj}, \quad j \in I.$$

Just as in the discrete-time case the explanation of the term equilibrium distribution is as follows. If $P\{X(0) = j\} = p_j$ for all $j \in I$, then, for any time $t > 0$, $P\{X(t) = j\} = p_j$ for all $j \in I$. The proof is nontrivial and will not be given.

Theorem 2.6.1   *Suppose that the continuous-time Markov chain $\{X(t), \ t \geq 0\}$ satisfies the Assumptions 2.5.2 and 2.6.1. Then:*

(a) The continuous-time Markov chain $\{X(t)\}$ has a unique equilibrium distribution $\{p_j, \ j \in I\}$. Moreover

$$p_j = \frac{\pi_j/v_j}{\sum\limits_{k \in I} \pi_k/v_k}, \quad j \in I, \tag{2.6.1}$$

where $\{\pi_j, \ j \in I\}$ is the unique equilibrium distribution of the embedded Markov chain $\{X_n\}$.

(b) Let $\{x_j\}$ be any solution to $v_j x_j = \sum_{k \neq j} x_k q_{kj}, \ j \in I$, with $\sum_j |x_j| < \infty$. Then, for some constant c, $x_j = cp_j$ for all $j \in I$.

Proof   We first verify that there is an one-to-one correspondence between the solutions of the two systems of linear equations

$$v_j x_j = \sum_{k \neq j} x_k q_{kj}, \quad j \in I$$

and

$$u_j = \sum_{k \in I} u_k p_{kj}, \quad j \in I.$$

This is an immediate consequence of the definition (2.5.1) of the p_{ij}'s. If $\{u_j\}$ is a solution to the second system with $\sum |u_j| < \infty$, then $\{x_j = u_j/v_j\}$ is a solution to the first system with $\sum |x_j| < \infty$, and conversely. This one-to-one correspondence and Theorem 2.3.9 imply the Theorem 2.6.1 provided we verify

$$\sum_{j \in I} \frac{\pi_j}{v_j} < \infty. \tag{2.6.2}$$

The proof that this condition holds is as follows. By Assumption 2.6.1, the process $\{X(t)\}$ regenerates itself each time the process makes a transition into state r. Let a cycle be defined as the time elapsed between two consecutive returns to state r. Using Wald's equation, it is readily seen that

$$E(\text{length of one cycle}) = \sum_{j \in I} E(\text{number of visits to state } j \text{ in one cycle}) \times \frac{1}{v_j}.$$

Thus, by part (c) of Lemma 2.3.10,

$$E(\text{length of one cycle}) = \frac{1}{\pi_r} \sum_{j \in I} \frac{\pi_j}{v_j}.$$

Since $E(\text{length of one cycle})$ is finite by Assumption 2.6.1 inequality (2.6.2) now follows by using that $\pi_r > 0$. This completes the proof.

A state j is said to be recurrent for the continuous-time Markov chain $\{X(t)\}$ only if $P\{\tau_j < \infty | X(0) = j\} = 1$, where τ_j is the first epoch 0 at which the process $\{X(t)\}$ makes a transition into state j. It is readily verified that a state j is recurrent for the continuous-time Markov chain $\{X(t)\}$ if and only if state j is recurrent for the embedded Markov chain $\{X_n\}$. In fact the state classification for discrete-time Markov chains carries over completely to the continuous-time case.

Let us now impose the following reward structure on the continuous-time Markov chain $\{X(t)\}$. A reward at rate $r(j)$ is earned whenever the process is in state j, while a lump reward of F_{jk} is earned each time the process jumps from state j to state k ($\neq j$). In addition to Assumption 2.6.1, we make the following assumption.

Assumption 2.6.2   *(a)* $\Sigma |r(j)| p_j < \infty$ *and* $\Sigma_{j \in I} p_j \Sigma_{k \neq j} q_{jk} |F_{jk}| < \infty$.
(b) For each nonrecurrent initial state $X(0) = i$, the total expected reward earned until the first epoch at which the process $\{X(t)\}$ makes a transition into the regeneration state r is finite.

For each $t > 0$, define the random variable $R(t)$ by

$$R(t) = \text{the total reward earned up to time } t.$$

We now come to a very useful ergodic theorem.

Theorem 2.6.2   *Suppose that the continuous-time Markov chain $\{X(t)\}$ satisfies the Assumptions 2.5.2, 2.6.1 and 2.6.2. Then, for each $X(0) = i$,*

$$\lim_{t \to \infty} \frac{R(t)}{t} = \sum_{j \in I} r(j) p_j + \sum_{j \in I} p_j \sum_{k \neq j} q_{jk} F_{jk} \quad \text{with probability 1.}$$

Proof   We first verify the result for a recurrent initial state $X(0) = s$. The process $\{X(t)\}$ regenerates itself each time the process makes a transition into state s. Let a cycle be defined as the time elapsed between two consecutive returns to state s. Using Wald's equation, it follows that

$$E(\text{length of one cycle}) = \sum_{j \in I} E(\text{number of visits to state } j \text{ in one cycle}) \times \frac{1}{\nu_j}$$

and so, by part (c) of Lemma 2.3.10

$$E(\text{length of one cycle}) = \frac{1}{\pi_s} \sum_{j \in I} \frac{\pi_j}{\nu_j}.$$

Since state s is recurrent, we have $\pi_s > 0$. Since condition (2.6.2) holds, it now follows that $E(\text{length of one cycle})$ is finite. Next apply the renewal-reward theorem in Section 1.3. This gives

$$\lim_{t \to \infty} \frac{R(t)}{t} = \frac{E(\text{reward earned during one cycle})}{E(\text{length of one cycle})} \tag{2.6.3}$$

with probability 1. Using again Wald's equation, we find

E(reward earned during one cycle)

$$= \sum_{j\in I} E(\text{number of visits to state } j \text{ in one cycle}) \times \left[\frac{r(j)}{v_j} + \sum_{k\neq j} p_{jk} F_{jk}\right].$$

Hence, by part (c) of Lemma 2.3.10 and expression (2.5.1)

$$E(\text{reward earned in one cycle}) = \sum_{j\in I} \frac{\pi_j}{\pi_s} \left[\frac{r(j)}{v_j} + \sum_{k\neq j} p_{jk} F_{jk}\right]$$

$$= \frac{1}{\pi_s} \sum_{j\in I} \frac{\pi_j}{v_j} \left[r(j) + \sum_{k\neq j} q_{jk} F_{jk}\right].$$

By substituting (2.6.1) into the expressions for the numerator and denominator of (2.6.3) we get the desired result when the initial state is recurrent. It remains to verify that the results also hold for a nonrecurrent initial state $X(0) = i$. This verification proceeds along the same lines as the proof of the result (2.3.8) in Theorem 2.3.4.

Interpretation of the equilibrium probabilities

A useful interpretation of the equilibrium probabilities p_j is provided by the following corollary.

Corollary 2.6.3 *Suppose that the continuous-time Markov chain $\{X(t)\}$ satisfies the Assumptions 2.5.2 and 2.6.1. Then*
(a) For each $X(0) = i$,

> *the long-run fraction of time the process is in state $k = p_k$*

with probability 1 for each $k \in I$.
(b) For each $X(0) = i$,

> *the long-run average number of transitions from state k to state j per unit time $= q_{kj} p_k$*

with probability 1 for all $j, k \in I$ with $k \neq j$.

Proof   Part (a) follows by applying Theorem 2.6.2 with $r(j) = 1$ for $j = k$, $r(j) = 0$ for $j \neq k$ and $F_{ij} = 0$ for all i, j. To find part (b) apply Theorem 2.6.2 with $r(i) = 0$ for all i, $F_{gh} = 1$ for $(g, h) = (k, j)$ and $F_{gh} = 0$ otherwise.

The first part of Corollary 2.6.3 states that $\lim_{t\to\infty}(1/t) \int_0^t I_j(u)\,du = p_j$ with probability 1, where the random variable $I_j(t)$ is defined by $I_j(t) = 1$ if $X(t) = j$

and $I_j(t) = 0$ otherwise. Using the bounded convergence theorem, it next follows that $\lim_{t\to\infty}(1/t)\int_0^t p_{ij}(u)\,du = p_j$ for all $i, j \in I$. In fact the stronger result

$$\lim_{t\to\infty} p_{ij}(t) = p_j \quad \text{for all } i, j \in I,$$

holds, see e.g. Chung (1967). The concept of periodicity and aperiodicity is not relevant for continuous-time Markov chains. The explanation is simple. Since the sojourn times in the states are exponentially distributed, we have $p_{ii}(t) \geq \exp(-v_i t) > 0$ for all t.

Since p_j gives the long-run fraction of time the process is in state j, it follows from the renewal-reward theorem in Section 1.3 that another representation for p_j is provided by

$$p_j = \frac{E(\text{amount of time the process is in state } j \text{ during one cycle})}{E(\text{length of one cycle})}$$

for any $j \in I$, where a cycle is defined as the time elapsed between two successive visits of the process $\{X(t)\}$ to some recurrent state s. Thus, using that p_s equals the ratio of $1/v_s$ and E(length of one cycle), we find the useful relation

E(amount of time the process is in state j between two successive
visits to state s) = $p_j/(v_s p_s)$, $j \in I$.   (2.6.4)

Computation of the equilibrium probabilities

The equilibrium probabilities p_j, $j \in I$ are the unique solution to the system of linear equations

$$v_j p_j = \sum_{k \neq j} q_{kj} p_k, \quad j \in I, \tag{2.6.5}$$

$$\sum_{j \in I} p_j = 1. \tag{2.6.6}$$

Corollary 2.6.3 is helpful in memorizing these equations. By physical considerations,

the long-run average number of transitions out of state j per unit time
= the long-run average number of transitions into state j per unit time

for each $j \in I$. Since v_j is the leaving rate out of state j and p_j is the long-run fraction of time the process is in state j, the long-run average number of transitions out of state j per unit time equals $v_j p_j$. On the other hand,

the long-run average number of transitions into j per unit time

$$= \sum_{k \neq j} \text{the long-run average number of transitions from } k \text{ to } j \text{ per unit time}$$

$$= \sum_{k \neq j} q_{kj} p_k.$$

The balance equations (2.6.5) are usually abbreviated as

$$\text{rate out of state } j = \text{rate into state } j.$$

More generally, for any set A of states with $A \neq I$,

$$\text{rate out of the set } A = \text{rate into the set } A. \tag{2.6.7}$$

In mathematical terms,

$$\sum_{j \in A} p_j \sum_{k \notin A} q_{jk} = \sum_{k \notin A} p_k \sum_{j \in A} q_{kj},$$

cf. also the discrete-time analogue (2.3.11). The principle (2.6.7) enables us to write down a recursive equation for the p_j's when

$$I = \{0, 1, \ldots, N\} \quad \text{and} \quad q_{ij} = 0 \quad \text{for } j \leq i - 2, \tag{2.6.8}$$

where $N \leq \infty$. Then, by taking $A = \{i, \ldots, N\}$ and applying the balance principle (2.6.7), we get

$$q_{i,i-1} p_i = \sum_{k=0}^{i-1} p_k \sum_{j=i}^{N} q_{kj}, \quad i = 1, \ldots, N.$$

This recursive relation can be used quite often in applications.

The same tools as discussed in the last part of Section 2.3.1 and in Section 2.3.3 for the discrete-time case can be used to calculate the equilibrium probabilities p_j. It is remarked that instead of solving the equilibrium equations (2.6.5) one may solve the equilibrium equations of the embedded Markov chain $\{X_n\}$ and next apply the relation (2.6.1).

Example 2.6.1 The M/M/1 queue (Example 2.5.1 continued).

In Section 2.5 we have already specified the transition rates of the continuous-time Markov chain $\{X(t)\}$ describing the number of customers in the system. Let us now make the assumption that the mean service time $1/\mu$ is smaller than the mean interarrival time $1/\lambda$. In other words,

$$\rho = \frac{\lambda}{\mu} < 1.$$

The dimensionless quantity ρ is often called the traffic intensity. The assumption of $\rho < 1$ implies that the expected time between two consecutive arrivals finding the system empty is finite (a formal proof of this result is contained in the analysis of Example 1.3.4 of Section 1.3). Thus Assumption 2.6.1 applies by taking the regeneration state r equal to state 0.

In general, to write down the balance equations, it may be helpful to use a *transition rate diagram*. The nodes of the diagram represent the states and the

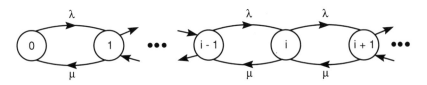

Figure 2.6.1   The transition rate diagram for the number of customers present

arrows in the diagram give the possible state transitions. An arrow from node i to node j is only drawn when the transition rate q_{ij} is positive, in which case the arrow is labelled with the value q_{ij}. In Figure 2.6.1 we display for the M/M/1 queue the transition rates of the $\{X(t)\}$ process.

Using the balance equation 'rate out of state i = rate into state i', we obtain the equilibrium equations

$$\lambda p_0 = \mu p_1,$$

$$(\lambda + \mu)p_i = \lambda p_{i-1} + \mu p_{i+1}, \quad i \geq 1.$$

These equations can easily be rewritten as

$$\mu p_i = \lambda p_{i-1}, \quad i \geq 1.$$

The latter equations can be directly written down by applying the balance principle (2.6.7) with $A = \{i, i+1, \ldots\}$. Iterating these equations yields $p_i = (\lambda/\mu)^i p_0$, $i \geq 1$. Using the normalizing equation $\Sigma_{i=0}^{\infty} p_i = 1$, we find $p_0 = 1 - \rho$. Hence we obtain the equilibrium probabilities

$$p_i = (1 - \rho)\rho^i, \quad i = 0, 1, \ldots . \tag{2.6.9}$$

From this result we obtain directly an expression for

L = the long-run average number of customers present.

More precisely, since the process $\{X(t)\}$ is regenerative, $\lim_{t\to\infty}(1/t)\int_0^t X(u)\,du$ is equal to a constant L with probability 1 with L having the above meaning. In view of the interpretation of the p_j's, it is intuitively clear that

$$L = \sum_{j=0}^{\infty} j p_j.$$

A rigorous proof of this result follows by assuming a cost at rate j whenever j customers are present and next applying the ergodic Theorem 2.6.2. Using (2.6.9), we obtain

$$L = \frac{\rho}{1 - \rho}.$$

Example 2.6.2 Response areas for emergency units (Example 2.5.2 continued)

For the given design of the two response areas, performance measures of interest are the fraction of alarms lost and the fraction of time each unit is busy. These performance measures can easily be expressed in terms of the equilibrium probabilities of the continuous-time Markov chain $\{X(t)\}$. To do so, we denote these equilibrium probabilities by $p(j, k)$ for $0 \le j, k \le 2$. Since $p(j, k)$ represents the long-run fraction of time the system is in state (j, k), it follows that

$$\text{the fraction of time that unit 1 is busy} = \sum_{j=1}^{2} \sum_{k=0}^{2} p(j, k).$$

A similar expression applies for the fraction of time that unit 2 is busy. To find the fraction of alarms from district $d(= 1, 2)$ that are lost, we use the property that 'Poisson arrivals see time averages'; see Theorem 1.7.1 in Section 1.7. Hence the long-run fraction of alarms from district d that find upon occurrence the system in state (j, k) equals the long-run fraction of time the system is in state (j, k). This gives for each district $d = 1, 2$,

$$\text{the fraction of alarms from district } d \text{ that are lost} = \sum_{j=1}^{2} \sum_{k=1}^{2} p(j, k).$$

It remains to calculate the equilibrium probabilities $p(j, k)$. To do so, it is helpful to display in Figure 2.6.2 the transition rates of the continuous-time Markov chain $\{X(t)\}$. Next, by applying the balance equation 'rate out of a state = rate into that

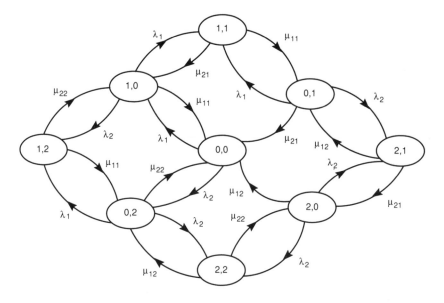

Figure 2.6.2   The transition rate diagram

state', we obtain the equations

$$(\lambda_1 + \lambda_2)p(0, 0) = \mu_{21}p(0, 1) + \mu_{22}p(0, 2) + \mu_{11}p(1, 0) + \mu_{12}p(2, 0),$$

$$(\lambda_1 + \lambda_2 + \mu_{21})p(0, 1) = \mu_{11}p(1, 1) + \mu_{12}p(2, 1),$$

$$(\lambda_1 + \lambda_2 + \mu_{22})p(0, 2) = \lambda_2 p(0, 0) + \mu_{11}p(1, 2) + \mu_{12}p(2, 2),$$

$$(\lambda_1 + \lambda_2 + \mu_{11})p(1, 0) = \lambda_1 p(0, 0) + \mu_{21}p(1, 1) + \mu_{22}p(1, 2),$$

$$(\mu_{11} + \mu_{21})p(1, 1) = \lambda_1 p(0, 1) + \lambda_1 p(1, 0),$$

$$(\mu_{11} + \mu_{22})p(1, 2) = \lambda_1 p(0, 2) + \lambda_2 p(1, 0),$$

$$(\lambda_1 + \lambda_2 + \mu_{12})p(2, 0) = \mu_{21}p(2, 1) + \mu_{22}p(2, 2),$$

$$(\mu_{12} + \mu_{21})p(2, 1) = \lambda_2 p(0, 1) + \lambda_1 p(2, 0),$$

$$(\mu_{12} + \mu_{22})p(2, 2) = \lambda_2 p(0, 2) + \lambda_2 p(2, 0).$$

Also, we have the normalizing equation $\Sigma_{j=0}^2 \Sigma_{k=0}^2 p(j, k) = 1$. These linear equations have to be numerically solved.

In the above analysis the assumption of exponentially distributed service times was made because it simplifies the analysis. In practice, however, this assumption will often not be satisfied. Therefore it is important to know whether the results obtained for exponentially distributed service times are approximately correct for the actual service-time distributions. The reader is asked to study this question in Exercise 2.19.

2.7 APPLICATIONS OF CONTINUOUS-TIME MARKOV CHAINS

In this section we give several other applications of the continuous-time Markov chain model. These applications concern a variety of practical problems and illustrate the wide applicability of the continuous-time Markov chain model.

Example 2.7.1 Erlang's loss model

Consider a communication system with c transmission channels at which messages are offered according to a Poisson process with rate λ. An arriving message that finds all c channels busy is lost and has no further influence on the system; otherwise, the message is assigned to a free channel and transmission starts immediately. The transmission times of the messages are independent and identically distributed random variables. Also, the arrival process and the transmission times are independent of each other. The goal is to find an expression for the long-run fraction of messages lost.

For the moment let us assume that the transmission times are exponentially distributed with mean $1/\mu$. To analyse the problem, define for each $t \geq 0$ the

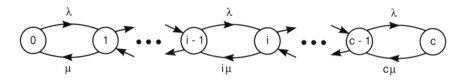

Figure 2.7.1   The transition diagram for the number of busy channels

random variable $X(t)$ by

$$X(t) = \text{the number of busy channels at time } t.$$

The continuous-time process $\{X(t), \ t \geq 0\}$ has the state space $I = \{0, 1, \ldots, c\}$. The process $\{X(t)\}$ allows for the characterization (2.5.2) and is thus a continuous-time Markov chain. The transition rate diagram is displayed in Figure 2.7.1. To see that

$$P\{X(t + \Delta t) = i - 1 | X(t) = i\} = i\mu\Delta t + o(\Delta t),$$

use the memoryless property of the exponential distribution and note that the minimum of i independent exponential services each having mean $1/\mu$ is exponentially distributed with mean $1/(i\mu)$.

The transition rates have the property expressed in (2.6.8). Hence a recursive equation can be given for the state probabilities p_i. Using the balance equation 'rate out of the set A = rate into the set A' for the set $A = \{i, \ldots, c\}$, we find

$$i\mu p_i = \lambda p_{i-1}, \quad 1 \leq i \leq c.$$

This equation and the normalizing equation $\Sigma_{i=0}^{c} p_i = 1$ yield

$$p_i = \frac{(\lambda/\mu)^i/i!}{\displaystyle\sum_{k=0}^{c}(\lambda/\mu)^k/k!}, \quad 0 \leq i \leq c. \tag{2.7.1}$$

To find the long-run fraction of messages lost, we use the property 'Poisson arrivals see time averages'. The long-run fraction of messages finding upon arrival all c channels busy is equal to the long-run fraction of time that all c channels are busy. The latter fraction is given by p_c. Thus we find

$$\text{the long-run fraction of messages lost} = \frac{(\lambda/\mu)^c/c!}{\displaystyle\sum_{k=0}^{c}(\lambda/\mu)^k/k!} \tag{2.7.2}$$

with probability 1.

The above formulas have been derived under the assumption that the transmission times are exponentially distributed. However, a famous insensitivity result from teletraffic theory tells us that the equilibrium distribution of the number of busy channels is also given by (2.7.1) when the transmission time has a general distribution with mean $1/\mu$. In other words, the equilibrium distribution of the number of busy channels requires the probability distribution of the transmission time only through its mean and is independent of the form of the distribution. This result was already conjectured by the telephone engineer A. K. Erlang in the early 1900s. The truncated Poisson distribution (2.7.1) is known as Erlang's loss distribution and the formula (2.7.2) is often called Erlang's loss formula. A rigorous proof of the insensitivity result was only given many years after Erlang made his conjecture, see e.g. Cohen (1976). The Erlang loss model is a very useful model that has a variety of applications.

The robustness of the present model is characteristic for many stochastic service systems in which queueing is not possible. It should be pointed out that opposite to stochastic loss systems are stochastic delay systems. In delay systems service requests that find upon arrival all servers busy are delayed until a free server becomes available. Stochastic delay systems typically do not have the property that the performance measures require the service time only through its first moment. For example, this is nicely demonstrated by the Pollaczek–Khintchine formula for the delay system dealt with in Example 2.9.1 in Section 2.9.

Example 2.7.2 Performance analysis for an unloader at a container terminal

Consider a container terminal with a finite number of N trailers which bring loads of containers from ships to a single unloader. The unloader can serve only one trailer at a time and the unloading time per trailer has an exponential distribution with mean $1/\mu_s$. A trailer leaves when it is unloaded and returns at the unloader with a next load of containers after an exponentially distributed trip time with mean $1/\lambda$. However, after a trailer is unloaded, the unloader needs an extra finishing time for the unloaded containers before the unloader is available to unload the next trailer. This finishing time has an exponential distribution with mean $1/\mu_f$. The unloading times, the trip times and the finishing times are assumed to be independent of each other.

We are interested in performance measures such as the average number of trailers that are unloaded per unit time and the probability that an arbitrary trailer has to wait. To find these performance measures, define for each $t \geq 0$,

$$X_1(t) = \text{the number of trailers present at the unloader at time } t,$$

$$X_2(t) = \begin{cases} 0 & \text{if the unloader is unloading or idle at time } t, \\ 1 & \text{if the unloader is in the finishing process.} \end{cases}$$

The process $\{X(t) = (X_1(t), X_2(t)), \ t \geq 0\}$ is a continuous-time Markov chain with state space $I = \{(i, j) | i = 0, \ldots, N, \ j = 0, 1\}$. The transition rate diagram in Figure 2.7.2 is easily verified by using the fact that the arrival rate of trailers at the unloader equals the sum of $N - i$ individual rates λ when i trailers are already present at the unloader.

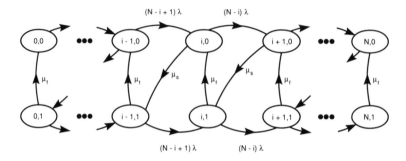

Figure 2.7.2   The transition rate diagram for the process $\{X(t)\}$

We denote the equilibrium probabilities of the process $\{X(t)\}$ by $p(i, j)$. Using the balance equation 'rate out of a state = rate into that state', we obtain the equations

$$N\lambda p(0, 0) = \mu_f p(0, 1),$$

$$\{\mu_s + (N - i)\lambda\}p(i, 0) = (N - i + 1)\lambda p(i - 1, 0) + \mu_f p(i, 1), \qquad 1 \leq i \leq N,$$

$$\{\mu_f + (N - i)\lambda\}p(i, i) = (N - i + 1)\lambda p(i - 1, 1) + \mu_s p(i + 1, 0), \qquad 0 \leq i \leq N,$$

with $p(-1, 1) = p(N + 1, 0) = 0$ by convention. Also, we have the normalizing equation

$$\sum_{i=0}^{N}\{p(i, 0) + p(i, 1)\} = 1.$$

A closer look at the balance equations reveals that they can recursively be solved. Starting with $\overline{p}(0, 0) = 1$, we successively compute $\overline{p}(0, 1)$, $\overline{p}(1, 0)$, $\overline{p}(1, 1)$, etc., and finally $\overline{p}(N - 1, 0)$, $\overline{p}(N, 0)$, $\overline{p}(N, 1)$. Next the desired probabilities $p(i, j)$ follow by normalizing the $\overline{p}(i, j)$'s.

We are now in a position to specify the following performance measures:

P_I = the long-run fraction of time the unloader is idle,

L_q = the long-run average number of trailers waiting to be unloaded,

λ_T = the long-run average number of trailers unloaded per unit time,

π_W = the long-run fraction of trailers that have to queue.

It is immediately seen that

$$P_I = p(0, 0) \quad \text{and} \quad L_q = \sum_{i=1}^{N}(i - 1)p(i, 0) + \sum_{i=1}^{N}ip(i, 1).$$

To find λ_T, note that λ_T can be seen as the long-run average number of transitions per unit time from some state $(i, 0)$ to state $(i - 1, 1)$. Thus

$$\lambda_T = \mu_s \sum_{i=1}^{N} p(i, 0). \qquad (2.7.3)$$

An alternative expression for λ_T is obtained by observing that

the long-run average number of trailers unloaded per unit time
= the long-run average number of trailers arriving at the unloader
per unit time.

Thus, λ_T can also be represented by the long-run average number of transitions per unit time from some state (i, δ) to state $(i + 1, \delta)$. This shows that

$$\lambda_T = \sum_{i=0}^{N} (N - i)\lambda[p(i, 0) + p(i, 1)]. \qquad (2.7.4)$$

The identity implied by (2.7.3) and (2.7.4) can be used as an extra accuracy check when solving the linear equations for the equilibrium probabilities.

The delay probability π_W is found by using the relation

$$\pi_W = \frac{\text{the long-run average number of delayed arrivals per unit time}}{\text{the long-run average number of arrivals per unit time}}.$$

The denominator of this ratio equals λ_T. By the same arguments as used to derive (2.7.4), the numerator equals

$$\sum_{i=1}^{N} (N - i)\lambda p(i, 0) + \sum_{i=0}^{N} (N - i)\lambda p(i, 1).$$

Thus

$$\pi_W = \frac{1}{\lambda_T} \left[\sum_{i=1}^{N} (N - i)\lambda p(i, 0) + \sum_{i=0}^{N} (N - i)\lambda p(i, 1) \right].$$

The above analysis was done under simplifying assumptions about the probability distributions of the trip time, the unloading time and the finishing time in order to keep the analysis tractable. In reality, exponential distributions cannot always be expected to occur. In particular, the assumption of exponentially distributed trip times is rather unrealistic. The continuous-time Markov chain analysis can be extended when assuming Erlangian or hyperexponential distributions. In Table 2.7.1 we give for some numerical examples the values of the performance measures P_1, λ_T, L_q and π_W for various distributions of the trip time and unloading time when the means are kept fixed. We assume the numerical data

$$N = 10, \quad E(T) = 50, \quad E(U) = 3 \quad \text{and} \quad E(F) = 0.5,$$

Table 2.7.1 Numerical results for the unloader

(c_T^2, c_U^2)	P_1	λ_T	L_q	π_W
(0.5, 0.5)	0.3714	0.180	0.481	0.5761
(1, 0.5)	0.3727	0.179	0.501	0.5841
(4, 0.5)	0.3751	0.179	0.537	0.5978
(0.5, 1)	0.3780	0.178	0.582	0.5680
(1, 1)	0.3786	0.178	0.590	0.5736
(4, 1)	0.3795	0.177	0.604	0.5835

where the generic variables T, U and F denote the trip time, the unloading time and the finishing time. The squared coefficient of variation c_T^2 of the trip time is varied as 0.5, 1 and 4, with $c_T^2 = 0.5$ corresponding to the E_2 distribution, $c_T^2 = 1$ to the exponential distribution and $c_T^2 = 4$ to an H_2 distribution with the gamma normalization (see Appendix B). The squared coefficient of variation c_U^2 of the unloading time has the two values 0.5 and 1 with $c_U^2 = 0.5$ corresponding to the E_2 distribution and $c_U^2 = 1$ to the exponential distribution. The numerical investigations indicate that some of the performance measures are fairly insensitive to more than the first moments of the various distributions. This is particularly true for P_1, λ_T and π_W. In other words, the analysis of the present problem may be based on the simplifying assumption of exponential distributions for practical purposes. This empirical finding is typical for many practical queueing systems with *finite-source* input.

Remark 2.7.1 Little's formula

Another important performance measure is

W_q = the long-run average delay in queue per trailer.

The delay in queue excludes unloading time. The quantity W_q can be directly related to the long-run average number of trailers waiting in queue by Little's formula. We have

$$L_q = \lambda^* W_q, \qquad (2.7.5)$$

where $\lambda^* = \lambda_T$ denotes the long-run average arrival rate of trailers. To see this relation, we impose an appropriate cost structure on the system and apply the general principle (cf. Section 1.6)

the long-run average cost incurred by the system per unit time
= (the long-run average number of arrivals per unit time)
× (the long-run average cost incurred per arrival). (2.7.6)

Assuming that for each arriving trailer the system incurs a cost at a rate of 1 while the trailer waits in queue, we have that the long-run average cost incurred per trailer equals W_q. On the other hand, the system incurs a cost at a rate of j when

j trailers are waiting in queue. Hence the long-run average cost incurred by the system per unit time equals L_q.

Example 2.7.3 Satellite capacity allocation to two competing user classes

Messages from two sources are sent to a satellite communication system which has s circuits for handling the submitted messages. The two sources 1 and 2 consist of M_1 and M_2 users respectively. Each user of type j generates messages for the satellite according to a Poisson process with rate λ_j whenever the user has no message in service at the satellite, and generates no new messages otherwise. The users act independently distributed of each other. The handling time of a message of type j is exponentially distributed with mean $1/\mu_j$. Each circuit is able to handle a message of any type but can transmit only one message at a time. No queueing is allowed in the satellite system. An arriving message gets immediate access to a free circuit when that message is admitted to the system. The following acceptance/rejection rule is used for the messages submitted by the two competing user classes. Messages of type 1 are always accepted whenever not all of the s circuits are occupied, whereas messages of type 2 are only accepted when less than L messages of type 2 are being processed and not all of the circuits are occupied. We are interested in the value of the control parameter L that minimizes a weighted sum of the rejection rates of the messages of the types 1 and 2 with c_1 and c_2 as the respective weights.

Assuming a given value of the control parameter L, define the stochastic process $\{(X_1(t), X_2(t)), t \geq 0\}$ by

$X_j(t) = $ the number of messages of type j in the satellite system at time t.

This process is readily seen to be a continuous-time Markov chain with finite state space $I = \{(i_1, i_2) | 0 \leq i_1 + i_2 \leq s, \ 0 \leq i_2 \leq L\}$. The equilibrium probabilities of this Markov chain are denoted by $p(i_1, i_2)$.

Denoting by $R_j(L)$ the long-run average number of messages of type j that are rejected per unit time, the design criterion is given by

$$R(L) = c_1 R_1(L) + c_2 R_2(L).$$

To evaluate this criterion, note that messages of type k arrive at the satellite system according to a Poisson process with the state-dependent rate $(M_k - i_k)\lambda_k$ for $k = 1, 2$ whenever the system is in state (i_1, i_2). Then, assuming that a fixed cost of 1 is incurred each time a message of type 1 is rejected, it follows from the ergodic Theorem 2.6.2 that

$$R_1(L) = \sum_{i_2=0}^{L} [M_1 - (s - i_2)]\lambda_1 p(s - i_2, i_2).$$

Similarly, we have

$$R_2(L) = \sum_{i_1=0}^{s-L}(M_2 - L)\lambda_2 p(i_1, L) + \sum_{i_2=0}^{L-1}(M_2 - i_2)\lambda_2 p(s - i_2, i_2).$$

Other interesting operating characteristics are the throughput $T(L)$ and the server occupancy $O(L)$, which are defined as the long-run average number of messages served per unit time and the long-run fraction of time a circuit is occupied. To find the average throughput, assume that a lump reward of 1 is earned each time the service completion of a message occurs. Then the long-run average cost per unit time gives the average throughput. Since messages of type k are completed at a rate of $i_k \mu_k$ whenever i_k messages of type k are present, it follows from the ergodic Theorem 2.6.2 that

$$T(L) = \sum_{i_1, i_2}(i_1\mu_1 + i_2\mu_2)p(i_1, i_2).$$

Also, noting that the server occupancy equals $(1/s)$ times the long-run average number of occupied circuits, it follows that the server occupancy is given by

$$O(L) = \frac{1}{s}\sum_{i_1, i_2}(i_1 + i_2)p(i_1, i_2).$$

It remains to evaluate the equilibrium probabilities $p(i_1, i_2)$. The familiar balance equation 'rate out of a state = rate into that state' yields for any state (i_1, i_2),

$$\{i_1\mu_1 + i_2\mu_2 + (M_1 - i_1)\lambda_1 + (M_2 - i_2)\lambda_2\delta(L - i_2)\}p(i_1, i_2)$$

$$= (M_1 - i_1 + 1)\lambda_1 p(i_1 - 1, i_2) + (M_2 - i_2 + 1)\lambda_2 p(i_1, i_2 - 1)$$

$$+ (i_1 + 1)\mu_1 p(i_1 + 1, i_2) + (i_2 + 1)\mu_2\delta(L - i_2 - 1)p(i_1, i_2 + 1),$$

when $i_1 + i_2 < s$ and $i_2 \leq L$; otherwise the left side of this equation should be changed as $(i_1\mu_1 + i_2\mu_2)p(i_1, i_2)$. Here $\delta(x) = 1$ for $x \geq 1$ and $\delta(x) = 0$ otherwise. Together with the normalizing equation these linear equations enable us to compute the probabilities $p(i_1, i_2)$.

Numerical illustration

As an illustration we consider the following numerical data

$$s = 10, \quad M_1 = 10, \quad M_2 = 10, \quad \lambda_1 = 3, \quad \lambda_2 = 1,$$
$$\mu_1 = 4, \quad \mu_2 = 1, \quad c_1 = 1, \quad \text{and} \quad c_2 = 1.$$

In Table 2.7.2 we give the values of the rejection rate $R(L)$, the throughput $T(L)$ and the server occupancy $O(L)$ for several values of the control parameter L. It

Table 2.7.2   The performance measures

L	R(L)	T(L)	O(L)
4	4.456	19.833	0.746
5	4.387	19.790	0.789
6	4.653	19.587	0.811

turns out from the calculations that the design criterion $R(L)$ assumes its minimum value for the L-policy with $L = 5$.

Insensitivity

The following remarks are in order with regard to the above model. First, in reality transmission times are not always expected to be exponentially distributed. To find out the effect of the distributional form of the transmission times on the performance measures, we numerically analysed the case of generalized Erlangian distributed transmission times. These additional numerical investigations indicated that under any L-policy the state probabilities and thus the performance measures depend on the transmission times of the types 1 and 2 only through their respective means $1/\mu_1$ and $1/\mu_2$. This empirical finding is not surprising since many stochastic service systems in which no queueing is possible have the property that the state probabilities are insensitive to the form of the service-time distributions and depend only on their means. Indeed, for the present model controlled by an L-policy this insensitivity property can be proved theoretically; see Van Dijk and Tijms (1986). For the special case of the L-policy with $L = s$ the insensitivity result is quite known and is sometimes referred to as the generalized Engset formula. Also, for the present finite-source population model with blocking it can be shown that the state probabilities are also correct when the think time of a user of type j until sending a new request is generally distributed with mean $1/\lambda_j$.

Second, the above discussion limited itself to the easily implementable L-policy, but other control rules are conceivable. The question of how to compute the optimal control rule among the class of all possible control rules will be addressed in the next chapter dealing with Markovian decision problems. It should be noted that the best L-policy is in general not optimal among the class of all possible control rules. However, we found that using the best L-policy rather than the overall optimal control rule often leads to only a small deviation from the optimal value of the design criterion. For example, for the above numerical data the rejection rate of 4.387 of the best L-policy is only 3.2% above the minimum rejection rate of 4.246. The minimum rejection rate is achieved by the following control rule. Each arriving message of type 1 is accepted whenever not all of the circuits are occupied. A message of type 2 finding upon arrival that i messages of type 1 are present, is accepted only when less than L_i messages of type 2 are in service and not all of the circuits are occupied. For the above data, $L_0 = L_1 = 6$, $L_2 = L_3 = L_4 = 5$, $L_5 = 4$, $L_6 = L_7 = 3$, $L_8 = 2$, $L_9 = 1$ and $L_{10} = 0$. Although for this more general control rule the insensitivity property is no longer exactly true, our numerical investigations indicate that the dependence on the distributional form of the service times is quite weak. For example, taking an Erlang-2 distribution for the service times of the requests of type 1 with the other data kept the same, the average rejection rate for the above L_i-policy equals 4.272, as opposed to 4.246 for the case of exponential services.

2.8   TRANSIENT ANALYSIS OF CONTINUOUS-TIME MARKOV CHAINS

In many practical situations one is not interested in the long-run behaviour of a stochastic system but in its transient behaviour. A typical example concerns airport runway operations. The demand profile for runway operations shows considerable variation over time with peaks at certain hours of the day. Equilibrium models are of no use in this kind of situation. The computation of transient solutions for Markov systems is a very important issue that arises in numerous problems in queueing, inventory and reliability.

This section deals with computational methods for the transient probabilities and first-passage time probabilities. Also, we discuss the computation of transient rewards (costs). Throughout this section it is assumed that the continuous-time Markov chain satisfies Assumption 2.5.2 requiring that the transition rates v_i, $i \in I$ are positive and bounded. In Section 2.8.1 we consider the computation of the transient probabilities $p_{ij}(t)$. A common approach is to derive Kolmogoroff's forward differential equations and solve them by a standard method from numerical analysis. An alternative and more transparent approach is based on the powerful probabilistic idea of uniformization. Section 2.8.2 deals with the computation of first-passage time probabilities and discusses both the approach of Kolmogoroff's backwards differential equation and the uniformization approach.

2.8.1 Transient probabilities

The transient probabilities $p_{ij}(t)$ of a continuous-time Markov chain $\{X(t)\}$ are defined by

$$p_{ij}(t) = P\{X(t) = j \mid X(0) = i\}, \quad i, j \in I \text{ and } t > 0.$$

We first discuss the useful technique of Kolmogoroff's forward differential equations.

Theorem 2.8.1 (Kolmogoroff's forward differential equations)   *Suppose that the continuous-time Markov chain* $\{X(t), \ t \geq 0\}$ *satisfies Assumption 2.5.2. Then, for any* $i \in I$,

$$p'_{ij}(t) = \sum_{k \neq j} q_{kj} p_{ik}(t) - v_j p_{ij}(t), \quad j \in I \text{ and } t > 0. \tag{2.8.1}$$

Proof   We sketch the proof only for the case of a finite state space I. The proof of the validity of the forward equations for the case of an infinite state space is very intricate. Fix $i \in I$ and $t > 0$. Let us consider what may happen in $(t, t + \Delta t]$ with Δt very small. The number of transitions in any finite time interval is finite with probability 1 and thus we can condition on the state that will occur at time t. We then find

$$p_{ij}(t + \Delta t) = P\{X(t + \Delta t) = j \mid X(0) = i\}$$

$$= \sum_{k \in I} P\{X(t + \Delta t) = j \mid X(0) = i, X(t) = k\} P\{X(t) = k \mid X(0) = i\}$$

$$= \sum_{k \in I} P\{X(t + \Delta t) = j \mid X(t) = k\} p_{ik}(t)$$

$$= \sum_{k \neq j} q_{kj} \Delta t p_{ik}(t) + (1 - v_j \Delta t) p_{ij}(t) + o(\Delta t),$$

using that a finite sum of $o(\Delta t)$ terms is again $o(\Delta t)$. Hence

$$\frac{p_{ij}(t + \Delta t) - p_{ij}(t)}{\Delta t} = \sum_{k \neq j} q_{kj} p_{ik}(t) - v_j p_{ij}(t) + \frac{o(\Delta t)}{\Delta t}.$$

Letting $\Delta t \to 0$ yields the desired result.

The linear differential equations (2.8.1) can only be explicitly solved in very special cases.

Example 2.6.1 (continued) The M/M/1 queue

Using the transition diagram in Figure 2.6.1, we find that Kolmogoroff's forward differential equations are as follows for the *M/M/1* queue:

$$p'_{ij}(t) = \mu p_{i,j+1}(t) + \lambda p_{i,j-1}(t) - (\lambda + \mu) p_{ij}(t), \quad i, j = 0, 1, \ldots \text{ and } t > 0$$

with $p_{i,-1}(t) = 0$. An explicit solution of these equations is given by

$$p_{ij}(t) = \frac{2}{\pi} \rho^{(j-i)/2} \int_0^\pi \frac{e^{-\mu t \gamma(y)}}{\gamma(y)} a_i(y) a_j(y) \, dy + \begin{cases} (1 - \rho)\rho^j, & \rho < 1 \\ 0, & \rho \geq 1 \end{cases}$$

for $i, j = 0, 1, \ldots$, where $\rho = \lambda / \mu$ and the functions $\gamma(y)$ and $a_k(y)$ are defined by

$$\gamma(y) = 1 + \rho - 2\sqrt{\rho} \cos(y) \quad \text{and} \quad a_k(y) = \sin(ky) - \sqrt{\rho} \sin[(k+1)y].$$

A proof of this explicit solution is not given here; see Morse (1955). The trigonometric integral representation for $p_{ij}(t)$ is very convenient for numerical computations. Integral representations can also be given for the first two moments of the number of customers in the system. The formulas will only be given for the case of $\rho < 1$. Denoting by $L(i, t)$ the number of customers in the system at time t when initially there are i customers present, we have

$$E[L(i, t)] = \frac{2}{\pi} \rho^{(1-i)/2} \int_0^\infty \frac{e^{-\mu t \gamma(y)}}{\gamma^2(y)} a_i(y) \sin(y) \, dy + \frac{\rho}{1 - \rho}$$

and

$$E[L^2(i, t)] = \frac{4(1 - \rho)}{\pi} \rho^{(1-i)/2} \int_0^\infty \frac{e^{-\mu t \gamma(y)}}{\gamma^3(y)} a_i(y) \sin(y) \, dy$$

$$+ 2\rho(1 - \rho)^{-2} - E[L(i, t)].$$

An important factor is the relaxation factor $\tau_c = [\mu(1 - \sqrt{\rho})^2]^{-1}$. This is the dominating factor for the time it takes until the system reaches statistical equilibrium. Empirical investigations by Odoni and Roth (1983) indicate that the relative difference between $E[L(0, t)]$ and its limiting value $\rho/(1 - \rho)$ is below $\alpha\%$ for $t \geq t_\alpha$, where $t_\alpha = 0.713 \tau_c \ln(100/\alpha)$ for $\alpha \geq 1$.

In general the linear differential equations (2.8.1) have to be solved numerically. A generally applicable method is the well-known Runge-Kutta method from numerical analysis. This method can also be applied when the transition rates of the continuous-time Markov chain are time-dependent. Another possibility to compute the $p_{ij}(t)$'s is to use the Fast Fourier Transform method when an explicit expression for the generating function $P(t; z) = \Sigma_{j \in I} p_{ij}(t) z^j$ can be derived. A simple and transparent probabilistic method can be given to compute the transient probabilities $p_{ij}(t)$ when the transition rates of the continuous-time Markov chain are time-independent. This method will be discussed next.

Uniformization method

To introduce this method, consider first the special case in which the leaving rates ν_i of the states are identical, say $\nu_i = \nu$ for all i. Then the transition epochs are generated by a Poisson process with rate ν. In this situation an expression for $p_{ij}(t)$ is directly obtained by conditioning on the number of Poisson events up to time t and using the n-step transition probabilities of the embedded Markov chain $\{X_n\}$ defined by (2.5.1). However, the leaving rates ν_i are in general not identical. Fortunately, there is a simple trick for reducing the case of non-identical leaving rates to the case of identical leaving rates. The uniformization method transforms the original continuous-time Markov chain with non-identical leaving rates into an equivalent stochastic process in which the transition epochs are generated by a Poisson process at a *uniform* rate and the transitions from state to state are described by a discrete-time Markov chain that allows for self-transitions leaving the state of the process unchanged.

To formulate the uniformization method, choose a finite number ν with

$$\nu \geq \nu_i \quad \text{for all } i \in I.$$

The state of the continuous-time Markov chain $\{X(t)\}$ just after a state transition is described by the discrete-time Markov chain $\{X_n\}$ whose one-step transition probabilities p_{ij} are given by (2.5.1). Define now $\{\overline{X}_n\}$ as the discrete-time Markov chain whose one-step transition probabilities $\overline{p}_{ij}$ are given by

$$\overline{p}_{ij} = \begin{cases} \dfrac{\nu_i}{\nu} p_{ij}, & j \neq i, \\ 1 - \dfrac{\nu_i}{\nu}, & j = i, \end{cases}$$

for any $i \in I$. Let $\{N(t), \ t \geq 0\}$ be a Poisson process with rate ν such that the process is independent of the discrete-time Markov chain $\{\overline{X}_n\}$. Define now the continuous-time stochastic process $\{\overline{X}(t), \ t \geq 0\}$ by

$$\overline{X}(t) = \overline{X}_{N(t)}, \quad t \geq 0. \tag{2.8.2}$$

In other words, the process $\{\overline{X}(t)\}$ makes state transitions at epochs generated by a Poisson process with rate ν and the state transitions are governed by the discrete-time Markov chain $\{\overline{X}_n\}$ with one-step transition probabilities $\overline{p}_{ij}$. Each time the Markov chain $\{\overline{X}_n\}$ is in state i, the next transition is the same as in the Markov chain $\{X_n\}$ with probability ν_i/ν and is a self-transition with probability $1 - \nu_i/\nu$. The transitions out of state i are in fact delayed by a time factor of ν/ν_i, while the time itself between two state transitions from state i is condensed by a factor of ν_i/ν. This heuristically explains why the continuous-time process $\{\overline{X}(t)\}$ is probabilistically identical to the original continuous-time Markov chain $\{X(t)\}$. A formal proof of this extremely useful result is given in the next theorem.

Theorem 2.8.2   *Suppose that the continuous-time Markov chain $\{X(t)\}$ satisfies Assumption 2.5.2. Then*

$$p_{ij}(t) = P\{\overline{X}(t) = j \mid \overline{X}(0) = i\} \quad \text{for all } i, j \in I \text{ and } t > 0.$$

Proof   For any $t > 0$, define the matrix $P(t)$ by $P(t) = (p_{ij}(t))$, $i, j \in I$. Denote by Q the matrix $Q = (q_{ij})$, $i, j \in I$, where the diagonal elements q_{ii} are defined by

$$q_{ii} = -\nu_i.$$

Then Kolmogoroff's forward differential equations can be written as $P'(t) = P(t)Q$ for any $t > 0$. It is left to the reader to verify that the solution of this system of differential equations is given by

$$P(t) = e^{tQ} = \sum_{n=0}^{\infty} \frac{t^n}{n!} Q^n, \quad t \geq 0. \tag{2.8.3}$$

The matrix $\overline{P} = (\overline{p}_{ij})$, $i, j \in I$, can be written as $\overline{P} = Q/\nu + I$, where I is the identity matrix. Thus

$$P(t) = e^{tQ} = e^{\nu t(\overline{P}-I)} = e^{\nu t \overline{P}}e^{-\nu t I} = e^{-\nu t}e^{\nu t \overline{P}} = \sum_{n=0}^{\infty} e^{-\nu t}\frac{(\nu t)^n}{n!}\overline{P}^n.$$

On the other hand, by conditioning on the number of Poisson events up to time t in the $\{\overline{X}(t)\}$ process, we have

$$P\{\overline{X}(t) = j | \overline{X}(0) = i\} = \sum_{n=0}^{\infty} e^{-\nu t}\frac{(\nu t)^n}{n!}\overline{p}_{ij}^{(n)},$$

where $\overline{p}_{ij}^{(n)}$ is the n-step transition probability of the discrete-time Markov chain $\{\overline{X}_n\}$. Together the latter two equations yield the desired result.

Corollary 2.8.3 *The probabilities $p_{ij}(t)$ are given by*

$$p_{ij}(t) = \sum_{n=0}^{\infty} e^{-\nu t}\frac{(\nu t)^n}{n!}\overline{p}_{ij}^{(n)}, \quad i, j \in I \text{ and } t > 0, \tag{2.8.4}$$

where the probabilities $\overline{p}_{ij}^{(n)}$ can be recursively computed from

$$\overline{p}_{ij}^{(n)} = \sum_{k \in I} \overline{p}_{ik}^{(n-1)}\overline{p}_{kj}, \quad n = 1, 2, \ldots \tag{2.8.5}$$

starting with $\overline{p}_{ii}^{(0)} = 1$ and $\overline{p}_{ij}^{(0)} = 0$ for $j \neq i$.

This probabilistic result is extremely useful for computational purposes. The series in (2.8.4) converges much faster than the series expansion (2.8.3). The computations required by (2.8.4) are simple and transparent. For fixed $t > 0$ the infinite series can be truncated beforehand using that

$$\sum_{n=M}^{\infty} e^{-\nu t}\frac{(\nu t)^n}{n!}\overline{p}_{ij}^{(n)} \leq \sum_{n=M}^{\infty} e^{-\nu t}\frac{(\nu t)^n}{n!}.$$

For a prespecified accuracy number $\varepsilon > 0$, we choose M such that the right side of this inequality is smaller than ε. Note that the Poisson probabilities can be recursively computed; see Remark 1.2.1 in Section 1.2. Also, the following remark may be helpful. For fixed initial state i, the recursion scheme (2.8.5) boils down to the multiplication of a vector with the matrix $\overline{P}$. In many applications the matrix $\overline{P}$ is sparse. Then the computational effort can be considerably reduced by using a data structure for sparse matrix multiplications.

Transient rewards

Let us assume that a reward at rate $r(j)$ is earned whenever the continuous-time Markov chain $\{X(t)\}$ is in state j, while a lump reward of F_{jk} is earned each time the process makes a transition from state j to state $k(\neq j)$. Define the random variable $R(t)$ by

$$R(t) = \text{the total reward earned up to time } t, \quad t \geq 0.$$

Lemma 2.8.4 *Suppose that the continuous-time Markov chain $\{X(t)\}$ satisfies Assumption 2.5.2. Then*

$$E[R(t)|X(0) = i] = \sum_{j \in I} r(j) E_{ij}(t) + \sum_{j \in I} E_{ij}(t) \sum_{k \neq j} q_{jk} F_{jk}, \quad t > 0,$$

where $E_{ij}(t)$ is the expected amount of time that the process $\{X(t)\}$ is in state j up to time t when the process starts in state i. For any $i, j \in I$,

$$E_{ij}(t) = \frac{1}{\nu} \sum_{k=1}^{\infty} e^{-\nu t} \frac{(\nu t)^k}{k!} \sum_{n=0}^{k-1} \overline{p}_{ij}^{(n)}, \quad t > 0.$$

Proof   The first term in the right side of the relation for the expected reward is obvious. To explain the second term, we use the property 'Poisson arrivals see time averages'. Fix $j, k \in I$ with $k \neq j$. Observe that the transitions out of state j occur according to a Poisson process with rate ν_j whenever the process $\{X(t)\}$ is in state j. Hence, using part (b) of Theorem 1.2.3 in Section 1.2, transitions from state j to state $k(\neq j)$ occur according to a Poisson process with rate $q_{jk}(= p_{jk}\nu_j)$ whenever the process $\{X(t)\}$ is in state j. Next, by applying part (a) of Theorem 1.7.1 in Section 1.7, it is readily seen that the expected number of transitions from state j to state k up to time t equals q_{jk} times the expected amount of time the process $\{X(t)\}$ is in state j up to time t.

Using the representation (2.8.4), we find

$$E_{ij}(t) = \int_0^t p_{ij}(u)\, du = \int_0^t \left[\sum_{n=0}^{\infty} e^{-\nu u} \frac{(\nu u)^n}{n!} \overline{p}_{ij}^{(n)} \right] du$$

$$= \sum_{n=0}^{\infty} \overline{p}_{ij}^{(n)} \int_0^t e^{-\nu u} \frac{(\nu u)^n}{n!}\, du.$$

Using the identity (1.2.3) in Section 1.2 and interchanging the order of summation, we get the desired result.

2.8.2 First-passage time probabilities

In this subsection it is assumed for ease that the state space I is finite. For any set C of states with $C \neq I$, define the first-passage time τ_C by

$\tau_C = $ the first epoch at which the process $\{X(t)\}$ makes a transition into a
  state of the set C.

Also, define the first-passage time probability $Q_{iC}(t)$ by

$$Q_{iC}(t) = P\{\tau_C > t \mid X(0) = i\} \quad \text{for } i \notin C \text{ and } t > 0.$$

Just as in the case of the transient probabilities $p_{ij}(t)$, the first-passage time probabilities $Q_{iC}(t)$ can be computed both through a system of linear differential equations and through the uniformization method. The probabilities $Q_{iC}(t)$ satisfy Kolmogoroff's backward differential equations

$$Q'_{iC}(t) = -v_i Q_{iC}(t) + \sum_{\substack{j \notin C \\ j \neq i}} q_{ij} Q_{jC}(t), \quad i \notin C \text{ and } t > 0. \tag{2.8.6}$$

These differential equations are easily derived. To find $Q_{iC}(t + \Delta t)$, we use a conditioning argument with respect to the possible transitions in $(0, \Delta t)$. This gives

$$Q_{iC}(t + \Delta t) = (1 - v_i \Delta t) Q_{iC}(t) + \sum_{\substack{j \notin C \\ j \neq i}} q_{ij} \Delta t \, Q_{jC}(t) + o(\Delta t).$$

Subtracting $Q_{iC}(t)$ from both sides of this equation, dividing by Δt and letting $\Delta t \to 0$, we obtain (2.8.6). The linear differential equations can be numerically solved by standards methods from numerical analysis.

Let us now make the assumption that $P\{\tau_C < \infty \mid X(0) = i\} = 1$ for all $i \notin C$. Then the mean first-passage times $m_{iC} = E\{\tau_C \mid X_0 = i\}$ can be computed from the system of linear equations

$$-v_i m_{iC} + \sum_{\substack{j \notin C \\ j \neq i}} q_{ij} m_{jC} = -1, \quad i \notin C.$$

To see this, integrate both sides of (2.8.6) from $t = 0$ to ∞, use the relation $m_{iC} = \int_0^\infty Q_{iC}(t) \, dt$ and note that $\int_0^\infty Q'_{iC}(t) \, dt = -1$. More generally, denoting the kth moment of τ_C by $m_{iC}^{(k)} = E[\tau_C^k \mid X(0) = i]$, it is readily verified from (2.8.6) and relation (A.10) in Appendix A that

$$-v_i m_{iC}^{(k)} + \sum_{\substack{j \notin C \\ j \neq i}} q_{ij} m_{jC}^{(k)} = -k m_{iC}^{(k-1)}, \quad i \notin C$$

for any $k \geq 2$. Thus the higher moments $m_{iC}^{(k)}$ can be computed by solving successively systems of linear equations. A quick approximation to the first-passage time distribution function is sometimes possible through a sum of exponential functions by matching a number of moments.

Next we show how to compute $Q_{iC}(t)$ by using the uniformization method. To do so, we first define for $i, j \notin C$ and $t > 0$ the taboo probability

$_C p_{ij}(t) =$ the probability that the process $\{X(t)\}$ will be in state j at time t without having visited the set C up to time t when the initial state $X(0) = i$.

Then, for any $i \notin C$,

$$Q_{iC}(t) = \sum_{j \notin C} c\,p_{ij}(t), \quad t > 0. \tag{2.8.7}$$

To find the $c\,p_{ij}(t)$'s, we use the uniformized process $\{\overline{X}(t)\}$ defined by (2.8.2). Denote by $c\overline{p}_{ij}^{(n)}$ the probability that the embedded Markov chain $\{X_n\}$ is in state j after n transitions without having visited the set C when the initial state is i. By conditioning on the number of Poisson events up to time t in the modified $\{\overline{X}(t)\}$ process, it follows that

$$c\,p_{ij}(t) = \sum_{n=0}^{\infty} e^{-\nu t} \frac{(\nu t)^n}{n!} \, c\overline{p}_{ij}^{(n)}, \quad i, j \notin C \text{ and } t > 0. \tag{2.8.8}$$

The taboo probabilities $c\overline{p}_{ij}^{(n)}$ are recursively computed from

$$c\overline{p}_{ij}^{(n)} = \sum_{k \notin C} c\overline{p}_{ik}^{(n-1)} \overline{p}_{kj}, \quad i, j \notin C \text{ and } n \geq 1, \tag{2.8.9}$$

starting with $c\overline{p}_{ij}^{(0)} = 1$ for $j = i$ and $c\overline{p}_{ij}^{(0)} = 0$ for $j \neq i$. Note that the uniformization algorithm for the first-passage time probabilities is basically the same as for the transient probabilities.

Expected reward earned until absorption

Suppose that a reward at rate $r(j)$ is earned whenever the $\{X(t)\}$ process is in state j and a lump reward of F_{jk} is earned each time the process makes a transition from state j to state k. Let C be a set of states such that $P\{\tau_C < \infty | X(0) = i\} = 1$ for all $i \notin C$. Define the random variable R_C by

$R_C = $ the total reward earned before the process $\{X(t)\}$ makes for the first time a transition into a state of the set C.

Then, for each $i \notin C$,

$$E[R_C | X(0) = i] = \frac{1}{\nu} \left[\sum_{j \notin C} r(j) \sum_{n=0}^{\infty} c\overline{p}_{ij}^{(n)} + \sum_{j \notin C} \sum_{n=0}^{\infty} c\overline{p}_{ij}^{(n)} \sum_{\substack{k \notin C \\ k \neq j}} q_{jk} F_{jk} \right].$$

It is left to the reader to verify that this expression follows from

$$E[R_C | X(0) = i] = \lim_{t \to \infty} \left[\sum_{j \notin C} r(j) \, cE_{ij}(t) + \sum_{j \notin C} cE_{ij}(t) \sum_{\substack{k \notin C \\ k \neq j}} q_{jk} F_{jk} \right],$$

where $cE_{ij}(t) = \int_0^t c\,p_{ij}(u)\,du$.

Example 2.8.1 Reliability of a 1-out-of-n system

A computer system uses one operating unit but has built in redundancy in the form of $N \geq 1$ standby units. The operating unit has an exponential lifetime with mean $1/\lambda$. If the operating unit fails it is replaced immediately by a standby unit if available. Each failed unit enters repair immediately and is again available after an exponentially distributed repair time with mean $1/\mu$. There are ample repair facilities so that any number of units can be in repair simultaneously. The repairs of the various units occur independently of each other, and also the repair times are independent of the operating times. It is assumed that a unit in standby status cannot fail. The system is down only if all $N+1$ units are down. We are interested in the system reliability which is defined as the probability that the system will not go down during $[0, t]$.

To analyse the system, define the random variable $X(t)$ by

$$X(t) = \text{the number of units in repair at time } t, \quad t \geq 0.$$

The process $\{X(t)\}$ is a continuous-time Markov chain with the finite state space $I = \{0, 1, \ldots, N+1\}$. The transition rate diagram is displayed in Figure 2.8.1.

The probability that the system will not go down during $[0, t]$ is given by the first-passage time probability $Q_{iC}(t)$ with $C = \{N+1\}$. Let us write $Q_i(t) = Q_{iC}(t)$ for $0 \leq i \leq N$. For the reliability problem, Kolmogoroff's backward differential equations become

$$Q_i'(t) = -(i\mu + \lambda)Q_i(t) + i\mu Q_{i-1}(t) + \lambda Q_{i+1}(t), \quad 0 \leq i \leq N \text{ and } t > 0$$

with the convention $Q_{-1}(t) = Q_{N+1}(t) = 0$. These differential equations cannot be solved explicitly except for the special case of $N = 1$ standby unit. Then, using Laplace transform theory (see Appendix C), it can be derived that

$$Q_0(t) = e^{-(\lambda+\mu/2)t} \left\{ \cosh\left(\frac{t}{2}\sqrt{4\lambda\mu + \mu^2}\right) \right.$$

$$\left. + \frac{2\lambda + \mu}{\sqrt{4\lambda\mu + \mu^2}} \sinh\left(\frac{t}{2}\sqrt{4\lambda\mu + \mu^2}\right) \right\}, \quad t \geq 0,$$

where $\cosh(z) = \frac{1}{2}(e^z + e^{-z})$ and $\sinh(z) = \frac{1}{2}(e^z - e^{-z})$. A similar expression applies to $Q_1(t)$. Also, we have for the case of $N = 1$ standby unit that $m_0 = 2/\lambda + \mu/\lambda^2$ in agreement with the result for the reliability problem in Appendix A.

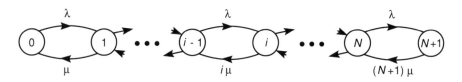

Figure 2.8.1   The transition rate diagram for the number of units down

In general the probability $Q_i(t)$ can only be evaluated numerically. To do this, a simple and transparent method is the uniformization method. This method requires the leaving rates

$$v_i = \begin{cases} i\mu + \lambda, & 0 \le i \le N, \\ (N+1)\mu, & i = N+1, \end{cases}$$

and the one-step transition probabilities

$$p_{ij} = \begin{cases} \lambda/(i\mu + \lambda), & j = i+1, & 0 \le i \le N, \\ i\mu/(i\mu + \lambda), & j = i-1, & 1 \le i \le N, \\ 1, & j = N, & i = N+1, \\ 0, & \text{otherwise.} \end{cases}$$

The uniformization rate v can be chosen as $v = \max(N\mu + \lambda, (N+1)\mu)$. The system reliability $Q_i(t)$ is then computed by the algorithm described by the relations (2.8.7)–(2.8.9).

Numerical considerations

In the analysis we have assumed that both the lifetime and the repair time are exponentially distributed. It is interesting to see what the effect is on the system reliability when we deviate from the assumption of exponentiality. To study the effect of the distributional forms of the lifetime and repair time on the system reliability, we assumed that the lifetime L of an operating unit has a Weibull distribution. The mean is normalized as $E(L) = 1$ and the squared coefficient of variation is varied as $c_L^2 = 0.5$, 1 and 2. Note that for $c_L^2 = 1$ the Weibull distribution reduces to the exponential distribution. The repair time R of a unit has either an exponential distribution or a deterministic distribution. Table 2.8.1 deals with the special case of $N = 1$ standby unit and gives for various examples the values of the mean $E(\tau)$, the coefficient of variation c_τ and the tail probabilities $Q_0(t) = P\{\tau > t\}$ for $t = 2, 5$ and 10. Here τ denotes the time until the first system failure. In all examples of Table 2.8.1 the mean repair time is taken as $E(R) = 1/10$. The results corresponding to the cases of exponential and deterministic repair times are denoted by 'exp' and 'det' in the table. For the examples in which the lifetimes and repair times are not both exponentially distributed, we have used computer simulation to obtain the values of the performance measures. In each

Table 2.8.1 The effect of the distributional forms on the system reliability

		$E(\tau)$	c_τ	$Q_0(2)$	$Q_0(5)$	$Q_0(10)$
$c_L^2 = 1$	exp	12.0	0.99	0.851	0.662	0.435
	det	11.6(0.3)	1.00(0.02)	0.840(0.007)	0.650(0.009)	0.424(0.010)
$c_L^2 = 1/2$	exp	27.1(0.5)	0.99(0.02)	0.942(0.005)	0.841(0.007)	0.699(0.009)
	det	32.5(0.6)	0.99(0.02)	0.947(0.005)	0.863(0.007)	0.737(0.009)
$c_L^2 = 2$	exp	6.7(0.2)	1.05(0.02)	0.708(0.009)	0.462(0.010)	0.232(0.009)
	det	6.0(0.2)	1.06(0.02)	0.682(0.009)	0.426(0.010)	0.196(0.008)

Table 2.8.2   The effect of extra standby units on the system reliability

	$E(R) = 1/5$					$E(R) = 1/10$			
	$E(\tau)$	c_τ	$Q_0(5)$	$Q_0(10)$	$Q_0(25)$	$E(\tau)$	c_τ	$Q_0(5)$	$Q_0(10)$	$Q_0(25)$
$N = 1$	7	0.98	0.493	0.238	0.027	12	0.99	0.662	0.435	0.124
$N = 2$	68	1.00	0.932	0.866	0.694	233	1.00	0.979	0.959	0.899
$N = 3$	984	1.00	0.995	0.990	0.975	6864	1.00	0.999	0.999	0.996

example the simulated values and the corresponding 95% confidence intervals are based on 10 000 independent runs; the notation 11.6(.3) denotes that the simulated value is 11.6 with [11.3, 11.9] as the 95% confidence interval. The numerical results in Table 2.8.1 show that the distributional form of the lifetime has a much larger effect on the system reliability than the distributional form of the repair time; for the important case of exponentially distributed lifetimes the system reliability is fairly insensitive to the distributional form of the repair time. In Table 2.8.2 we give some numerical results indicating the effect of an extra standby unit on the system reliability. Assuming that both the lifetimes and the repair times are exponentially distributed, the number N of standby units is varied as 1, 2 and 3. The mean lifetime $E(L)$ is taken equal to 1 and the average repair time $E(R)$ has the two values $1/5$ and $1/10$. The numerical results in Table 2.8.2 show that the effect of an extra standby unit is considerably larger than one would intuitively expect. Another remarkable finding is that in the examples considered the coefficient of variation c_τ is very close to 1, suggesting that the first-passage time τ is approximately exponentially distributed. Indeed, our numerical investigations indicate that

$$P\{\tau > t\} \approx e^{-t/E(\tau)} \quad \text{for all } t \geq 0 \qquad (2.8.9)$$

is an excellent approximation when the mean failure rate $1/E(L)$ is sufficiently small compared with the mean repair rate $1/E(R)$, as will be the case in most practical applications. This empirical finding is in agreement with the theoretical result that under general conditions the time until the first occurrence of a 'rare' event in a regenerative stochastic process is approximately exponentially distributed; see also Example 1.4.2 in Section 1.4. The importance of a result such as (2.8.9) is that it provides quantitative insight.

As a final remark, the above discussion shows that numbers are often indispensable for gaining system understanding, which is ultimately the primary purpose of stochastic modelling. Both analytical methods and computer simulation might be useful for that purpose.

2.9   PHASE METHOD

The phase method makes it possible to use the continuous-time Markov chain approach for a wide variety of practical probability problems in which the underlying probability distributions are not necessarily exponential. The method goes essentially back to A. K. Erlang, who did pioneering work in stochastic processes at the beginning of the twentieth century. In his analysis of telephone problems Erlang devised the trick of considering the duration of a call as the sum of a number of consecutive phases that are sequentially processed. Thus Erlang used the idea of approximating a positive random variable by a sum of independent exponentials with the same means. This explains the name of Erlang distribution. More generally, the probability distribution of any positive random variable can be arbitrarily closely approximated by a mixture of Erlangian distributions with the same scale parameters. The theoretical basis for the use of mixtures of Erlangian distributions is provided by the following theorem.

Theorem 2.9.1   *Let $F(t)$ be the probability distribution function of a positive random variable. For fixed $\Delta > 0$ define the probability distribution function $F_\Delta(x)$ by*

$$F_\Delta(x) = \sum_{j=1}^{\infty} p_j(\Delta) \left\{ 1 - \sum_{k=0}^{j-1} e^{-x/\Delta} \frac{(x/\Delta)^k}{k!} \right\}, \quad x \geq 0, \tag{2.9.1}$$

where $p_j(\Delta) = F(j\Delta) - F((j-1)\Delta)$, $j = 1, 2, \ldots$. Then

$$\lim_{\Delta \to 0} F_\Delta(x) = F(x)$$

for any continuity point x of $F(t)$.

Proof   For fixed Δ, $x > 0$, let $U_{\Delta,x}$ be a Poisson distributed random variable with

$$P\{U_{\Delta,x} = k\Delta\} = e^{-x/\Delta} \frac{(x/\Delta)^k}{k!}, \quad k = 0, 1, \ldots .$$

It is immediately verified that

$$E(U_{\Delta,x}) = x \quad \text{and} \quad \sigma^2(U_{\Delta,x}) = x\Delta.$$

Let $g(t)$ be a bounded function. We now prove that

$$\lim_{\Delta \to 0} E[g(U_{\Delta,x})] = g(x) \tag{2.9.2}$$

for any continuity point x of $g(t)$. To see this, fix $\varepsilon > 0$ and a continuity point x of $g(t)$. Then there exists a number $\delta > 0$ such that $|g(t) - g(x)| \leq \varepsilon/2$ for all t with $|t - x| \leq \delta$. Also, let $M > 0$ be such that $|g(t)| \leq M/2$ for all t. Then

$$|E[g(U_{\Delta,x})] - g(x)| \leq \sum_{k=0}^{\infty} |g(k\Delta) - g(x)| P\{U_{\Delta,x} = k\Delta\}$$

$$\leq \frac{\varepsilon}{2} + M \sum_{k:|k\Delta-x|>\delta} P\{U_{\Delta,x} = k\Delta\}$$

$$= \frac{\varepsilon}{2} + M P\{|U_{\Delta,x} - E(U_{\Delta,x})| > \delta\}.$$

By Tschebyshev's inequality, $P\{|U_{\Delta,x} - E(U_{\Delta,x})| > \delta\} \le x\Delta/\delta^2$. For Δ small enough, we have $Mx\Delta/\delta^2 \le \frac{1}{2}\varepsilon$. This proves the relation (2.9.2). Next, we apply (2.9.2) with $g(t) = F(t)$. Hence, for any continuity point x of $F(t)$,

$$F(x) = \lim_{\Delta \to 0} E[F(U_{\Delta,x})] = \lim_{\Delta \to 0} \sum_{k=0}^{\infty} F(k\Delta) e^{-x/\Delta} \frac{(x/\Delta)^k}{k!}$$

$$= \lim_{\Delta \to 0} \sum_{k=0}^{\infty} e^{-x/\Delta} \frac{(x/\Delta)^k}{k!} \sum_{j=1}^{k} p_j(\Delta),$$

where the latter equality uses that $F(0) = 0$. Interchanging the order of summation, we next obtain

$$F(x) = \lim_{\Delta \to 0} \sum_{j=1}^{\infty} p_j(\Delta) \sum_{k=j}^{\infty} e^{-x/\Delta} \frac{(x/\Delta)^k}{k!}$$

yielding the desired result.

The proof of Theorem 2.9.1 shows that the result also holds when $F(t)$ has a positive mass at $t = 0$. We should then add the term $F(0)$ to the right side of (2.9.1). Roughly stated, Theorem 2.9.1 tells us that the probability distribution of any positive random variable can be arbitrarily closely approximated by a mixture of Erlangian distributions with the *same* scale parameters. The fact that the Erlangian distributions from the mixture have identical scale parameters simplifies the construction of an appropriate continuous-time Markov chain in specific applications. In practice it is not always obvious how to choose a mixture that is sufficiently close to the distribution considered. One often confines to a mixture of two Erlangian distributions by matching the first two moments of the distribution considered, see also Appendix B.

The power of the phase method is illustrated with the following example.

Example 2.9.1 The M/G/1 queue

Customers arrive at a single-server station according to a Poisson process with rate λ. The service times of the customers are independent and identically distributed random variables and are also independent of the arrival process. The single server can handle only one customer at a time and customers are served in order of arrival. The performance measure of interest is the probability distribution of the waiting time of a customer.

Denoting by the generic variable S the service time of a customer, it is assumed that

$$P\{S \le x\} = \sum_{j=1}^{\infty} \beta_j \left(1 - \sum_{k=0}^{j-1} e^{-\mu x} \frac{(\mu x)^k}{k!} \right), \qquad x \ge 0, \qquad (2.9.3)$$

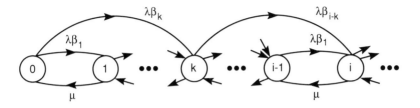

Figure 2.9.1   The transition diagram for the number of uncompleted phases

where $\beta_j \geq 0$ and $\Sigma_{j=1}^{\infty}\beta_j = 1$. Thus we can think of the service of a customer as follows. With probability β_j the customer has to go through j service phases before its service is completed, where the phases are sequentially processed and are independently and exponentially distributed with common mean $1/\mu$. This interpretation enables us to define the following continuous-time Markov chain. For any $t > 0$, let

$X(t) = $ the number of uncompleted service phases present at time t.

The process $\{X(t)\}$ is a continuous-time Markov chain with infinite state space $I = \{0, 1, \ldots\}$. Its transition rate diagram is displayed in Figure 2.9.1. It will be seen below that the waiting-time distribution is obtained from the equilibrium distribution of the process $\{X(t)\}$ and the property 'Poisson arrivals see time averages'. To ensure that the process $\{X(t)\}$ has an equilibrium distribution, we need the assumption

$$\rho = \lambda E(S) < 1.$$

This is a natural assumption since $\lambda E(S)$ represents the expected amount of work that is offered to the server in one unit of time. It is stated without proof that the assumption $\rho < 1$ implies that Assumption 2.6.1 is satisfied. Denote by $\{f_j, \; j = 0, 1, \ldots\}$ the equilibrium distribution of the continuous-time Markov chain $\{X(t)\}$. Note that the process $\{X(t)\}$ has the property stated in (2.6.8). Thus a recursive equation can be given for the f_j's. Applying the balance equation 'rate out of the set A = rate into the set A' with $A = \{i, i+1, \ldots\}$, we see from Figure 2.9.1 that

$$\mu f_i = \sum_{k=0}^{i-1} f_k \lambda \sum_{j=i-k}^{\infty} \beta_j, \quad i = 1, 2, \ldots . \tag{2.9.4}$$

This recursion provides an effective method to compute the f_j's.

Let us now turn to the waiting-time distribution. For any $n \geq 1$, define the random variable D_n by

$D_n = $ the delay in queue of the nth customer.

The delay in queue excludes the service time. For ease let us assume that the system is empty at epoch 0 and so $D_1 = 0$. We are interested in the long-run fraction of customers whose delay in queue does not exceed a given value x. For $n = 1, 2, \ldots$, define the indicator variable $I_n(x)$ by

$$I_n(x) = \begin{cases} 1 & \text{if } D_n \leq x, \\ 0 & \text{otherwise.} \end{cases}$$

The process $\{I_n(x), \; n \geq 1\}$ is a discrete-time regenerative process. It regenerates itself each time an arriving customer finds the server idle. The expected value of the number of customers arriving between two regeneration epochs is finite. Hence, by Theorem 1.3.2 in Section 1.3, there is a constant $W_q(x)$ such that

$$\lim_{n \to \infty} \frac{1}{n} \sum_{k=1}^{n} I_k(x) = W_q(x) \quad \text{with probability 1.}$$

In words, $W_q(x)$ is the long-run fraction of customers whose delay in queue does not exceed the value x.

Theorem 2.9.2   *The waiting-time distribution function $W_q(x)$ is given by*

$$W_q(x) = \sum_{j=0}^{\infty} f_j \left(1 - \sum_{k=0}^{j-1} e^{-\mu x} \frac{(\mu x)^k}{k!} \right), \quad x \geq 0. \tag{2.9.5}$$

Proof   A formal proof is as follows. By the bounded convergence theorem,

$$\lim_{n \to \infty} \frac{1}{n} \sum_{k=1}^{n} E[I_k(x)] = W_q(x).$$

Denote by $\pi_j^{(n)}$ the probability that there are j uncompleted service phases in the system just prior to the nth arrival. Under the condition that an arriving customer finds j uncompleted service phases present, the delay in queue of the customer is the sum of j independent exponentials with the same means $1/\mu$ and hence has an Erlang-j distribution. Thus, denoting by Z_j an Erlang-j distributed random variable with mean j/μ, we find by conditioning that

$$E[I_k(x)] = \pi_0^{(k)} + \sum_{j=1}^{\infty} \pi_j^{(k)} P\{Z_j \leq x\}.$$

Using expression (1.2.1) in Section 1.2, it follows that

$$E[I_k(x)] = \sum_{j=0}^{\infty} \pi_j^{(k)} \left\{ 1 - \sum_{m=0}^{j-1} e^{-\mu x} \frac{(\mu x)^m}{m!} \right\}, \quad k \geq 1. \tag{2.9.6}$$

Next we apply the property 'Poisson arrivals see time averages'. By part (b) of Theorem 1.7.1 in Section 1.7, the long-run (expected) fraction of arrivals who see

j uncompleted service phases equals f_j. Thus

$$\lim_{n\to\infty} \frac{1}{n} \sum_{k=1}^{n} \pi_j^{(k)} = f_j, \quad j = 0, 1, \ldots . \tag{2.9.7}$$

Together (2.9.6) and (2.9.7) yield the result (2.9.5).

We have deliberately given a rigorous derivation of the result (2.9.5). Usually the following informal reasoning is used. First it is argued that a customer arriving when the system has reached statistical equilibrium finds j uncompleted service phases present with probability f_j in view of the property of 'Poisson arrivals see time averages'. Next it is argued that the conditional delay in queue of a customer who finds upon arrival j uncompleted phases present has an Erlang-j distribution. Together the two arguments yield (2.9.5). This heuristic reasoning will be frequently used in Chapter 4. The proof of the result (2.9.5) demonstrates how the heuristic derivation can be made rigorously.

The expression (2.9.5) for $W_q(x)$ is very useful for computational purposes. For numerical calculations it is recommended to rewrite (2.9.5) as

$$W_q(x) = 1 - \sum_{k=0}^{\infty} e^{-\mu x} \frac{(\mu x)^k}{k!} \left[1 - \sum_{j=0}^{k} f_j \right], \quad x \geq 0 \tag{2.9.8}$$

by interchanging the order of summation. The series representation (2.9.8) converges faster than the series (2.9.5). Using the explicit result (2.6.9), we have for the special case of the *M/M/1* queue that

$$W_q(x) = 1 - \rho e^{-\mu(1-\rho)x}, \quad x \geq 0.$$

Denote by W_q the long-run average delay in queue per customer. Using that the long-run fraction of customers finding upon arrival j uncompleted service phases present is equal to f_j, it is intuitively obvious that

$$W_q = \sum_{j=1}^{\infty} \frac{j}{\mu} f_j.$$

A rigorous proof of this result is rather subtle and will not be given here.

An explicit expression for W_q can be derived. To do so, we need the generating function

$$F(z) = \sum_{i=0}^{\infty} f_i z^i, \quad |z| \leq 1.$$

Multiplying both sides of (2.9.4) by z^i and summing over i, we find after an interchange of the order of summation

$$\mu F(z) = \mu f_0 + \frac{\lambda z F(z)}{1-z} \{1 - \beta(z)\},$$

where $\beta(z) = \Sigma_{j=1}^{\infty} \beta_j z^j$. Hence

$$F(z) = \frac{\mu f_0 (1 - z)}{\mu (1 - z) - \lambda z \{1 - \beta(z)\}}. \tag{2.9.9}$$

We are now in a position to derive one of the most famous results in queueing theory.

Theorem 2.9.3 (Pollaczek–Khintchine formula)   *For the M/G/1 queue, the long-run average delay in queue per customer is given by*

$$W_q = \frac{1}{2}(1 + c_S^2)\frac{\rho E(S)}{1 - \rho} \tag{2.9.10}$$

where $c_S^2 = \sigma^2(S)/E^2(S)$ denotes the squared coefficient of variation of the service time.

Proof   We first show that the relation $F(1) = 1$ implies that

$$f_0 = 1 - \rho. \tag{2.9.11}$$

By letting $z \to 1$ in both sides of (2.9.9) and using L'Hôpital's rule, we find $f_0 = 1 - \lambda \Sigma j \beta_j / \mu$ showing that $f_0 = 1 - \rho$. Next we apply the relation $F'(1) = \Sigma_{j=1}^{\infty} j f_j$. Using twice L'Hôpital's rule, we obtain from (2.9.9) that

$$\sum_{j=1}^{\infty} j f_j = \frac{\lambda \beta''(1) + 2\lambda \beta'(1)}{2\mu(1 - \rho)}.$$

The numerator of the right side becomes $\lambda \Sigma_j j(j + 1)\beta_j = \lambda \mu^2 E(S^2)$, using that the second moment of an Erlang-j distribution with scale parameter μ is given by $j(j + 1)/\mu^2$. Since $W_q = (1/\mu)\Sigma_{j=1}^{\infty} j f_j$, the formula (2.9.10) now follows.

The Pollaczek–Khintchine formula uses the service time S only through $E(S)$ and $E(S^2)$. Since any service-time distribution can be arbitrarily closely approximated by a service-time distribution of the form (2.9.3), it is reasonable to expect that the Pollaczek–Khintchine formula holds for a general service-time distribution. This is indeed true. A rigorous proof of this result can be given by using the theory of weak convergence.

Importance of the Pollaczek–Khintchine formula

The Pollaczek–Khintchine formula gives not only an explicit expression for W_q, but more importantly it gives useful qualitative insights as well. It shows that the average delay per customer in the $M/G/1$ queue depends on the service-time distribution only through the first two moments. Through the factor $\frac{1}{2}(1 + c_S^2)$ the Pollaczek–Khintchine formula specifies how the average delay decreases when the variability in the service time is reduced while the average arrival rate and the average service time are kept fixed. Note the representation $W_q = \frac{1}{2}(1 + c_S^2)W_q(\exp)$, where $W_q(\exp)$ is the average delay for the special case of exponential services ($c_S^2 = 1$) with the same mean $E(S)$. In particular, the average delay for deterministic services is one half of the average delay for exponential services. Also, the Pollaczek–Khintchine formula shows that the average delay explodes when the average arrival rate becomes very close to the average service rate. To be specific, the slope of increase of W_q as function of the offered load ρ is proportional to $(1 - \rho)^{-2}$, as follows by differentiation of W_q. As an illustration of this qualitative result, a small increase in the arrival rate λ when the load $\rho = 0.9$ causes an increase in the average delay 25 times greater than it would cause when the load $\rho = 0.5$. This non-intuitive finding shows the danger of designing a stochastic system with a high utilization level, since then a small increase in the traffic input may cause a dramatic degradation in system performance.

EXERCISES

2.1 To meet the demand for water in a region, it is decided that a dam should be built in a river. For the dam capacity there is an option of either 2 units or 3 units. The probability distribution of the number of units of water that flow into the dam during each week is given by $p_0 = \frac{1}{8}$, $p_1 = \frac{1}{4}$, $p_2 = \frac{1}{2}$ and $p_3 = \frac{1}{8}$. If the supply of water by the river exceeds the remaining capacity of the dam, the excess water is lost. The demand for water is 2 units in each week where it is assumed that the water has to be provided at the beginning of each week. If the dam does not contain enough water to satisfy the demand, the shortage will be fulfilled at a cost of 10 per unit shortage. The weekly depreciation costs of the dam have the values of 2 and 2.5 for the respective capacities of 2 and 3 units. Determine for which dam capacity the long-run average costs per week are minimal. (*Answer*: capacity 3 with average weekly costs of 6.56.)

2.2 A factory has a buffer with a capacity of 4 m^3 for temporarily storing waste produced by the factory. Each week the factory produces k m^3 waste with a probability p_k, where $p_0 = \frac{1}{8}$, $p_1 = \frac{1}{2}$, $p_2 = \frac{1}{4}$ and $p_3 = \frac{1}{8}$. If the amount of waste produced in one week exceeds the remaining capacity of the buffer, the excess is specially removed at a cost of 30 per m^3. At the end of each week there is a regular opportunity to remove waste from the storage buffer at a fixed cost of 25 and a variable cost of 5 per m^3. The following policy is used. If at the end of the week the storage buffer contains more than 2 m^3 the buffer is emptied; otherwise no waste is removed. Determine the long-run average cost per week (*Answer*: 17.77.)

2.3 A machine has two critical parts that are subject to failure. The machine can continue to operate if one part has failed. Only in the case where both parts are no longer intact does a repair need to be done. A repair takes one day and after a repair both parts are intact again. At the beginning of each day the machine is examined to determine whether a repair is required or not. If at the beginning of a day both parts are intact, then at the end of the day exactly one part will be intact with probability 0.25 and both parts will be intact with probability 0.5. In case only one part is intact at the beginning of a day, this part will have failed at the end of the day with probability 0.5. Each repair costs 50, while for each day the machine is running a reward of 100 is earned. Determine the long-run average reward per day (*Answer*: 62.5.)

2.4 Every time unit a job is offered at a work station. The work station has no buffer to store temporarily jobs. An arriving job is rejected when the work station is busy. The processing time of a job has a Coxian-2 distribution. Show how to calculate the long-run fraction of jobs rejected.

2.5 Consider a periodic review inventory system in which every period a replenishment order is placed for the demand that occurred in the last period. The lead time of a replenishment order is either $1/4$ period or $1 1/2$ periods. If the lead time of the current replenishment order is α periods, then the lead time of the next replenishment order will also be α periods with probability $q(\alpha)$ for $\alpha = 1/4, 1 1/2$. For the numerical data $q(1/4) = 0.8$ and $q(1 1/2) = 0.5$, calculate the long-run fraction of replenishment orders crossing in time. (*Answer*: 1/7).

2.6 At the beginning of each day a batch of containers arrives at a stackyard having only capacity to store N containers. The batch size has a discrete probability distribution $\{q_k, k \geq 1\}$. If the stackyard has not sufficient space to store the whole batch, the batch as a whole is taken elsewhere. Each accepted container stays an exponentially distributed time at the stackyard, where the holding times of the various containers are independent of each other. How to calculate the long-run fraction of containers that are taken elsewhere?

2.7 An electrician is offered with probability p a job at the beginning of each day. For each offer the electrician can exactly estimate how many days of work the job involves. The offer is only accepted if the number of days of work still to do does not exceed the critical value M after acceptance of the order. The probability of getting an offer involving k days of work is a_k, $k = 1, 2, \ldots$. A profit of $p(k)$ is made for each accepted offer that involves k days of work. An idle cost of c is incurred for each day that the electrician has no work to do. Use Markov chain analysis to find the long-run net profit per day as function of the control parameter M.

2.8 At the beginning of each day a crucial piece of an electronic equipment is inspected and then classified as being in one of the working conditions $i = 1, \ldots, N$. Here the working condition i is better than the working condition $i + 1$. If the working condition is $i = N$ the piece must be replaced by a new one and such an enforced replacement takes two days. If the working condition is i with $i < N$ there is a choice between replacing preventively the piece by a new one and letting the piece operate for the present day. A preventive replacement takes one day. A new piece has working condition $i = 1$. A piece whose present working condition is i has the next day the working condition j with known probability q_{ij} where $q_{ij} = 0$ for $j < i$. The following replacement rule is used. The current piece is only replaced by a new one when its working condition is greater than the critical value m where m is a given integer with $1 \leq m < N$. Determine the fraction of days that the equipment is inoperative and determine the fraction of replacements occurring in the failure state N.

2.9 Consider Example 2.4.2. Use the results in Lemma 1.2.4 in Section 1.2 to show that an exact expression for the long-run average number of messages in the buffer is given by

$$L_q = \sum_{j=0}^{\infty} \pi_j \sum_{k=0}^{\infty} e^{-\lambda} \frac{\lambda^k}{k!} \sum_{i=1}^{\min(k,N)} \frac{k+1-i}{k+1}$$

$$+ \sum_{j=c+1}^{N} \pi_j \sum_{k=0}^{\infty} e^{-\lambda} \frac{\lambda^k}{k!} \left\{ j - c + \sum_{i=0}^{\min(k,N-j+c)} \frac{k+1-i}{k+1} \right\}.$$

Denote by W_q the long-run average waiting time in the buffer per accepted message. Argue that $L_q = \lambda(1 - P_{\text{rej}})W_q$, where P_{rej} denotes the long-run fraction of messages that are rejected.

2.10 Consider a communication channel at which messages arrive according to a Poisson process with rate λ. The messages are temporarily stored in a finite buffer to await transmission. Each message that finds upon arrival the buffer full is lost. The transmission time of each message is a constant slot length of one time unit. The beginnings of the time slots provide the only opportunity to start the transmission of a message. Only one message can be transmitted at a time. The transmission of a message is successful with probability f, otherwise the transmission has to be retried in the next time slot. Show how to calculate the long-run fraction of messages lost.

2.11 At a telephone exchange calls arrive according to a Poisson process with rate λ. The calls are first put in an infinite-capacity buffer before they can be processed further. The buffer is periodically scanned every T time units and only at those scanning epochs calls in the buffer are allocated to free servers. There are c servers and each server can handle only one call at a time. The service times of the calls are independent random variables having a common exponential distribution with mean $1/\mu$. Setup the equilibrium equations of a suitably chosen discrete-time Markov chain in order to determine the average number of calls in the buffer. Also, relate the average delay in the buffer per call to the average number of calls in the buffer.

2.12 Consider a stochastically failing equipment with two identical components that operate independently of each other. The lifetime in days of each component has a discrete probability distribution $\{p_j, j = 1, \ldots, M\}$. A component being in the failure state at the beginning of a day is replaced instantaneously. It may be economical to replace preventively the other working component at the same time the failed component has to be replaced. The cost of replacing only one component is K_1, while the cost of replacing simultaneously both components equals K_2 with $0 < K_2 < 2K_1$. The control rule is as follows. Replace a component upon failure or reaching the age of R days, whichever occurs first. If a component is replaced and the other component is still working, the other component is preventively replaced when it has been in use for r or more days. Show how to obtain the long-run average cost per day for a given (r, R) rule with $1 \leq r < R$.

2.13 Suppose that a conveyer belt is running at a uniform speed and transporting items on individual carriers equally spaced along the conveyer. There are N work stations placed in order along the conveyer. Each time unit an item for processing arrives and is handled by the first work station that is idle. Any station can process only one item at a time and has no storage capacity. An item that finds all of the N work stations busy is lost. The processing time of an item at station i has an Erlang-r_i distribution with mean m_i, $i = 1, \ldots, N$. The work stations are numbered $i = 1, \ldots, N$ according to the order they are placed along the conveyer. Give a Markov chain analysis aimed at the computation of the loss probability. For the model with $N = 2$ work stations, solve for the two cases:

(a) the processing times at the stations 1 and 2 are exponentially distributed with respective means $m_1 = 0.75$ and $m_2 = 1.25$ (*answer*: 0.0467);
(b) the processing times at the stations 1 and 2 are Erlang-3 distributed with respective means $m_1 = 0.75$ and $m_2 = 1.25$ (*answer*: 0.0133).

(This problem is motivated by Gregory and Litton (1975).)

2.14 A single communication channel is shared for transmitting voice packets and data packets. The channel can transmit only one packet at a time and the transmission time of each packet is the same constant. The time is divided into slots of size corresponding to the transmission time. Transmission of a packet can only start at the beginning of a time slot. Within each time slot voice packets and data packets arrive according to a general input process with $p(i_1, i_2)$ denoting the joint probability of i_1 arrivals of voice packets and i_2 arrivals of data packets. The arrivals in the successive slots are independent of each other. The voice and data packets have to wait in buffers of respective sizes N_1 and N_2 (the waiting positions are exclusive to any packet in transmission). A packet finding upon arrival that its buffer is full is lost. The transmission scheme uses randomization in the situation that both voice and data packets are present when the transmission channel becomes free for a next transmission. In those situations a voice packet is always chosen with the same probability π. The goal is to find the value of π for which the long-run average queue size of data packets is minimal subject to the constraint that the long-run average queue size of voice packets does not exceed a prespecified level α. Compute the optimal value of π for the numerical data $N_1 = 5$, $N_2 = 7$, $p(0, 0) = 0.36$, $p(0, 1) = p(1, 0) = 0.24$, $p(1, 1) = 0.16$ and $\alpha = 0.5$. (*Answer:* $\pi = 0.8169$.)

2.15 At a single-server station potential customers arrive according to a Poisson process with rate λ. A customer joins the system with probability $1/(n + 1)$ when the customer finds upon arrival n other customers present. The service times of the customers are independent random variables having a common exponential distribution with mean $1/\mu$. Verify that the limiting distribution of the number of customers in the system is a Poisson distribution with mean λ/μ. Also, determine the average number of customers per unit time who actually join the system.

2.16 A common car-service between cities in Israel is called 'Sheroot'. Passengers who wish to use this service to travel from city A to city B arrive at a cab station at city A according to a Poisson process with rate λ and take their seat in a 7-seat Sheroot-cab when available. Passengers who find no cab available leave and go to a competitor. As soon as the cab is full it leaves for city B. After a cab leaves the station it takes an exponential time with mean $1/\beta$ until a new cab becomes available. Find the probability that an arriving customer finds a cab and does not have to wait until the cab leaves. What is the long-run average waiting time per customer who finds upon arrival a cab?

2.17 Messages arrive at a communication channel according to a Poisson process with rate λ. The message length is exponentially distributed with mean $1/\mu$. An arriving message finding that the line is idle is provided with service immediately; otherwise the message waits until access to the line can be given. The communication line is only able to submit one message at a time, but has available two possible transmission rates σ_1 and σ_2 with $0 < \sigma_1 < \sigma_2$. Thus the transmission time of a message is exponentially distributed with mean $1/(\sigma_i \mu)$ when the transmission rate σ_i is used. It is assumed that $\lambda/(\sigma_2 \mu) < 1$. At any time the transmission line may switch from one rate to the other. The transmission rate is controlled by a single-critical-number rule. The transmission rate σ_1 is used whenever less than R messages are present, while otherwise the faster transmission rate σ_2 is used. The following costs are involved. There is a holding cost at rate hj whenever j messages are in the system. An operating cost at rate $r_i > 0$ is incurred when the line is transmitting a message using rate σ_i, while an operating cost at rate $r_0 \geq 0$ is incurred when the line is idle. Derive a recursion scheme for the computation of the limiting distribution of the number of messages present and give an expression for the long-run average cost per unit time. Write a computer program for the calculation of the value of R which minimizes the average cost and solve for the numerical data $\lambda = 0.8$, $\mu = 1$, $\sigma_1 = 1$, $\sigma_2 = 1.5$, $h = 1$, $r_0 = 0$, $r_1 = 5$ and $r_2 = 25$. (*Answer:* $R = 10$ and the minimal average cost is 7.66.)

2.18 In an inventory system for a single product the depletion of stock is due to demand and deterioration. The demand process for the product is a Poisson process with rate λ. The lifetime of each unit product is exponentially distributed with mean $1/\mu$. The stock control is exercised as follows. Each time the stock drops to zero an order for Q units is placed. The lead time of each order is negligible. Determine the average stock and the average number of orders placed per unit time.

2.19 Consider Example 2.6.2 and denote by S_{ij} the time needed to serve an alarm for district j by unit i. Assume that S_{ij} has a Coxian-2 distribution for all i, j. Show how to calculate the performance measures π_L = the fraction of alarms that is lost and P_i = the fraction of time that unit i is busy. Letting m_{ij} and c_{ij}^2 denote the mean and the squared coefficient of variation of S_{ij}, assume the numerical data $\lambda_1 = 0.25$, $\lambda_2 = 0.25$, $m_{11} = 0.75$, $m_{12} = 1.25$, $m_{21} = 1.25$ and $m_{22} = 1$. Write a compute program to verify the following numerical results:

(i) $\pi_L = 0.0704$, $P_1 = 0.2006$, $P_2 = 0.2326$ when $c_{ij}^2 = \frac{1}{2}$ for all i, j;

(ii) $\pi_L = 0.0708$, $P_1 = 0.2004$, $P_2 = 0.2324$ when $c_{ij}^2 = 1$ for all i, j;

(iii) $\pi_L = 0.0718$, $P_1 = 0.2001$, $P_2 = 0.2321$ when $c_{ij}^2 = 4$ for all i, j.

Here the values $c_{ij}^2 = \frac{1}{2}$, 1 and 4 correspond to the E_2 distribution, the exponential distribution and the H_2 distribution with balanced means.

2.20 Consider Example 2.7.2 and suppose now that there are two unloaders each having its own queue. An arriving trailer joins the unloader with the shortest queue and randomly picks an unloader when both queues are equally long. No jockeying between the queues is possible. Extend the analysis in Example 2.7.2.

2.21 An assembly line for a certain product has two stations in series. Each station has only room for a single unit of the product. If the assembly of a unit is completed at station 1, it is forwarded immediately to station 2 provided station 2 is idle; otherwise the unit remains in station 1 until station 2 becomes free. Units for assembly arrive at station 1 according to a Poisson process with rate λ, but a newly arriving unit is only accepted by station 1 when no other unit is present in station 1. Each unit rejected is handled elsewhere. The assembly times at the stations 1 and 2 are exponentially distributed with respective means $1/\mu_1$ and $1/\mu_2$. Show how to calculate the fraction of units that are not accepted by station 1. Also, use Little's formula to determine the average time spent by an accepted unit in the assembly line.

2.22 Consider a single unloader system at which trains arrive, bringing coal from various mines. There are N trains involved in the coal transport. The coal unloader can handle only one train at a time and the unloading time per train has an exponential distribution with mean $1/\mu$. The unloader is subject to breakdowns when the unloader is in operation. The operating time of an unloader has an exponential distribution with mean $1/\eta$ and the time to repair a broken unloader is exponentially distributed with mean $1/\xi$. The unloading of the train that is in service when the unloader breaks down is resumed as soon as the repair of the unloader is completed. An unloaded train returns to the mines for another trainload of coal. The time for a train to complete a trip from the unloader to the mines and back is assumed to have an exponential distribution with mean $1/\lambda$. Show how to calculate performance measures such as the average number of trains at the unloader (L), the average time a train has to wait until unloading is completed (W), and the average number of trains unloaded per unit time (T). Write a computer program and solve for the numerical data $N = 5$, $\mu = 2.5$, $\eta = 0.10$, $\xi = 0.35$ and $\lambda = 0.2$. (*Answer: L = 0.765, W = 0.903* and *T = 0.847.*) Extend your analysis in order to investigate the effect of the simplifying assumption of exponentially distributed trip times on the performance measures. (This problem is based on Chelst *et al* (1981).)

2.23 Consider a data communication system with a finite-capacity buffer for temporarily storing messages. The messages arrive according to a Poisson process with rate λ. The transmission time of each message is exponentially distributed with mean $1/\mu$. The system can transmit only one message at a time. The buffer has room for only N messages excluding the message in transmission (if any). The following protocol for the admission of messages is used. As soon as the buffer is full, the arrival process of messages is stopped until the number in the buffer has fallen to the resume level R with $0 \le R < N$. Use Markov chain analysis to compute the long-run fraction of messages blocked.

2.24 Suppose we wish to determine the capacity of a stackyard at which containers arrive according to a Poisson process with a rate of $\lambda = 1$ per hour. A container finding upon arrival that the yard is full is brought elsewhere. The time that a container is stored in the yard is exponentially distributed with a mean $1/\mu = 10$ hours. Determine the required capacity of the yard so that no more than 1% of the arriving containers find the yard full. (*Answer*: capacity 18.) How does the answer change when the time that a container is stored in the yard is uniformly distributed between 5 and 15 hours?

2.25 Consider the continuous-review (s, Q) inventory model with Poisson demand, where demand occurring while the system is out of stock is lost. Assume that the order quantity $Q = 1$, that is, a replenishment order for one unit is placed each time the inventory position decreases to the reorder point s. The lead times of the replenishment orders are independent and identically distributed random variables. Establish an equivalence with Erlang's loss model in order to conclude that the long-run fraction of time the stock on hand equals j is given by

$$\frac{(\lambda\tau)^{s+1-j}/(s+1-j)!}{\sum_{k=0}^{s+1}(\lambda\tau)^k/k!} \quad \text{for } j = 0, 1, \ldots, s+1,$$

where λ is the average demand rate and τ is the average replenishment lead time.

2.26 Long-parkers and short-parkers arrive at a parking place for cars according to independent Poisson processes with respective rates of $\lambda_1 = 4$ and $\lambda_2 = 6$ per hour. The parking place has room for $N = 10$ cars. Each arriving car which finds all places occupied goes elsewhere. The parking time of long-parkers is uniformly distributed between 1 and 2 hours, while the parking time of short-parkers has a uniform distribution between 20 and 60 minutes. Calculate the probability that upon arrival a car finds all parking places occupied. (*Answer*: 0.215.)

2.27 Suppose you have two groups of servers each without waiting room. The first group consists of c_1 identical servers each having an exponential service rate μ_1 and the second group consists of c_2 identical servers each having an exponential service rate μ_2. Customers for group i arrive according to a Poisson process with rate λ_i ($i = 1, 2$). A customer who finds upon arrival that all servers in his group are busy is served by a server in the other group provided one is free, otherwise the customer is lost. Show how to calculate the long-run fraction of customers who are lost.

2.28 Consider a conveyor system at which items for processing arrive according to a Poisson process with rate λ. The service requirements of the items are independent random variables having a common exponential distribution with mean $1/\mu$. The conveyor system has to work stations 1 and 2 that are placed according to this order along the conveyor. Work station i consists of s_i identical service channels, each having a constant processing rate of σ_i ($i = 1, 2$), that is, an item processed at work station i has an average processing time of $1/\sigma_i\mu$. Both work stations have no storage capacity and each service channel can handle only one item at a time. An arriving item is processed by the first work station in which a service channel is free and is lost when no service channel is available at either of the stations. Show how to calculate the fraction of items that is lost and solve for the numerical data $\lambda = 10$, $\mu = 1$, $\sigma_1 = 2$, $\sigma_2 = 1.5$, $s_1 = 5$ and $s_2 = 5$. (*Answer*: 0.0306.) Verify experimentally that the lost probability is nearly insensitive to the distributional form of the service requirement (e.g. compute the loss probability 0.0316 for the above numerical data when the service requirement has an H_2 distribution with balanced means and a squared coefficient of variation of 4).

2.29 Consider a stochastic service system with Poisson arrivals at rate λ and two different groups of servers, where each arriving customer requires simultaneously a server from both groups. An arrival not finding that both groups have a free server is lost and has no further influence on the system. The ith group consists of s_i identical servers ($i = 1, 2$) and each server can handle only one customer at a time. An entering customer occupies the two assigned servers from the groups 1 and 2 during independently exponentially distributed times with respective means $1/\mu_1$ and $1/\mu_2$. Show how to calculate the loss probability and solve for the numerical data $\lambda = 1$, $1/\mu_1 = 2$, $1/\mu_2 = 5$, $s_1 = 5$ and $s_2 = 10$. (*Answer*: 0.0464.) Verify experimentally that the loss probability is nearly insensitive to the distributional form of the service times (e.g. compute the loss probability 0.0470 for the above data when the service time in group 1 has an E_2 distribution and the service time in group 2 has an H_2 distribution with balanced means and a squared coefficient of variation of 4).

2.30 Consider the following stochastic service model having applications in both message storage and data multiplexing problems. Customers of the types 1 and 2 arrive at a shared resource with c units according to independent Poisson processes with respective rates λ_1 and λ_2. A customer of type i requires b_i resource units and is rejected anyway when less than b_i units are free. An accepted customer of type i has an exponentially distributed residency time with mean $1/\mu_i$ during which all of the b_i assigned resource units are kept occupied so that the b_i units are relinquished simultaneously when the customer departs. The complete sharing policy is followed under which an arriving customer of type i is rejected only if less than b_i resource units are free. Show how to calculate the fraction of customers of type i lost. Solve for the numerical data $c = 50$, $b_1 = 3$, $b_2 = 2$, $\lambda_1 = 10$, $\lambda_2 = 5$, $1/\mu_1 = 1$ and $1/\mu_2 = 1$. (*Answer*: the respective loss probabilities of type 1 and type 2 customers are 0.0863 and 0.0551.) Verify experimentally that in the present model the state probabilities are insensitive to the distributional form of residency times. (This problem is taken from Kaufman (1981).)

2.31 Suppose a production facility has M operating machines and a buffer of B standby machines. Machines in operation are subject to breakdowns. The running times of the operating machines are independent of each other and have a common exponential distribution with mean $1/\lambda$. An operating machine that breaks down is replaced by a standby machine if one is available. A failed machine immediately enters repair. There are ample repair facilities so that any number of machines can be repaired simultaneously. The repair time of a failed machine is assumed to have an exponential distribution with mean. $1/\mu$. For given values of μ, λ and M, demonstrate how to calculate the minimum buffer size B in order to achieve that the long-run fraction of time, that less than M machines are operating, is no more than a specific value β. Do you expect the answer to depend on the specific form of the repair-time distribution?

2.32 An operating system has $r + s$ identical units where r units must be operating and s units are in preoperation (warm standby). A unit in operation has a constant failure rate of λ, while a unit in preoperation has a constant failure rate of β with $\beta < \lambda$. Failed units enter a repair facility which is able to repair simultaneously at most c units. The repair of a failed unit has an exponential distribution with mean $1/\mu$. An operating unit that fails is replaced immediately by a unit from the warm standby if one is available. The operating system goes down when less than r units are in operation. Show how to calculate the probability distribution function of the time until the system goes down for the first time when all of the $r + s$ units are in a good condition at time 0.

2.33 Suppose a communication system has c transmission channels at which messages arrive according to a Poisson process with rate λ. Each message that finds upon arrival all of the c channels busy is lost; otherwise the message is randomly assigned to one of the free channels. The transmission length of an accepted message has an exponential distribution with mean $1/\mu$. However, each separate channel is subject to a randomly changing environment that influences the transmission rate of the channel. Independently of each other the channels alternate between periods of good condition and periods of bad condition. These alternating periods are independent of each other and have exponential distributions with respective means $1/\gamma_g$ and $1/\gamma_b$. The transmission rate of a channel being in good (bad) condition is $\sigma_g(\sigma_b)$. Setup the balance equations for the calculation of the fraction of messages that is rejected. Noting that $\sigma = (\sigma_b\gamma_g + \sigma_g\gamma_b)/(\gamma_g + \gamma_b)$ is the average transmission rate used by a channel, make some numerical comparisons with the case of a fixed transmission rate σ.

2.34 Messages arrive at a node in a communication network according to a Poisson process with rate λ. Each arriving message is temporarily stored in a infinite-capacity buffer until it can be transmitted. The messages have to be routed over one of two communication lines each with a different transmission time. The transmission time over the ith communication line is exponentially distributed with mean $1/\mu_i$ ($i = 1, 2$), where $1/\mu_1 < 1/\mu_2$ and $\mu_1 + \mu_2 > \lambda$. The faster communication line is always available for service, but the slower line will be used only when the number of messages in the buffer exceeds some critical level. Each line is only able to handle one message at a time and provides non-preemptive service. With the goal of minimizing the average sojourn time (including transmission time) of a message in the system, the following control rule with switching level L is used. The slower line is turned on for transmitting a message when the number of messages in the system exceeds the level L and is turned off again when it completes a transmission and the number of messages left behind is at or below L. Show how to calculate the average sojourn time of a message in the system. (This problem is taken from Lin and Kumar (1984).)

2.35 Two communication lines in a packet switching network share a finite storage space for incoming messages. Messages of types 1 and 2 arrive at the storage area according to two independent Poisson processes with respective rates λ_1 and λ_2. A message of type j is destined for communication line j and its transmission time is exponentially distributed with mean $1/\mu_j$, $j = 1, 2$. A communication line is only able to transmit one message at a time. The storage space consists of M buffer places. Each message requires exactly one buffer place and occupies the buffer place until its transmission time has been completed. A number of N_j buffer places is always reserved for messages of type j and a number of N_0 buffer places is to be used by messages of both types, where $N_0 + N_1 + N_2 = M$. An arriving message of type j enters the system only when a buffer place is available for the message; otherwise, the message is rejected. Discuss how to calculate the optimal values of N_0, N_1 and N_2 when the goal is to minimize the total rejection rate of both types of messages. Write a computer program and solve for the numerical data $M = 15$, $\lambda_1 = \lambda_2 = 1$ and $\mu_1 = \mu_2 = 1$ (*Answer:* $N_0 = 7$ and $N_1 = N_2 = 4$ with a total rejection rate of 0.207.) (This problem is based on Kamoun and Kleinrock (1980).)

BIBLIOGRAPHIC NOTES

Many good textbooks on stochastic processes are available and most of them treat the topic of Markov chains. Our favourite books include Cox and Miller (1965), Karlin and Taylor (1975) and Ross (1983), each offering an excellent introduction to Markov chain theory. A very fundamental treatment of denumerable Markov chains can be found in the book of Chung (1967). An excellent book on Markov chains with a general state space is Meyn and Tweedie (1993). The books of Bartholomew (1973) and Bartlett (1978) provide interesting applications of Markov chains to the fields of social sciences and natural sciences. The concept of the embedded Markov chain and its application in Example 3.1.2 are due to Kendall (1953). The numerical method for solving the balance equations for an infinite-state Markov chain is adapted from Tijms and Van de Coevering (1991), see also Takahashi and Takami (1976). The powerful technique of equating the flow out of an (aggregated) state to the flow into that state has a long history in teletraffic analysis. The method of phases using fictitious stages with exponentially distributed lifetimes has its origin in the pioneering work of Erlang on stochastic processes in the early 1900s and is discussed in full generality in Cox and Miller (1965); an extension of this method is the supplementary variable technique for constructing Markovian processes in continuous-time when the lifetime variables have continuous, non-exponential distributions; see also the book of Kosten (1973). Schassberger (1973) exploited the phase method by representing the probability distribution of a non-negative random variable by the limit of a sequence of mixtures of Erlangian distributions with the same scale parameters. The uniformization technique for continuous-time Markov chains goes at least back to Jensen (1953) and is quite useful for both analytical and computational purposes; see also Grassmann (1977). De Souza e Silva and Gail (1986) describe a useful extension of the uniformization technique to compute the transient probability distribution of the sojourn time in a given set of states. Many applications of Markov chains are to be found in operations research and computer science. The buffer design problem in Example 2.4.2 is based on Chu (1970). The resource allocation problem for urban emergency services in Example 2.5.2 is taken from Carter *et al.* (1972). Example 2.7.2 dealing with unloading of trailers at a stackyard is based on practical work of Van Hee (1984). The resource allocation problem for communication networks in Example 2.7.3 is adapted from Foschini *et al.* (1981). Insensitivity is a fundamental concept in stochastic service systems with losses. A general discussion of the insensitivity phenomenon in stochastic networks can be found in Kelly (1979, 1991) and Van Dijk (1993).

REFERENCES

Bartholomew, D.J. (1973). *Stochastic Models for Social Processes*, 2nd edn, Wiley, New York.
Bartlett, M.S. (1978). *An Introduction to Stochastic Processes*, 3rd edn, Cambridge University Press, Cambridge.

Carter, G., Chaiken, J.M., and Ignall, E.J. (1972). 'Response areas for two emergency units', Operat. Res. **20**, 571–594.

Chelst, K., Tilles, A.Z. and Pipis, J.S. (1981). 'A coal unloader: a finite queueing system with breakdowns', *Interfaces*, **11**, no. 5, 12–24.

Chu, W.W. (1970). 'Buffer behaviour for Poisson arrivals and multiple synchronous constant output', IEEE Trans. Comput., **19**, 530–534.

Chung, K.L. (1967). *Markov Chains with Stationary Transition Probabilities*, 2nd edn, Springer-Verlag, Berlin.

Cohen, J.W. (1976). *On Regenerative Processes in Queueing Theory*, Lecture Notes in Mathematical Economics and Mathematical Systems, Vol. 121, Springer-Verlag, Berlin.

Cox, D.R. and Miller, H.D. (1965). *The Theory of Stochastic Processes*, Chapman and Hall, London.

De Soua e Silva, E., and Gail, H.R. (1986). 'Calculating cumulative operational time distributions of repairable computer systems', *IEEE Trans. Comput.*, **35**, 322–332.

Federgruen, A. and Tijms, H.C. (1978). 'The optimality equation in average cost denumerable state semi-Markov decision problems, recurrency conditions and algorithms', *J. Appl. Prob.*, **15**, 356–373.

Foschini, G.J., Gopinath, B and Hayes, J.F. (1981). 'Optimum allocation of servers to two types of competing customers', *IEEE Trans Commun.*, **29**, 1051–1055.

Fox, B. and Landi, D.M. (1968). 'An algorithm for identifying the ergodic subchains and transient states of a stochastic matrix', *Commun. ACM*, **11**, 619–621.

Grassmann, W.K. (1977). 'Transient solutions in Markovian queueing systems', Comput. Operat. Res., **4**, 47–53.

Gregory, G. and Litton, C.D. (1975). 'A conveyor model with exponential service times', *Int. J. Prod. Res.*, **13**, 1–7.

Jensen, A. (1953). 'Markov chains as an aid in the study of Markoff process', *Skand. Aktuarietidskr.*, **36**, 87–91.

Kamoun, F., and Kleinrock, L. (1980). 'Analysis of a shared finite storage in a computer network node environment under general traffic conditions', *IEEE Trans. Commun.*, **28**, 992–1003.

Karlin, S., and Taylor, H.M. (1975). *A First Course in Stochastic Processes*, 2nd edn, Academic Press, New York.

Kaufman, J.S. (1981). 'Blocking in a shared resource environment', *IEEE Trans. Commun.*, **29**, 1474–1481.

Kelly, F.P. (1979). *Reversibility and Stochastic Networks*, Wiley, New York.

Kelly, F.P. (1991). 'Loss Networks', *Ann. Appl. Prob.*, **1**, 319–378.

Kendall, D.G. (1953). 'Stochastic processes occurring in the theory of queues and their analysis by the method of the embedded Markov chain', *Ann. Math. Statist.*, **24**, 338–354.

Kosten, L. (1973). *Stochastic Theory of Service Systems*, Pergamon Press, London.

Lin, W., and Kumar, P. (1984). 'Optimal control of a queueing system with two heterogeneous servers', *IEEE Trans. Automat. Contr.* **AC-29**, 696–703.

Markov, A.A. (1906). 'Extension of the law of large numbers to dependent events (in Russian)', *Bull. Soc. Phys. Math. Kazan*, **15**, 255–261.

Meyn, S.P. and Tweedie, R. (1993). '*Markov Chains and Stochastic Stability*', Springer-Verlag, Berlin.

Morse, P.M. (1955). 'Stochastic properties of waiting lines', *Operat. Res.*, **3**, 255–261.

Odoni, A.R., and Roth, E. (1983). 'An empirical investigation of the transient behaviour of stationary queueing systems', *Operat. Res.*, **31**, 432–455.

Ross, S.M. (1983), Stochastic Processes, Wiley, New York.

Schassberger, R. (1973). *Wartenschlangen* (in German), Springer-Verlag, Berlin.

Takahashi, Y. and Takami, Y. (1976). 'A numerical method for the steady state probabilities of a GI/G/c queueing system in a general class', *J. Operat. Soc. Japan*, **18**, 147–157.

Tijms, H.C. and Van de Coevering, M.C.T. (1991). 'A simple numerical approach for infinite-state Markov chains', *Prob. Eng. Inform. Sci.*, **5**, 285–295.

Van Dijk, N.M. and Tijms, H.C. (1986). 'Insensitivity in two-node blocking network models with applications', in: *Teletraffic Analysis and Computer Performance Evaluation*, eds O.J. Boxma, J.W. Cohen and H.C. Tijms, pp. 329–340, North-Holland, Amsterdam.

Van Dijk, N.M. (1993). *Queueing Networks and Product Form*, Wiley, Chichester.

Van Hee (1984). 'Models underlying decision support systems for port terminal planning', *Wissenschaftliche Zeitschrift Technische Hochschule Leipzig*, **8**, 161–170.

CHAPTER 3

Markovian Decision Processes and their Applications

3.0 INTRODUCTION

In the previous chapter we saw that in the analysis of many operational systems the concepts of a state of a system and a state transition are of basic importance. For dynamic systems with a *given* probabilistic law of motion, the simple Markov model is often appropriate. However, in many situations with uncertainty and dynamism the state transitions can be controlled by taking a sequence of actions. The Markov decision model is a versatile and powerful tool for analysing probabilistic sequential decision processes with an infinite planning horizon. This model is an outgrowth of the Markov model and dynamic programming. The latter concept, being developed by Bellman in the early 1950s, is a computational approach for analysing sequential decision processes with a finite planning horizon. The basic ideas of dynamic programming are states, the principle of optimality and functional equations. In fact dynamic programming is a recursion procedure for calculating optimal value functions from a functional equation. This functional equation reflects the principle of optimality, stating that an optimal policy has the property that whatever the initial state and initial decision are, the remaining decisions must constitute an optimal policy with regard to the state resulting from the first transition. This principle is always valid when the number of states and the number of actions are finite. At much the same time as Bellman (1957) popularized dynamic programming, Howard (1960) used basic principles from Markov chain theory and dynamic programming to develop a policy-iteration algorithm for solving probabilistic sequential decision processes with an infinite planning horizon. In the two decades following the pioneering work of Bellman and Howard, the theory of Markov decision processes has expanded at a fast rate and a powerful technology has developed. However, in that period relatively little effort was put in applying the quite useful Markov decision model to practical problems. The Markov decision model has many potential applications in inventory control, maintenance, manufacturing and telecommunications among others. It is to be believed that the Markov decision model will see many significant applications when this versatile model becomes more familiar to engineers, operations research analysts, computer science people and others. To this purpose, we focus in this chapter on the basic concepts and algorithms from Markov decision theory and illustrate the wide applicability of the Markov decision model to a variety of problems. In our discussion we confine ourselves to the optimality criterion of the long-run average cost per unit time. For most applications of Markov decision theory this criterion is believed to be more appropriate than the alternative criterion of the total expected discounted costs. The average cost criterion is particularly appropriate when many state transitions occur within a relatively short time. For example, this is typically the case in stochastic control problems in telecommunication applications.

This chapter is organized as follows. In Section 3.1 we discuss the basic concepts of the discrete-time Markov decision model. In this model the times between the decision epochs are constant. The policy-iteration algorithm is given in Section 3.2. In Section 3.3 we present a linear progamming formulation of the Markov decision model and discuss how to handle probabilistic constraints on the long-run state-action frequencies. The policy-iteration algorithm and the linear programming formulation, being related to each other, both require the solving of a system of simultaneous linear equations in each iteration step. In Section 3.4 we discuss the alternative method of value iteration which avoids the computationally burdensome solving of systems of linear equations but involves only recursive computations. The value-iteration method endowed with quickly converging lower and upper bounds on the minimal costs is usually the most effective method for solving Markov decision problems with a large number of states. Also, in Section 3.4 we give a modified value-iteration method with a dynamic relaxation factor to speed up convergence. The semi-Markov decision model is the subject of Section 3.5. In this model the times between the decision epochs are not constant but random. The algorithms for the discrete-time Markov decision model will be extended to the semi-Markov decision model. A data transformation is given by which a semi-Markov decision model can be transformed into an equivalent discrete-time Markov decision model. Consequently, value iteration applies as well to the semi-Markov model. Also, for semi-Markov decision processes with exponentially distributed transition times, the effectiveness of value iteration may be further enhanced by introducing fictitious decision epochs for the purpose of creating sparse matrices of transition probabilities. In the final Section 3.6 we show how the flexibility of policy iteration together with an embedding technique can be used to develop tailor-made algorithms for practical applications having a specific structure. Also, this section discusses a heuristic approach for handling multi-dimensional Markov decision applications.

3.1 DISCRETE-TIME MARKOV DECISION PROCESSES

In Section 2.1 we have considered a dynamic system that evolves over time according to a *fixed* probabilistic law of motion satisfying the Markovian assumption. This assumption states that the next state to be visited depends only on the present state of the system. In this chapter we deal with a dynamic system evolving over time where the probabilistic law of motion can be controlled by taking decisions. Also, costs are incurred (or rewards are earned) as a consequence of the decisions that are sequentially made when the system evolves over time. An *infinite planning horizon* is assumed and the goal is to find a control rule which minimizes the *long-run average cost per unit time*.

A typical example of a controlled dynamic system is an inventory system with stochastic demands where the inventory position is periodically reviewed. The decisions taken at the review times consist of ordering a certain amount of the product depending on the inventory position. The economic consequences of the decisions are reflected in ordering, inventory and shortage costs.

We now introduce the Markov decision model. Consider a dynamic system which is reviewed at equidistant points of time $t = 0, 1, \dots$. At each review the system is classified into one of a possible number of states and subsequently a decision has to be made. The set of possible states is denoted by I. For each state $i \in I$, a set $A(i)$ of decisions or actions is given. The state space I and the action sets $A(i)$ are assumed to be *finite*. The economic consequences of the decisions taken at the review times (decision epochs) are reflected in costs. This controlled dynamic system is called a *discrete-time Markov decision model* when the following Markovian property is satisfied. If at a decision epoch the action a is chosen in state i, then regardless of the past history of the system, the following happens:

(a) An immediate cost $c_i(a)$ is incurred.

(b) At the next decision epoch the system will be in state j with probability $p_{ij}(a)$ where

$$\sum_{j \in I} p_{ij}(a) = 1, \quad i \in I.$$

Note that the one-step costs $c_i(a)$ and the one-step transition probabilities $p_{ij}(a)$ are assumed to be time homogeneous. In specific problems the 'immediate' costs $c_i(a)$ will often represent the expected cost incurred until the next decision epoch when action a is chosen in state i. Also, it should be emphasized that the choice of the state space and of the action sets usually depends on the cost structure of the specific problem considered. For example, in a production/inventory problem involving a fixed setup cost for restarting production after an idle period, the state description should include a state variable indicating whether the production facility is on or off. Many practical control problems can be modelled as a Markov decision process by choosing appropriately the state space and action sets. Before we develop the required theory for the average cost criterion, we give two examples.

Example 3.1.1 A maintenance problem

At the beginning of each day a piece of equipment is inspected to reveal its actual working condition. The equipment will be found in one of the working conditions $i = 1, \dots, N$, where the working condition i is better than the working condition

$i + 1$. The equipment deteriorates in time. If the present working condition is i and no repair is done, then at the beginning of the next day the equipment has working condition j with probability q_{ij}. It is assumed that the equipment cannot improve on its own. In other words, $q_{ij} = 0$ for $j < i$ and $\Sigma_{j \geq i} q_{ij} = 1$. The working condition $i = N$ represents a malfunction that requires a repair taking two days. For the intermediate states i with $1 < i < N$ there is a choice between preventively repairing the equipment and letting the equipment operate for the present day. A preventive repair takes only one day. A repaired system has the working condition $i = 1$. We wish to determine a maintenance rule which minimizes the long-run fraction of time the system is in repair.

This problem can be put in the framework of a discrete-time Markov decision model. To do this, note first that, by assuming a cost of 1 for each day the system is in repair, the long-run average cost per day represents the long-run fraction of days the system is in repair. Also, since a repair for working condition N takes two days and in the discrete-time Markov decision model the state of the system has to be defined at the beginning of each day, we need an auxiliary state for the situation in which a repair is in progress. Thus the set of possible states of the system is chosen as

$$I = \{1, 2, \ldots, N, N + 1\}.$$

State i with $1 \leq i \leq N$ corresponds to the situation in which an inspection reveals working condition i, while state $N + 1$ corresponds to the situation in which a repair is in progress already for one day. Denote the two possible actions by

$$a = \begin{cases} 1 & \text{if the system is repaired,} \\ 0 & \text{otherwise.} \end{cases}$$

The set of possible actions in state i is chosen as

$$A(1) = \{0\}, \quad A(i) = \{0, 1\} \text{ for } 1 < i < N, \quad A(N) = A(N + 1) = \{1\}.$$

The one-step transition probabilities $p_{ij}(a)$ are given by

$$p_{i1}(1) = 1 \quad \text{for } 1 < i < N,$$
$$p_{N,N+1}(1) = 1, \quad p_{N+1,1}(1) = 1,$$
$$p_{ij}(0) = q_{ij} \quad \text{for } 1 \leq i < N \text{ and } j \geq i,$$

and the other $p_{ij}(a) = 0$. Further, the one-step costs $c_i(a)$ are given by

$$c_i(1) = 1 \quad \text{and} \quad c_i(0) = 0.$$

Stationary policies

We now introduce some concepts that will be needed in the algorithms to be described in the next sections. A *rule or policy* for controlling the system is a prescription for taking actions at each decision epoch. In principle a control rule may be quite complicated in the sense that the prescribed actions may depend on the whole history of the system. However, in view of the above Markovian assumptions and the fact that the planning horizon is infinitely long, it will intuitively be clear that we need only to consider the so-called *stationary* policies. A stationary policy R is a rule that always prescribes a single action R_i whenever the system is found in state i at a decision epoch. Thus in Example 3.1.1 the rule R prescribing a repair only when the system has a working condition of at least 5 is given by $R_i = 0$ for $1 \leq i < 5$ and $R_i = 1$ for $5 \leq i \leq N + 1$.

Define for $n = 0, 1, \ldots$,

$$X_n = \text{the state of the system at the } n\text{th decision epoch.}$$

Under a given stationary policy R, we have

$$P\{X_{n+1} = j | X_n = i\} = p_{ij}(R_i),$$

regardless of the past history of the system up to time n. Hence under a given stationary policy R the stochastic process $\{X_n\}$ is a discrete-time Markov chain with one-step transition probabilities $p_{ij}(R_i)$. This Markov chain incurs a cost $c_i(R_i)$ each time the system visits state i. Thus we can invoke results from Markov chain theory to specify the long-run average cost per unit time under a given stationary policy.

Average cost for a given stationary policy

We first introduce some notation. Assuming that a given stationary policy R is used, denote the n-step transition probabilities of the corresponding Markov chain $\{X_n\}$ by

$$p_{ij}^{(n)}(R) = P\{X_n = j | X_0 = i\}, \quad i, j \in I \text{ and } n = 1, 2, \ldots .$$

Note that $p_{ij}^{(1)}(R) = p_{ij}(R_i)$. By Chapman and Kolmogoroff's equations (2.1.3) in Section 2.1,

$$p_{ij}^{(n)}(R) = \sum_{k \in I} p_{ik}^{(n-1)}(R) p_{kj}(R_k), \quad n = 2, 3, \ldots . \tag{3.1.1}$$

Also, define the expected cost function $V_n(i, R)$ by

$$V_n(i, R) = \text{the total expected costs over the first } n \text{ decision epochs when the}$$
$$\text{initial state is } i \text{ and policy } R \text{ is used.}$$

Obviously, we have

$$V_n(i, R) = \sum_{t=0}^{n-1} \sum_{j \in I} p_{ij}^{(t)}(R) c_j(R_j), \tag{3.1.2}$$

where $p_{ij}^{(0)}(R) = 1$ for $j = i$ and $p_{ij}^{(0)}(R) = 0$ for $j \neq i$. Next we define the average cost function $g_i(R)$ by

$$g_i(R) = \lim_{n \to \infty} \frac{1}{n} V_n(i, R), \quad i \in I.$$

This limit exists by Theorem 2.2.2 in Section 2.2. The long-run average expected cost per unit time is independent of the initial state i when it is assumed that the Markov chain $\{X_n\}$ corresponding to policy R has no two disjoint closed sets (recall that a set C of states is closed if $p_{ij}(R) = 1$ for $i \in C$ and $j \notin C$). In the unichain case we write

$$g_i(R) = g(R), \quad i \in I.$$

By Theorem 2.3.4 in Section 2.3 we have

$$g(R) = \sum_{j \in I} c_j(R_j) \pi_j(R), \tag{3.1.3}$$

where $\{\pi_j(R), \ j \in I\}$ is the unique equilibrium distribution of the Markov chain $\{X_n\}$. The $\pi_j(R)$'s are the unique solution to the system of linear equations

$$\pi_j(R) = \sum_{k \in I} p_{kj}(R_k) \pi_k(R), \quad j \in I, \tag{3.1.4}$$

$$\sum_{j \in I} \pi_j(R) = 1.$$

Moreover, for any $j \in I$,

$$\pi_j(R) = \lim_{m \to \infty} \frac{1}{m} \sum_{n=1}^{m} p_{ij}^{(n)}(R) \quad \text{for all } i \in I. \tag{3.1.5}$$

Last but not least, we have that $g(R)$ is not only an expected value. Also, with probability 1,

the long-run *actual* average cost per unit time $= g(R)$

independently of the initial state. The strong result that the long-run average cost per unit time is equal to a constant $g(R)$ with probability 1 does not necessarily hold when the Markov chain $\{X_n\}$ associated with policy R has two disjoint closed sets. In the multichain situation with a transient initial state the long-run average cost per unit time is a random variable whose actual value depends on the recurrent class in which the Markov chain is eventually absorbed. However, the multichain situation will rarely occur. In practical applications each relevant stationary policy typically has the property that the associated Markov chain has no two disjoint closed sets. In the following we will usually assume that this property is satisfied.

Average cost optimal policy

A stationary policy R^* is said to be average cost optimal if

$$g_i(R^*) \leq g_i(R)$$

for each stationary policy R uniformly in the initial state i. It is stated without proof that an average cost optimal stationary policy R^* always exists. Moreover, policy R^* is not only optimal among the class of stationary policies but it is also optimal among the class of all conceivable policies. The interested reader is referred to the book of Derman (1970) for a proof.

In most applications it is computationally not feasible to find an average cost optimal policy by computing the average cost for each stationary policy separately. For example in the case that the state space has N states and each action set consists of two actions, the number of possible stationary policies is 2^N, and this number grows quickly beyond any practical bound. However, an efficient algorithm can be given that constructs a sequence of improved policies until an average cost optimal policy is found.

In improving a given policy R, a key role is played by the so-called *relative values* of the various states when policy R is used. The relative values indicate the transient effect of the starting states on the total expected costs under the given policy. In what follows it will be seen that the average cost per unit time and the relative values of the various states can be calculated simultaneously by solving a system of linear equations.

Relative values associated with a given policy R

It may be helpful to give first a heuristic discussion of the relative values of a given policy R before presenting a rigorous treatment. It is assumed that the Markov chain $\{X_n\}$ associated with policy R has no two disjoint closed sets. Then $g_i(R) = g(R)$ independently of the initial state $i \in I$. The starting point is the obvious relation $\lim_{n \to \infty} V_n(i, R)/n = g(R)$ for all i, where $V_n(i, R)$ denotes the total expected costs over the first n decision epochs when the initial state is i and policy R is used. This relation motivates the heuristic assumption that bias values $v_i(R)$, $i \in I$, exist such that, for each $i \in I$,

$$V_n(i, R) \approx ng(R) + v_i(R) \quad \text{for } n \text{ large.} \tag{3.1.6}$$

Note that $v_i(R) - v_j(R) \approx V_n(i, R) - V_n(j, R)$ for n large, so that $v_i(R) - v_j(R)$ measures the difference in total expected costs when starting in state i rather than in state j, given that policy R is followed. Next we heuristically argue that the average cost $g(R)$ and the relative values $v_i(R)$, $i \in I$, satisfy a simultaneous system of linear equations. To do so, note the recursion equation

$$V_n(i, R) = c_i(R_i) + \sum_{j \in I} p_{ij}(R_i) V_{n-1}(j, R), \quad n \geq 1 \text{ and } i \in I.$$

This equation follows by conditioning on the next state that occurs when action $a = R_i$ is made in state i when n decision epochs are to go. A cost $c_i(R_i)$ is incurred at the first decision epoch and the total expected cost over the remaining $n - 1$ decision epochs is $V_{n-1}(j, R)$ when the next state is j. By substituting the above asymptotic expansion in the recursion equation, we find, after cancelling out common terms,

$$g(R) + v_i(R) \approx c_i(R_i) + \sum_{j \in I} p_{ij}(R_i)v_j(R), \quad i \in I.$$

These equations represent the value-determination equations for policy R.

A rigorous way of introducing the relative values associated with a given stationary policy R is to consider the costs incurred until the first return to some regeneration state for policy R. We choose some state r such that for each initial state the Markov chain $\{X_n\}$ associated with policy R will visit state r with probability 1 after finitely many transitions. Thus we can define, for each state $i \in I$,

$T_i(R) = $ the expected time until the first return to state r when starting in state i and using policy R.

In particular, letting a cycle be the time elapsed between two consecutive visits to the regeneration state r under policy R, we have that $T_r(R)$ is the expected length of a cycle. Also, define for each $i \in I$

$K_i(R) = $ the expected costs incurred until the first return to state r when starting in state i and using policy R.

We use the convention that $K_i(R)$ includes the cost incurred when starting in state i but excludes the cost incurred when returning to state r. By the theory of renewal-reward processes, the average cost per unit time equals the expected costs incurred in one cycle divided by the expected length of one cycle and so

$$g(R) = \frac{K_r(R)}{T_r(R)}.$$

Next we define the relative values

$$w_i(R) = K_i(R) - g(R)T_i(R), \quad i \in I. \tag{3.1.7}$$

Note, as a consequence of (3.1.7), the normalization

$$w_r(R) = 0.$$

In the next theorem we prove that the average cost per unit time and the relative values can be calculated simultaneously by solving a system of linear equations.

Theorem 3.1.1 *Let R be a given stationary policy such that the associated Markov chain $\{X_n\}$ has no two disjoint closed sets. Then*

(a) *The average cost $g(R)$ and the relative values $w_i(R)$, $i \in I$, satisfy the following system of linear equations in the unknowns g and v_i, $i \in I$:*

$$v_i = c_i(R_i) - g + \sum_{j \in I} p_{ij}(R_i)v_j, \quad i \in I. \tag{3.1.8}$$

(b) *Let the numbers g and v_i, $i \in I$, be any solution to (3.1.8). Then*

$$g = g(R)$$

and, for some constant c,

$$v_i = w_i(R) + c \quad \text{for all } i \in I.$$

(c) *Let s be an arbitrarily chosen state. Then the linear equations (3.1.8) together with the normalization equation $v_s = 0$ have a unique solution.*

Proof (a) By conditioning on the next state following the initial state i, it can be seen that

$$T_i(R) = 1 + \sum_{j \neq r} p_{ij}(R_i)T_j(R), \quad i \in I,$$

$$K_i(R) = c_i(R_i) + \sum_{j \neq r} p_{ij}(R_i)K_j(R), \quad i \in I.$$

This implies that

$$K_i(R) - g(R)T_i(R) = c_i(R_i) - g(R) + \sum_{j \neq r} p_{ij}(R_i)\{K_j(R) - g(R)T_j(R)\}.$$

Hence, using that $w_r(R) = 0$, we find

$$w_i(R) = c_i(R) - g(R) + \sum_{j \in I} p_{ij}(R_i)w_j(R), \quad i \in I.$$

(b) Let $\{g, v_i\}$ be any solution to (3.1.8). We first verify by induction that the following identity holds for each $m = 1, 2, \ldots$:

$$v_i = \sum_{t=0}^{m-1} \sum_{j \in I} p_{ij}^{(t)}(R)c_j(R_j) - mg + \sum_{j \in I} p_{ij}^{(m)}(R)v_j, \quad i \in I, \tag{3.1.9}$$

where $p_{ij}^{(0)}(R) = 1$ for $j = i$ and $p_{ij}^{(0)}(R) = 0$ for $j \neq i$. Clearly, (3.1.9) is true for $m = 1$. Suppose now that (3.1.9) is true for $m = n$. Substituting the equations (3.1.8) in the right side of (3.1.9) with $m = n$, it follows that for each $i \in I$,

$$v_i = \sum_{t=0}^{n-1} \sum_{j \in I} p_{ij}^{(t)}(R)c_j(R_j) - ng + \sum_{j \in I} p_{ij}^{(n)}(R)\left\{c_j(R_j) - g + \sum_{k \in I} p_{jk}(R_j)v_k\right\}$$

$$= \sum_{t=0}^{n} \sum_{j \in I} p_{ij}^{(t)}(R) c_j(R_j) - (n+1)g + \sum_{k \in I} \left\{ \sum_{j \in I} p_{ij}^{(n)}(R) p_{jk}(R_j) \right\} v_k$$

where the latter equality involves an interchange of the order of summation. Next, using (3.1.1), we get (3.1.9) for $m = n + 1$ which completes the induction step.

Using the relation (3.1.2) for the total expected costs over the first m decision epochs, we can rewrite (3.1.9) in the more convenient form

$$v_i = V_m(i, R) - mg + \sum_{j \in I} p_{ij}^{(m)}(R) v_j, \quad i \in I. \tag{3.1.10}$$

Since $V_m(i, R)/m \to g(R)$ as $m \to \infty$ for each i, the result $g = g(R)$ follows by dividing both sides of (3.1.10) by m and letting $m \to \infty$. To prove the second part of assertion (b), let $\{g, v_i\}$ and $\{g', v'_i\}$ be any two solutions to (3.1.8). Since $g = g' = g(R)$, it follows from the representation (3.1.10) that

$$v_i - v'_i = \sum_{j \in I} p_{ij}^{(m)}(R) \{v_j - v'_j\}, \quad i \in I \text{ and } m \geq 1.$$

By summing both sides of this equation over $m = 1, \ldots, n$ and next dividing by n, it follows after an interchange of the order of summation that

$$v_i - v'_i = \sum_{j \in I} \left\{ \frac{1}{n} \sum_{m=1}^{n} p_{ij}^{(m)}(R) \right\} (v_j - v'_j), \quad i \in I \text{ and } n \geq 1.$$

Next, by letting $n \to \infty$ and using (3.1.5), we obtain

$$v_i - v'_i = \sum_{j \in I} \pi_j(R)(v_j - v'_j), \quad i \in I.$$

The right side of this equation does not depend on i. Assertion (b) now follows by taking the particular solution $v'_i = w_i(R)$, $i \in I$.

(c) Let $\{g, v_i\}$ be any solution to (3.1.8). Since $\sum_j p_{ij}(R_i) = 1$ for each $i \in I$, it follows that for any constant γ the numbers g and $v'_i = v_i + \gamma$, $i \in I$, satisfy (3.1.8) as well. Hence the equations (3.1.8) together with $v_s = 0$ for some state s have a solution. In view of assertion (b), this solution must be unique.

Economic interpretation of the relative values

For any solution $\{g(R), v_i(R)\}$ to the value-determination equations (3.1.8), the numbers $v_i(R)$, $i \in I$, are called the relative values of the various starting states when policy R is used. An explanation of this nomenclature is the following. Assuming that the Markov chain $\{X_n\}$ associated with rule R is aperiodic, we have, for any two states i, $j \in I$,

$v_i(R) - v_j(R) =$ the difference in total expected costs over an infinitely
long period of time by starting in state i rather than in
state j when using policy R.

In other words, $v_i(R) - v_j(R)$ is the maximum amount that a rational person is willing to pay to start the system in state j rather than in state i when the system is controlled by rule R. This interpretation is an easy consequence of (3.1.10). Using the assumption that the Markov chain $\{X_n\}$ is aperiodic, we find that $\lim_{m\to\infty} p_{ij}^{(m)}(R)$ exists. Moreover this limit is independent of the initial state i. Thus, by (3.1.10),

$$v_i(R) - v_j(R) = \lim_{m \to \infty} \{V_m(i, R) - V_m(j, R)\},$$

yielding the desired result since $V_m(i, R)$ represents the total expected costs over the first m decision epochs when the initial state is i and policy R is used.

3.2 POLICY-ITERATION ALGORITHM

For ease of presentation we will discuss the policy-iteration algorithm under the following assumption.

Unichain Assumption 3.2.1 *For each stationary policy the associated Markov chain $\{X_n\}$ has no two disjoint closed sets.*

This assumption is satisfied in most applications.

The relative values associated with a given policy R provide a tool for constructing a new policy $\overline{R}$ whose average cost is no more than that of the current policy R. It is insightful to give first a heuristic motivation for the policy-improvement step of Howard's policy-iteration algorithm.

Motivation for the policy-improvement step

The intuitive idea behind the procedure for improving a given policy is to consider the following difference in costs:

$\Delta(i, a, R) =$ the difference in total expected costs over an infinitely long period
of time by taking first action a and next using policy R rather than
using policy R from the beginning onwards when the initial state
is i.

This difference is equal to zero when action $a = R_i$ is chosen. We wish to make the difference in costs as negative as possible. The difference $\Delta(i, a, R)$ is given by

$$\Delta(i, a, R) = \lim_{n \to \infty} \left[c(i, a) + \sum_{j \in I} p_{ij}(a) V_{n-1}(j, R) - \left\{ c(i, R_i) \right. \right.$$

$$+ \sum_{j \in I} p_{ij}(R_i) V_{n-1}(j, R) \bigg\} \bigg]$$

provided that $\lim_{n \to \infty} \{V_n(i, R) - ng(R)\} = v_i(R)$ exists for all $i \in I$. A sufficient condition for the existence of this limit is the aperiodicity of the Markov chain $\{X_n\}$ associated with policy R. Then we find

$$\Delta(i, a, R) = c(i, a) + \sum_{j \in I} p_{ij}(a) v_j(R) - \bigg\{ c(i, R_i) + \sum_{j \in I} p_{ij}(R_i) v_j(R) \bigg\}$$

$$= c(i, a) + \sum_{j \in I} p_{ij}(a) v_j(R) - g(R) - v_i(R).$$

Thus for each state i we look for an action a which makes

$$c(i, a) - g(R) + \sum_{j \in I} p_{ij}(a) v_j(R)$$

as small as possible. This quantity is called the policy-improvement quantity. Its minimum over $a \in A(i)$ is always less than or equal to $v_i(R)$.

The above heuristic discussion on the policy-improvement step will now be formalized. The next theorem plays a crucial role in what follows. The theorem is stated in general terms.

Theorem 3.2.1   *Let g and v_i, $i \in I$, be given numbers. Suppose that the policy $\overline{R}$ has the property*

$$c_i(\overline{R}_i) - g + \sum_{j \in I} p_{ij}(\overline{R}_i) v_j \leq v_i \quad \text{for each } i \in I. \tag{3.2.1}$$

Then the long-run average cost of policy R satisfies

$$g_i(\overline{R}) \leq g, \quad i \in I. \tag{3.2.2}$$

The theorem is also true when the inequality signs (3.2.1) and (3.2.2) are reversed.

Proof   We first give an intuitive explanation of the theorem and next we formalize this explanation. Suppose that a control cost of $c_i(a) - g$ is incurred each time the action a is chosen in state i, while a terminal cost of v_j is incurred when the control of the system is stopped and the system is left behind in state j. Then (3.2.1) states that controlling the system for one step according to rule $\overline{R}$ and stopping next is preferable to stopping directly when the initial state is i. Since this property is true for each initial state, a repeated application of this property yields that controlling the system for m steps according to rule $\overline{R}$ and stopping after that is preferable to stopping directly. Thus, using the notation (3.1.2) with R replaced by $\overline{R}$, we have, for each initial state $i \in I$,

$$V_m(i, \overline{R}) - mg + \sum_{j \in I} p_{ij}^{(m)}(\overline{R}) v_j \leq v_i \quad \text{for } m = 1, 2, \dots . \tag{3.2.3}$$

Dividing both sides of (3.2.3) by m and letting $m \to \infty$, it follows that $g(\overline{R}) - g \leq 0$ as was to be proved. It is easy to formalize the above arguments. In the same way as we derived (3.1.9), we obtain (3.2.3) by a repeated substitution of the inequality (3.2.1). Finally, it will be clear from the above proof that the theorem remains true when the inequality signs in (3.2.1) and (3.2.2) are reversed.

In order to improve a given policy R whose average cost $g(R)$ and relative values $v_i(R)$, $i \in I$, have been computed, we apply the above theorem with $g = g(R)$ and $v_i = v_i(R)$, $i \in I$. Thus, by constructing a new policy $\overline{R}$ such that, for each state $i \in I$,

$$c_i(\overline{R}_i) - g(R) + \sum_{j \in I} p_{ij}(\overline{R}_i)v_j \leq v_i, \tag{3.2.4}$$

we obtain an improved rule $\overline{R}$ according to $g(\overline{R}) \leq g(R)$. In constructing such an improved policy $\overline{R}$ it is important to realize that for each state i *separately* an action $\overline{R}_i$ satisfying (3.2.4) can be determined. A particular way to find for state $i \in I$ an action $\overline{R}_i$ satisfying (3.2.4) is to minimize

$$c_i(a) - g(R) + \sum_{j \in I} p_{ij}(a)v_j(R) \tag{3.2.5}$$

with respect to $a \in A(i)$. Noting that (3.2.5) equals $v_i(R)$ for $a = R_i$, it follows that (3.2.4) is satisfied for the action $\overline{R}_i$ which minimizes (3.2.5) with respect to $a \in A(i)$. We are now in a position to formulate the following algorithm.

Policy-iteration algorithm

Step 0 (initialization). Choose a stationary policy R.

Step 1 (value-determination step). For the current rule R, compute the unique solution $\{g(R), v_i(R)\}$ to the following system of linear equations:

$$v_i = c_i(R_i) - g + \sum_{j \in I} p_{ij}(R_i)v_j, \quad i \in I$$

$$v_s = 0,$$

where s is an arbitrarily chosen state.

Step 2 (policy-improvement step). For each state $i \in I$, determine an action a_i yielding the minimum in

$$\min_{a \in A(i)} \left\{ c_i(a) - g(R) + \sum_{j \in I} p_{ij}(a)v_j(R) \right\}.$$

The new stationary policy $\overline{R}$ is obtained by choosing $\overline{R}_i = a_i$ for all $i \in I$ with the convention that $\overline{R}_i$ is chosen equal to the old action R_i when this action minimizes the policy-improvement quantity.

Step 3 (convergence test). If the new policy $\overline{R}$ equals the old policy R, the algorithm is stopped with policy R. Otherwise, go to step 1 with R replaced by $\overline{R}$.

The policy-iteration algorithm converges after a finite number of iterations to an average cost optimal policy. We defer the proof to the appendix at the end of this section.

The policy-iteration algorithm is empirically found to be a remarkably robust algorithm that converges very fast in specific problems. The number of iterations is practically independent of the number of states and varies typically between 3 and 15 (say). Also, it can be roughly stated that the average costs of the policies generated by policy iteration converge at least exponentially fast to the minimum costs, with the greatest improvements in the first few iterations.

Remark 3.2.1 The average cost optimality equation

Since the policy-iteration algorithm converges after finitely many iterations, there exist numbers g^* and v_i^*, $i \in I$, such that

$$v_i^* = \min_{a \in A(i)} \left\{ c_i(a) - g^* + \sum_{j \in I} p_{ij}(a)v_j^* \right\}, \quad i \in I. \tag{3.2.6}$$

This functional equation is called the average cost optimality equation. Using Theorem 3.2.1, we can directly verify that any stationary policy R^* such that the action $R^*(i)$ minimizes the right side of (3.2.6) for all $i \in I$ is average cost optimal. To see this, note that

$$v_i^* = c_i(R_i^*) - g^* + \sum_{j \in I} p_{ij}(R_i^*)v_j^*, \quad i \in I \tag{3.2.7}$$

and

$$v_i^* \leq c(i, a) - g^* + \sum_{j \in I} p_{ij}(a)v_j^*, \quad a \in A(i) \text{ and } i \in I. \tag{3.2.8}$$

The equality (3.2.7) and Theorem 3.2.1 imply that $g(R^*) = g^*$. Let $\overline{R}$ be any stationary policy. Taking $a = \overline{R}(i)$ in (3.2.8) for all $i \in I$ and applying Theorem 3.2.1, we find $g(\overline{R}) \geq g^*$. In other words, $g(R^*) \leq g(\overline{R})$ for any stationary policy $\overline{R}$. This shows not only that policy R^* is average cost optimal but shows also that the constant g^* in (3.2.6) is uniquely determined as the minimal average cost per unit time. It is stated without proof that the function v_i^*, $i \in I$, in (3.2.6) is uniquely determined up to an additive constant.

Next the policy-iteration algorithm is applied to compute an average cost optimal policy for the control problem in Example 3.1.1.

Example 3.1.1 (continued) A maintenance problem

It is assumed that the number of possible working conditions equals $N = 5$. The deterioration probabilities q_{ij} are given in Table 3.2.1. The policy-iteration

Table 3.2.1 The deteriorating probabilities q_{ij}

i \ j	1	2	3	4	5
1	0.75	0.20	0.05	0	0
2	0	0.50	0.20	0.20	0.10
3	0	0	0.50	0.25	0.25
4	0	0	0	0.30	0.70

algorithm is initialized with the policy which prescribes the repair action $a = 1$ in each state except state 1.

Iteration 1

Step 1. For the current policy $R^{(1)}$ whose actions are

$$R_1^{(1)} = 0, \quad R_2^{(1)} = R_3^{(1)} = R_4^{(1)} = R_5^{(1)} = R_6^{(1)} = 1,$$

the average cost $g(R^{(1)})$ and the relative values $v_i(R^{(1)})$ can be computed as the unique solution to the linear equations

$$v_1 = 0 - g + 0.75v_1 + 0.20v_2 + 0.05v_3,$$
$$v_k = 1 - g + v_1, \quad k = 2, 3, 4,$$
$$v_5 = 1 - g + v_6, \quad v_6 = 1 - g + v_1, \quad v_6 = 0,$$

where $s = 6$ is taken for the normalizing equation $v_s = 0$. The solution of these linear equations is given by

$$g(R^{(1)}) = 0.2, \quad v_1(R^{(1)}) = -0.8, \quad v_2(R^{(1)}) = 0, \quad v_3(R^{(1)}) = 0,$$
$$v_4(R^{(1)}) = 0, \quad v_5(R^{(1)}) = 0.8, \quad v_6(R^{(1)}) = 0.$$

Step 2. The calculations for the policy-improvement step for policy $R^{(1)}$ are displayed in Table 3.2.2. Note that for state 3 the action $a = 1$ prescribed by the current policy $R^{(1)}$ is one of the minimizing actions and thus the new action in state 3 is taken as $a = 1$. From Table 3.2.2 we find the new policy $R^{(2)}$ whose actions are

$$R_1^{(2)} = R_2^{(2)} = 0, \quad R_3^{(2)} = R_4^{(2)} = R_5^{(2)} = R_6^{(2)} = 1.$$

Table 3.2.2 The policy-improvement step for policy $R^{(1)}$

State i	Action a	Test quantity $c_i(a) - g(R^{(1)}) + \Sigma_{j=1}^{6} p_{ij}(a) v_j(R^{(1)})$	
2	0	$0 - 0.2 + (0.50)(0) + (0.20)(0) + (0.20)(0) + (0.10)(0.8)$	$= -0.12$
	1	$1 - 0.2 + (1.0)(-0.8)$	$= 0$
3	0	$0 - 0.2 + (0.50)(0) + (0.25)(0) + (0.25)(0.8)$	$= 0$
	1	$1 - 0.2 + (1.0)(-0.8)$	$= 0$
4	0	$0 - 0.2 + (0.30)(0) + (0.70)(0.8)$	$= 0.36$
	1	$1 - 0.2 + (1.0)(-0.8)$	$= 0$

Table 3.2.3   The policy-improvement step for policy $R^{(2)}$

State i	Action a	Test quantity $c_i(a) - g(R^{(2)}) + \sum_{j=1}^{6} p_{ij}(a) v_j(R^{(2)})$	
2	0	$0 - 0.172 + (0.50)(-0.178) + (0.20)(0) + (0.20)(0)$	
		$+ (0.10)(0.828)$	$= -0.178$
	1	$1 - 0.172 + (1.0)(-0.828)$	$= 0$
3	0	$0 - 0.172 + (0.50)(0) + (0.25)(0) + (0.25)(0.828)$	$= 0.035$
	1	$1 - 0.172 + (1.0)(-0.828)$	$= 0$
4	0	$0 - 0.172 + (0.30)(0) + (0.70)(0.828)$	$= 0.408$
	1	$1 - 0.172 + (1.0)(-0.828)$	$= 0$

Step 3. The new policy $R^{(2)}$ is different from the previous policy $R^{(1)}$ and hence a next iteration is performed.

Iteration 2
Step 1. For the current policy $R^{(2)}$ the average cost $g(R^{(2)})$ and the relative values $v_i(R^{(2)})$ can be computed as the unique solution to the linear equations

$$v_1 = 0 - g + 0.75v_1 + 0.20v_2 + 0.05v_3,$$
$$v_2 = 0 - g + 0.50v_2 + 0.20v_3 + 0.20v_4 + 0.10v_5,$$
$$v_k = 1 - g + v_1, \qquad k = 3, 4,$$
$$v_5 = 1 - g + v_6, \qquad v_6 = 1 - g + v_1, \qquad v_6 = 0.$$

The solution of these linear equations is

$$g(R^{(2)}) = 0.172, \quad v_1(R^{(2)}) = -0.828, \quad v_2(R^{(2)}) = -0.178,$$
$$v_3(R^{(2)}) = 0, \quad v_4(R^{(2)}) = 0, \quad v_5(R^{(2)}) = 0.828, \quad v_6(R^{(2)}) = 0.$$

Step 2. The improvement step for policy $R^{(2)}$ is shown in Table 3.2.3. From Table 3.2.3 we find the new policy $R^{(3)}$ whose actions are

$$R_1^{(3)} = R_2^{(3)} = 0, \quad R_3^{(3)} = R_4^{(3)} = R_5^{(3)} = R_6^{(3)} = 1.$$

Step 3. The new policy $R^{(3)}$ is the same as the previous policy $R^{(2)}$ and is thus average cost optimal. The minimum fraction of days the equipment is in repair equals 0.172.

Remark 3.2.2 Deterministic state transitions

For the case of deterministic state transitions the computational burden of the policy-iteration algorithm can be reduced considerably. Instead of solving a system of linear equations at each step, the average cost and relative values can be obtained from recursive calculations. The reason for this is that under each stationary policy the process moves cyclically among the recurrent states. The simplified policy-iteration calculations for deterministic state transitions are as follows:

(a) Determine for the current policy R the cycle of recurrent states among which the process cyclically moves.
(b) The cost rate $g(R)$ equals the sum of one-step costs in the cycle divided by the number of states in the cycle.
(c) The relative values for the recurrent states are calculated recursively, in reverse direction to the natural flow around the cycle, after assigning a value 0 to one recurrent state.
(d) The relative values for transient states are computed first for states which reach the cycle in one step, then for states which reach the cycle in two steps, and so forth.

It is worthwhile to point out that the simplified policy-iteration algorithm may be an efficient technique to compute a minimum-cost-to-time circuit in a deterministic network.

Appendix. The finite convergence of the policy-iteration algorithm

We first establish a lexicographical ordering for the average cost and the relative values associated with the policies that are generated by the algorithm. For that purpose we need to standardize the relative cost functions since a relative cost function is not uniquely determined. Let us (re)number the possible states as $i = 0, \ldots, N$. In view of the fact that the relative values of a given policy are unique up to an additive constant, the sequence of policies generated by the algorithm does not depend on the particular choice of the relative cost function for a given policy. For each stationary policy Q, we now consider the particular relative cost function $w_i(Q)$ defined by (3.1.7) where the regeneration state r is chosen as the *largest* state in $I(Q)$. The set $I(Q)$ is defined by

$$I(Q) = \{\text{states that are recurrent under policy } Q\}.$$

Let R and $\overline{R}$ be immediate successors in the sequence of policies generated by the algorithm. Suppose that $\overline{R} \neq R$. We assert that either

(a) $g(\overline{R}) < g(R)$

or

(b) $g(\overline{R}) = g(R)$ and $w_i(\overline{R}) \leq w_i(R)$ for all $i \in I$ with equality for each $i \in I(\overline{R})$ and strict inequality for at least one state $i \notin I(\overline{R})$.

That is each iteration either reduces the cost rate or else reduces the relative value of a transient state. Since the number of possible stationary policies is finite, this assertion implies that the algorithm converges after finitely many iterations. To prove the assertion, the starting point is the relation

$$c_i(\overline{R}_i) - g(R) + \sum_{j \in I} p_{ij}(\overline{R}_i) w_j(R) \leq w_i(R), \quad i \in I, \qquad (3.2.9)$$

with strict inequality only for those states i with $\overline{R}_i \neq R_i$. This relation is an immediate consequence of the construction of policy $\overline{R}$. By Theorem 3.2.1 and (3.2.9), we have $g(\overline{R}) \leq g(R)$. We now sharpen this result as follows. It holds that $g(\overline{R}) < g(R)$ only if the strict inequality holds in (3.2.9) for some state i that is recurrent under the new policy $\overline{R}$. To prove this, multiply both sides of (3.2.9) by $\pi_i(\overline{R})$ and sum over i. Then, using $\pi_i(\overline{R}) \geq 0$ for all i with the strict inequality sign only for $i \in I(\overline{R})$, we find after an interchange of the order of summation that

$$\sum_{i \in I} \pi_i(\overline{R})c_i(\overline{R}) - g(R) + \sum_{j \in I}\left\{\sum_{i \in I} \pi_i(\overline{R})p_{ij}(\overline{R}_i)\right\} w_j(R) \leq \sum_{i \in I} \pi_i(\overline{R})w_i(R)$$

with strict inequality only if there is strict inequality in (3.2.9) for some $i \in I(\overline{R})$. Using (3.1.3) and (3.1.4) with R replaced by $\overline{R}$, it follows that the above inequality is equivalent to

$$g(\overline{R}) - g(R) + \sum_{j \in I} \pi_j(\overline{R})w_j(R) \leq \sum_{i \in I} \pi_i(\overline{R})w_i(R)$$

with strict inequality only if there is strict inequality in (3.2.9) for some $i \in I(\overline{R})$. This yields the desired sharpening of Theorem 3.2.1. Thus we find either (a) $g(\overline{R}) < g(R)$ or (b) $g(\overline{R}) = g(R)$ with strict equality in (3.2.9) for all $i \in I(\overline{R})$. To complete the verification of the above assertion, suppose that case (b) applies. Then, by the convention made in the policy-improvement step,

$$\overline{R}_i = R_i, \quad i \in I(\overline{R}).$$

Thus the set $I(\overline{R})$ of states must be closed under policy R. Since policy R has no two disjoint closed sets, we can next conclude that

$$I(R) = I(\overline{R}).$$

For each policy Q the relative values $w_i(Q)$ for $i \in I(Q)$ do not depend on the actions Q_i for $i \notin I(Q)$. This follows easily from definition (3.1.7) and the fact that the process always stays in the set of recurrent states once it is in that set. Since the regeneration state r for the relative cost function is chosen as the largest state in $I(Q)$, this particular regeneration state is the same for the policies R and $\overline{R}$. Thus

$$w_i(R) = w_i(\overline{R}), \quad i \in I(\overline{R}).$$

The remainder of the proof is now easy. Proceeding in the same way as in the derivation of (3.1.9), we find by iterating the inequality (3.2.9) that

$$w_i(R) \geq V_m(i, \overline{R}) - mg(R) + \sum_{j \in I} p_{ij}^{(m)}(\overline{R})w_j(R), \quad i \in I \text{ and } m \geq 1, \quad (3.2.10)$$

where the strict inequality sign holds for each i with $\overline{R}_i \neq R_i$. Since it is assumed that $g(\overline{R}) = g(R)$, it follows from (3.1.10) with R replaced by $\overline{R}$ that

$$V_m(i, \overline{R}) - mg(R) + \sum_{j \in I} p_{ij}^{(m)}(\overline{R}) w_j(R)$$

$$= w_i(\overline{R}) + \sum_{j \in I} p_{ij}^{(m)}(\overline{R})\{w_j(R) - w_j(\overline{R})\}, \quad i \in I \text{ and } m \geq 1.$$

Hence (3.2.10) can be written as

$$w_i(R) \geq w_i(\overline{R}) + \sum_{j \in I} p_{ij}^{(m)}(\overline{R})\{w_j(R) - w_j(\overline{R})\}, \quad i \in I \text{ and } m \geq 1,$$

where the strict inequality sign holds for each i with $R_i \neq \overline{R}_i$. Noting that $w_j(R) = w_j(\overline{R})$ for j recurrent under $\overline{R}$ and $p_{ij}^{(m)}(\overline{R}) \to 0$ as $m \to \infty$ for j transient under $\overline{R}$, it next follows that $w_i(R) \geq w_i(\overline{R})$ with strict inequality for each i with $R_i \neq \overline{R}_i$. This completes the proof.

3.3   LINEAR PROGRAMMING FORMULATION

The policy-iteration algorithm solves the average-cost optimality equation (3.2.6) in a finite number of steps by generating a sequence of improved policies. Another convenient way of solving the optimality equation is the application of a linear programming formulation for the average cost case. The linear programming formulation to be given below allows the Unichain Assumption 3.2.1 in Section 3.2 to be weakened as follows.

Weak Unichain Assumption 3.3.1   *For each average cost optimal stationary policy the associated Markov chain $\{X_n\}$ has no two disjoint closed sets.*

This assumption allows non-optimal policies to have multiple disjoint closed sets. The Unichain Assumption 3.2.1 may be too strong for some applications; for example, in inventory problems with strictly bounded demands it may be possible to construct stationary policies with disjoint ordering regions such that the levels between which the stock fluctuates remain dependent on the initial level. However, the Weak Unichain Assumption will practically always be satisfied in real-world applications. For the weak unichain case, the minimal average cost per unit time is independent of the initial state and, moreover, the average cost optimality equation (3.2.6) applies and determines uniquely g^* as the minimal average cost per unit time, (for a proof see Denardo and Fox (1968)). This reference also gives the following linear programming algorithm for the computation of an average cost optimal policy.

Linear programming algorithm

Step 1. Apply the simplex method to compute an optimal basic solution (x_{ia}^*) to the linear program

$$\text{Minimize} \sum_{i \in I} \sum_{a \in A(i)} c_i(a) x_{ia}$$

subject to

$$\sum_{a \in A(j)} x_{ja} - \sum_{i \in I} \sum_{a \in A(i)} p_{ij}(a) x_{ia} = 0, \quad j \in I \qquad (3.3.1)$$

$$\sum_{i \in I} \sum_{a \in A(i)} x_{ia} = 1, \quad x_{ia} \geq 0 \quad \text{for all } i, a.$$

Step 2. Start with the nonempty set

$$I_0 := \left\{ i \;\middle|\; \sum_{a \in A(i)} x_{ia}^* > 0 \right\}$$

and, for any state $i \in I_0$, set the decision

$$R_i^* := a \quad \text{for some } a \text{ such that } x_{ia}^* > 0.$$

Step 3. If $I_0 = I$, then the algorithm is stopped with the average cost optimal policy R^*. Otherwise, determine some state $i \notin I_0$ and action $a \in A(i)$ such that $p_{ij}(a) > 0$ for some $j \in I_0$, set $R_i^* := a$ and $I_0 := I_0 \cup \{i\}$, and repeat step 3.

The linear program (3.3.1) can heuristically be explained by interpreting the variables x_{ia} as

x_{ia} = the long-run fraction of decision epochs at which the system is in state i and action a is made.

The objective of the linear program is the minimization of the long-run average cost per unit time, while the first set of constraints represent the balance equations requiring that for any state $j \in I$ the long-run average number of transitions from state j per unit time must be equal to the long-run average number of transitions into state j per unit time. The last constraint obviously requires that the sum of the fractions x_{ia} must be equal to 1.

Next we sketch a proof that the above linear programming algorithm leads to an average cost optimal policy when the Weak Unichain Assumption 3.3.1 is satisfied. Our starting point is the average cost optimality equation (3.2.6). Since this equation is solvable, the linear inequalities

$$g + v_i - \sum_{j \in I} p_{ij}(a) v_j \leq c_i(a), \quad i \in I \text{ and } a \in A(i), \qquad (3.3.2)$$

must have a solution. It follows from Theorem 3.2.1 that any solution $\{g, v_i\}$ to these inequalities satisfies $g \leq g_i(R)$ for any $i \in I$ and any policy R. Hence we can conclude that for any solution $\{g, v_i\}$ to the linear inequalities (3.3.2) holds that $g \leq g^*$ with g^* being the minimal average cost per unit time. Hence, using the fact that relative values v_i^*, $i \in I$, exist such that $\{g^*, v_i^*\}$ constitutes a solution to (3.3.2), the linear program

Maximize g

subject to

$$g + v_i - \sum_{j \in I} p_{ij}(a) v_j \leq c_i(a) \quad \text{for } i \in I \tag{3.33}$$

and $a \in A(i)$, g, v_i unrestricted

has the minimal average cost g^* as the optimal objective-function value. Next observe that the linear program (3.3.1) is the dual of the primal linear program (3.3.3). By the dual theorem of linear programming the primal and dual linear programs have the same optimal objective-function value. Hence the minimal objective-function value of the linear program (3.3.1) yields the minimal average cost g^*. Next we show that an optimal basic solution (x_{ia}^*) to the linear program (3.3.1) induces an average cost optimal policy. To do so, define the set

$$S_0 = \left\{ i \,\middle|\, \sum_{a \in A(i)} x_{ia}^* > 0 \right\}.$$

Then the set S_0 is closed under any policy R such that the action $a = R_i$ satisfies $x_{ia}^* > 0$ for all $i \in S_0$. To see this, suppose that $p_{ij}(R_i) > 0$ for some $i \in S_0$ and $j \notin S_0$. Then the first set of constraints of the linear program (3.3.1) implies that $\Sigma_a x_{ja}^* > 0$ contradicting that $j \notin S_0$. Next consider the set I_0 as constructed in the linear programming algorithm. Let R^* be a policy such that the actions R_i^* for $i \in I_0$ are chosen according to the algorithm. It remains to verify that $I_0 = I$ and that policy R^* is average cost optimal. To do so, let $\{g_i^*, v_i^*\}$ be the particular optimal basic solution to the primal linear program (3.3.3) such that this basic solution is complementary to the optimal basic solution (x_{ia}^*) of the dual linear program (3.3.1). Then, by the complementary slackness property of linear programming,

$$g^* + v_i^* - \sum_{j \in I} p_{ij}(R_i^*) v_j^* = c_i(R_i^*) \quad \text{for } i \in S_0.$$

The term $\Sigma_{j \in I} p_{ij}(R_i^*) v_j^*$ can be replaced by $\Sigma_{j \in S_0} p_{ij}(R_i^*) v_j^*$ for $i \in S_0$, since the set S_0 is closed under policy R^*. Thus, by Theorem 3.2.1, we can conclude that $g_i(R^*) = g^*$ for all $i \in S_0$. The states in $I_0 \backslash S_0$ are transient under policy R^* and are ultimately leading to a state in S_0. Hence $g_i(R^*) = g^*$ for all $i \in I_0$. To prove that $I_0 = I$, assume to the contrary that $I_0 \neq I$. By the construction of I_0, the set $I \backslash I_0$ is closed under any policy. Let R_0 be any average cost optimal policy. Define the policy R_1 by

$$R_1(i) = \begin{cases} R^*(i) & \text{for } i \in I_0, \\ R_0(i) & \text{for } i \in I \backslash I_0. \end{cases}$$

Since $I \backslash I_0$ and I_0 are both closed sets under policy R_1, we have constructed an average cost optimal policy with two disjoint closed sets. This contradicts the Weak

Unichain Assumption 3.3.1. Hence $I_0 = I$. This completes the proof that the linear programming algorithm leads to an average cost optimal policy.

Comparison of linear programming and policy iteration

In comparing the linear programming and policy-iteration formulations for the Markovian decision problem, the following remarks are in order. The linear programming formulation has the advantage that sophisticated linear programming codes with the additional option of sensitivity analysis are widely available. The policy-iteration formulation usually involves the writing of its own code. The number of iterations required by the simplex method depends heavily on the specific problem considered, whereas the policy-iteration algorithm requires typically only a very small number of iterations regardless of the problem size. In general a statement about the relative efficiency of the two methods seems difficult to make. On the one hand, the policy-iteration algorithm usually requires a much smaller number of iterations than the simplex method. However, on the other hand, the computational burden per iteration is greater for the policy-iteration method than for the simplex method. Unlike in linear programming, in policy iteration a system of linear equations must be solved anew at each iteration. It is interesting to note that the policy-iteration algorithm may be interpreted as a modified linear programming algorithm in which pivot operations are performed simultaneously on several variables, rather than a pivot operation on a single variable as in the simplex method. Also, there is considerable flexibility in changing the decisions in an iteration of the policy-iteration algorithm. It will be seen later in Section 3.6 that this flexibility may enable us to formulate very efficient, tailor-made algorithms in applications having a specific structure. In general both policy iteration and linear progamming are unattractive for solving large-scale Markovian decision problems, since both methods require per iteration the solving of a system of linear equations whose size equals the number of states. Another computational method based on successive substitutions will be discussed in the next section; this method is in general the best method for solving large-scale Markov decision problems.

We illustrate the linear programming formulation to the Markovian decision problem dealt with in Example 3.1.1. The specification of the basic elements of the Markovian decision model for this problem is given in Section 3.1.

Example 3.1.1 (continued) A maintenance problem

The linear programming formulation for this problem is

$$\text{Minimize} \sum_{i=2}^{N+1} x_{i1}$$

subject to

$$x_{10} - \left(q_{11}x_{10} + \sum_{i=2}^{N-1} x_{i1} + x_{N+1,1} \right) = 0,$$

$$x_{j0} + x_{j1} - \sum_{i=1}^{j} q_{ij} x_{i0} = 0, \quad 2 \leq j \leq N - 1,$$

$$x_{N1} - \sum_{i=1}^{N-1} q_{iN} x_{i0} = 0,$$

$$x_{N+1,1} - X_{N1} = 0,$$

$$x_{10} + \sum_{i=2}^{N-1} (x_{i0} + x_{i1}) + x_{N1} + x_{N+1,1} = 1,$$

$$x_{10}, x_{i0}, x_{i1}, x_{N1}, x_{N+1,1} \geq 0.$$

For the numerical data given in Table 3.2.1 in Section 3.2, the linear program has the optimal basic solution

$$x_{10}^* = 0.5917, \quad x_{20}^* = 0.2367, \quad x_{31}^* = 0.0769,$$

$$x_{41}^* = 0.0473, \quad x_{51}^* = x_{61}^* = 0.0237.$$

This yields an average cost optimal policy $R^* = (0, 0, 1, 1, 1, 1)$ with the minimal average cost $\Sigma_{i=2}^{N+1} x_{i1}^* = 0.172$ in agreement with the results obtained by the policy-iteration algorithm.

Linear programming and probabilistic constraints

The linear programming formulation may often handle conveniently Markovian decision problems with probabilistic constraints. In many practical applications constraints are imposed on certain state frequencies. For example, in inventory problems when shortage costs are difficult to estimate, probabilistic constraints may be placed on the probability of shortage or the fraction of demand that cannot be met directly from stock on hand. Similarly, in a maintenance problem involving a randomly changing state a constraint may be placed on the frequency at which a certain inoperative state occurs.

The following illustrative example taken from Wagner (1975) shows that for control problems with probabilistic constraints it may be optimal to choose the decisions in a random way rather than in a deterministic way. Suppose the daily demand D for some product is described by the probability distribution

$$P\{D = 0\} = P\{D = 1\} = \frac{1}{6}, \quad P\{D = 2\} = \frac{2}{3}.$$

The demands on the successive days are independent of each other. At the beginning of each day it has to be decided how much to order of the product. The delivery of any order is instantaneous. The variable ordering cost of each unit is $c > 0$. Any unit that is not sold at the end of the day becomes obsolete and must be discarded. The decision problem is to minimize the average ordering cost per day, subject to the constraint that the fraction of the demand to be met is at least $1/3$. This probabilistic constraint is satisfied when using the policy of ordering one unit every day, a policy which has an average cost of c per day. However, this deterministic control rule is not optimal, as can be seen by considering the randomized control rule under which every day no unit is ordered with probability $4/5$ and two units are ordered with probability $1/5$. Under this randomized rule the probability that the daily demand is met equals $(4/5)(1/6)+(1/5)(1) = 1/3$ and the average ordering cost per day equals $(4/5)(0) + (1/5)(2c) = (2/5)c$. It is readily seen that the above randomized rule is optimal.

So far we have considered only stationary policies under which the actions are chosen deterministically. A policy π is called a stationary *randomized* policy when it is described by a probability distribution $\{\pi_a(i), a \in A(i)\}$ for each state $i \in I$. Under policy π action $a \in A(i)$ is chosen with probability $\pi_a(i)$ whenever the process is in state i. If $\pi_a(i)$ is 0 or 1 for every i and a, the stationary randomized policy π reduces to the familiar stationary policy choosing the actions in a deterministic way. For any policy π, let the state-action frequencies $f_{ia}(\pi)$ be defined by

$f_{ia}(\pi) = $ the long-run fraction of decision epochs at which the process is in
state i and action a is chosen when policy π is used.

Consider now a Markovian decision problem in which the goal is to minimize the long-run average cost per unit time subject to the following linear constraints on the state-action frequencies

$$\sum_{i \in I} \sum_{a \in A(i)} \alpha_{ia}^{(s)} f_{ia}(\pi) \leq \beta^{(s)} \quad \text{for } s = 1, \dots, L,$$

where $\alpha_{ia}^{(s)}$ and $\beta^{(s)}$ are given constants. It is assumed that the constraints allow for a feasible solution. In case the Unichain Assumption 3.2.1 of Section 3.2 is satisfied, it can be shown that an optimal policy may be obtained by solving the following linear program (see Derman (1970) and Hordijk and Kallenberg (1984)),

$$\text{minimize} \sum_{i \in I} \sum_{a \in A(i)} c_i(a) x_{ia}$$

subject to

$$\sum_{a \in A(j)} x_{ja} - \sum_{i \in I} \sum_{a \in A(i)} p_{ij}(a) x_{ia} = 0, \quad j \in I,$$

$$\sum_{i \in I} \sum_{a \in A(i)} x_{ia} = 1,$$

$$\sum_{i \in I} \sum_{a \in A(i)} \alpha_{ia}^{(s)} x_{ia} \leq \beta^{(s)}, \quad s = 1, \dots, L,$$

$$x_{ia} \geq 0, \quad i \in I \text{ and } a \in A(i).$$

Denoting by $\{x_{ia}^*\}$ an optimal basic solution to this linear program and letting the set $S_0 = \{i \mid \Sigma_a x_{ia}^* > 0\}$, an optimal stationary randomized policy π^* is given by

$$\pi_a^*(i) = \begin{cases} x_{ia}^*/\Sigma_a x_{ia}^*, & a \in A(i) \text{ and } i \in S_o, \\ \text{arbitrary}, & \text{otherwise}. \end{cases}$$

Here it is pointed out that the Unichain Assumption 3.2.1 is essential for guaranteeing the existence of an optimal stationary randomized policy.

A heuristic approach for handling probabilistic constraints is the Lagrange-multiplier method by which method only stationary non-randomized policies are produced. To describe this method, assume a single probabilistic constraint

$$\sum_{i \in I} \sum_{a \in A(i)} \alpha_{ia} f_{ia}(\pi) \leq \beta$$

on the state-action frequencies. Here it is assumed that $\alpha_{ia} \geq 0$ for all i and a. In the Lagrange-multiplier method the above constraint is eliminated by putting it into the criterion function by means of a Lagrange multiplier λ; that is, the goal function is changed from $\Sigma_{i,a} c_i(a) x_{ia}$ to $\Sigma_{i,a} c_i(a) x_{ia} + \lambda(\Sigma_{i,a} \alpha_{ia} x_{ia} - \beta)$. The Lagrange multiplier may be interpreted as the cost to each unit that is used from some resource. Thus, for a given value of the Lagrange multiplier $\lambda \geq 0$, we consider the unconstrained Markov decision problem with one-step costs

$$c_i^\lambda(a) = c_i(a) + \lambda \alpha_{ia}$$

and one-step transition probabilities $p_{ij}(a)$ as before. Solving this unconstrained Markov decision problem yields an optimal stationary policy $R(\lambda)$ that prescribes always the same action $R_i(\lambda)$ whenever the system is in state i. Let $\beta(\lambda)$ be the constraint level associated with policy $R(\lambda)$, that is

$$\beta(\lambda) = \sum_{i \in I} \alpha_{i,R_i(\lambda)} f_{i,R_i(\lambda)}(R(\lambda)).$$

Note that $f_{i,R_i(\lambda)}(R(\lambda))$ is just the steady-state probability $\pi_i(R(\lambda))$ giving the frequency at which the system visits state i under policy $R(\lambda)$. Since the number of stationary policies is finite, there will be gaps in the range of values of $\beta(\lambda)$ when the Lagrange multiplier λ is varied. Actually, by a standard result in parametric linear programming, there are a finite number of values $\lambda_0 < \lambda_1 < \ldots < \lambda_{m-1} < \lambda_m$ with $\lambda_0 = 0$ and $\lambda_m = \infty$ such that for any $0 \leq j \leq m - 1$ a same optimal stationary policy $R(\lambda) = R(\lambda_j)$ applies for all $\lambda \in [\lambda_j, \lambda_{j+1})$. Also, we have that $\beta(\lambda)$ is piecewise constant and non-increasing in $\lambda \geq 0$. The heuristic Lagrangian approach searches now for the stationary policy $R(\lambda_k)$ corresponding to the first interval $[\lambda_k, \lambda_{k+1})$ for which

$$\beta(\lambda_k) \leq \beta.$$

The gaps in the constraint levels of the stationary policies $R(\lambda)$ explain partly why the average cost of the stationary policy obtained by the Lagrangian approach will in general be larger than the average cost of the stationary randomized policy resulting from the linear programming formulation. Also, it should be pointed out that there is no guarantee that the stationary policy $R(\lambda_k)$ obtained by the Lagrangian approach is the best policy among all stationary policies satisfying the probabilistic constraint, although in most practical situations this may be expected to be the case. In spite of the possible pitfalls of the Lagrangian approach, this approach may be quite useful in practical applications having a specific structure (cf. Example 3.5.1 in Section 3.5).

3.4   VALUE-ITERATION ALGORITHM

The policy-iteration algorithm and the linear programming formulation both require that in each iteration a system of linear equations of the same size as the state space is solved. In general, this will be computationally burdensome for a large state space and makes these algorithms computationally unattractive for large-scale Markov decision problems. In this section we discuss an alternative algorithm which avoids solving systems of linear equations but uses instead the recursive solution approach from dynamic programming. This method is the *value-iteration algorithm* which computes recursively a sequence of value functions approximating the minimal average cost per unit time. The value functions provide lower and upper bounds on the minimal average cost and under a certain aperiodicity condition these bounds approximate arbitrarily closely the minimal cost rate. The value-iteration algorithm endowed with these lower and upper bounds is in general the best computational method for solving large-scale Markov decision problems. This is even true in spite of the fact that the value-iteration algorithm does not has the robustness of the policy-iteration algorithm. It turns out that in value iteration the number of iterations is typically problem dependent. Another important advantage of value iteration is that it is usually easy to write an own code for specific applications. By exploiting the structure of the particular application one usually avoids computer memory problems that may be encountered when using policy iteration. Value iteration is not only a powerful method for controlled Markov chains, but it is also a useful tool to compute bounds on performance measures in a single Markov chain.

The value-iteration algorithm computes recursively for $n = 1, 2, \ldots$ the value function $V_n(i)$ from

$$V_n(i) = \min_{a \in A(i)} \left\{ c_i(a) + \sum_{j \in I} p_{ij}(a) V_{n-1}(j) \right\}, \quad i \in I, \qquad (3.4.1)$$

starting with an arbitrarily chosen function $V_0(i)$, $i \in I$. The quantity $V_n(i)$ can be interpreted as the minimal total expected costs with n periods left to the time horizon when the current state is i and a terminal cost of $V_0(j)$ is incurred when the system ends up at state j (for a proof see Denardo (1982) and Derman (1970)). This interpretation suggests that for large n the one-step difference $V_n(i) - V_{n-1}(i)$

will come very close to the minimal average cost per unit time, while the stationary policy whose actions minimize the right side of (3.4.1) for all i will be very close in costs to the minimal average costs. However, these matters appear to be rather subtle for the average cost criterion due to the effect of possible periodicities in the underlying decision processes. To state this precisely, let

$$M_n = \max_{j \in I}\{V_n(j) - V_{n-1}(j)\}, \quad n = 1, 2, \ldots,$$

$$m_n = \min_{j \in I}\{V_n(j) - V_{n-1}(j)\}, \quad n = 1, 2 \ldots . \tag{3.4.2}$$

Theorem 3.4.1 *Suppose that the Weak Unichain Assumption 3.3.1 of Section 3.3 holds. For any $n \geq 1$, let $R(n)$ be a stationary policy whose actions minimize the right side of the value-iteration equation (3.4.1) for all i. Then*

$$m_n \leq g^* \leq g_i(R(n)) \leq M_n, \quad i \in I \text{ and } n \geq 1, \tag{3.4.3}$$

where g^* denotes the minimal average cost per unit time. Moreover, the sequence $\{m_n, \ n \geq 1\}$ is nondecreasing and the sequence $\{M_n, \ n \geq 1\}$ is non-increasing.

Proof   By Theorem 3.2.1, it suffices to verify that

$$c_i(R_i(n)) + \sum_{j \in I} p_{ij}(R_i(n))V_{n-1}(j) \leq V_{n-1}(i) + M_n, \quad i \in I, \tag{3.4.4}$$

in order to get $g_i(R(n)) \leq M_n$ for all i. Similarly, by verifying that

$$c_i(R_i) + \sum_{j \in I} p_{ij}(R_i)V_{n-1}(j) \geq V_{n-1}(i) + m_n, \quad i \in I, \tag{3.4.5}$$

for any stationary policy R, it follows from Theorem 3.2.1 that $g_i(R) \geq m_n$ for all $i \in I$ and so $g^* \geq m_n$. To prove (3.4.4), note that, by the definition of policy $R(n)$,

$$V_n(i) = c_i(R_i(n)) + \sum_{j \in I} p_{ij}(R_i(n))V_{n-1}(j), \quad i \in I. \tag{3.4.6}$$

Writing $V_n(i) = V_{n-1}(i) + V_n(i) - V_{n-1}(i)$ and using that $V_n(i) - V_{n-1}(i) \leq M_n$, the inequality (3.4.4) follows. To prove (3.4.5), note that for any stationary policy R

$$c_i(R_i) + \sum_{j \in I} p_{ij}(R_i)V_{n-1}(j) \geq V_n(i), \quad i \in I. \tag{3.4.7}$$

Replacing $V_n(i)$ by $V_{n-1}(i) + V_n(i) - V_{n-1}(i)$ and using $V_{n-1}(i) - V_n(i) \geq m_n$, we obtain the inequality (3.4.5).

It remains to prove that $M_{k+1} \leq M_k$ and $m_{k+1} \geq m_k$ for any $k \geq 1$. Taking $n = k$ in (3.4.6) and taking $n = k + 1$ and $R = R(k)$ in (3.4.7), it follows that

$$V_{k+1}(i) - V_k(i) \leq \sum_{j \in I} p_{ij}(R_i(k))\{V_k(j) - V_{k-1}(j)\}, \quad i \in I. \tag{3.4.8}$$

Similarly, by taking $n = k + 1$ in (3.4.6) and taking $n = k$ and $R = R(k + 1)$ in (3.2.7), we find

$$V_{k+1}(i) - V_k(i) \geq \sum_{j \in I} p_{ij}(R_i(k + 1))\{V_k(j) - V_{k-1}(j)\}, \quad i \in I. \qquad (3.4.9)$$

From (3.4.8), it follows that $V_{k+1}(i) - V_k(i) \leq M_k$, $i \in I$ and so $M_{k+1} \leq M_k$. Similarly, we obtain from (3.4.9) that $m_{k+1} \geq m_k$.

In formulating the value-iteration algorithm it is no restriction to assume that each one-step cost $c_i(a)$ is positive. Otherwise, add a sufficiently large constant to each $c_i(a)$ and note that by this transformation the average cost of any policy changes by the same amount. Then, by choosing $V_0(i)$ such that $0 \leq V_0(i) \leq \min_a c_i(a)$ for all $i \in I$, we have $V_1(i) \geq V_0(i)$ for all i, implying that each term of the nondecreasing sequence $\{m_n, n \geq 1\}$ is non-negative. Using this we next obtain from Theorem 3.4.1 that

$$\frac{M_n - m_n}{m_n} \leq \varepsilon \text{ implies } 0 \leq \frac{g(R(n)) - g^*}{g^*} \leq \varepsilon.$$

In other words, the average cost of the policy $R(n)$ obtained at the nth iteration cannot deviate more than $100\varepsilon\%$ from the theoretically minimal average cost when $(M_n - m_n)/m_n \leq \varepsilon$. In practical applications one is usually satisfied with a policy whose average cost is sufficiently close to the minimal average cost.

Value-iteration algorithm

Step 0. Choose $V_0(i)$ with $0 \leq V_0(i) \leq \min_a c_i(a)$ for all $i \in I$. Let $n := 1$.
Step 1. Compute the value function $V_n(i)$, $i \in I$, from

$$V_n(i) = \min_{a \in A(i)} \left\{ c_i(a) + \sum_{j \in I} p_{ij}(a)V_{n-1}(j) \right\}$$

and determine $R(n)$ as a stationary policy whose actions minimize the right side of this equation for all $i \in I$.
Step 2. Compute the bounds

$$m_n = \min_{j \in I}\{V_n(j) - V_{n-1}(j)\} \quad \text{and} \quad M_n = \max_{j \in I}\{V_n(j) - V_{n-1}(j)\}.$$

The algorithm is stopped with policy $R(n)$ when $0 \leq M_n - m_n \leq \varepsilon m_n$ where ε is a prespecified tolerance number (for example $\varepsilon = 0.001$). Otherwise go to step 3.
Step 3. $n := n + 1$ and go to step 1.

The value-iteration algorithm has in general not the robustness of the policy-iteration algorithm. The number of iterations required by the value-iteration algorithm is typically problem dependent and will usually increase when the number of states becomes larger. Also, the tolerance number ε in the stopping criterion and the starting vector $V_0(i)$, $i \in I$, will affect the number of iterations required.

The remaining question is whether the algorithm will be stopped after finitely many iterations. In other words, the question is whether the upper and lower bounds M_n and m_n converge to the same limits. In general M_n and m_n need not have the same limits, as the following example demonstrates. Consider the trivial Markov decision problem with the two states 1 and 2 and a single action a_0, where the one-step costs and the one-step transition probabilities are given by $c_1(a_0) = 1$, $c_2(a_0) = 0$, $p_{12}(a_0) = p_{21}(a_0) = 1$ and $p_{11}(a_0) = p_{22}(a_0) = 0$. Hence the system cycles between the states 1 and 2. It is easily verified that $V_{2k}(1) = V_{2k}(2) = k$, $V_{2k-1}(1) = k$ and $V_{2k-1}(2) = k - 1$ for all $k \geq 1$. Hence $m_n = 0$ and $M_n = 1$ for all n, implying that the sequences $\{m_n\}$ and $\{M_n\}$ have different limits. The reason for the oscillating behaviour of $V_n(i) - V_{n-1}(i)$ is the periodicity of the Markov chain describing the state of the system.

The next theorem gives sufficient conditions for the convergence of the value-iteration algorithm. The proof of this theorem is deferred to the Appendix of this Section.

Theorem 3.4.2   *Suppose that the Weak Unichain Assumption 3.3.1 holds and that for each average cost optimal stationary policy the associated Markov chain $\{X_n\}$ is aperiodic. Then there are finite constants $\alpha > 0$ and $0 < \beta < 1$ such that*

$$|M_n - m_n| \leq \alpha\beta^n, \quad n \geq 1.$$

In particular, $\lim_{n\to\infty} M_n = \lim_{n\to\infty} m_n = g^*$.

Data transformation

The periodicity issue can be circumvented by a *perturbation* of the one-step transition probabilities. The perturbation technique is based on the following two observations. First, a recurrent state allowing for a direct transition to itself must be aperiodic. Second, the relative frequencies at which the states of a Markov chain are visited do not change when at each transition epoch the Markov chain is allowed only with a constant probability $\tau > 0$ to make a transition according to the original law of motion and is forced to make a self-transition otherwise. In other words, if the one-step transition probabilities p_{ij} of a Markov chain $\{X_n\}$ are perturbed as $\overline{p}_{ij} = \tau p_{ij}$ for $j \neq i$ and $\overline{p}_{ii} = \tau p_{ii} + 1 - \tau$ for some constant τ with $0 < \tau < 1$, the perturbed Markov chain $\{\overline{X}_n\}$ with one-step transition probabilities $\overline{p}_{ij}$ is aperiodic and has the same equilibrium probabilities as the original Markov chain $\{X_n\}$; see also the proof of Theorem 2.3.3 in Section 2.3. Thus a Markov decision model involving periodicities may be perturbed as follows. Choosing some constant τ with $0 < \tau < 1$, the state space, the action sets, the one-step costs and the one-step transition probabilities of the perturbed Markov decision model are defined by

$$\bar{I} = I,$$

$$\overline{A}(i) = A(i) \qquad\qquad \text{for } i \in \overline{I},$$

$$\overline{c}_i(a) = c_i(a) \qquad\qquad \text{for } i \in \overline{I} \text{ and } a \in \overline{A}(i), \qquad\qquad (3.4.10)$$

$$p_{ij}(a) = \begin{cases} \tau p_{ij}(a) & \text{for } j \neq i, \ i \in \overline{I} \text{ and } a \in \overline{A}(i), \\ \tau p_{ii}(a) + 1 - \tau & \text{for } j = i, \ i \in \overline{I} \text{ and } a \in \overline{A}(i). \end{cases}$$

For each stationary policy, the associated Markov chain $\{\overline{X}_n\}$ in the perturbed model is aperiodic. It is not difficult to verify that for each stationary policy the average cost per unit time in the perturbed model is the same as that in the original model. For the unichain case this is an immediate consequence of the representation (3.1.3) for the average cost and the fact that for each stationary policy the Markov chain $\{\overline{X}_n\}$ has the same equilibrium probabilities as the Markov chain $\{X_n\}$ in the original model. For the multichain case a similar argument can be used to show that the two models are in fact equivalent. Thus the value-iteration algorithm can be applied to the perturbed model in order to solve the original model. In specific problems involving periodicities the 'optimal' value of τ is usually not clear beforehand; empirical investigations indicate that $\tau = \frac{1}{2}$ is usually a satisfactory choice.

Modified value-iteration algorithm with a dynamic relaxation factor

The convergence rate of the value-iteration algorithm may be accelerated by using a relaxation factor, such as in successive overrelaxation for solving a single system of linear equations. Then at the nth iteration a new approximation to the value function $V_n(i)$ is obtained by using both the previous values $V_{n-1}(i)$ and the residuals $V_n(i) - V_{n-1}(i)$. It is possible to select dynamically a relaxation factor and thus avoid the experimental determination of the best value of a fixed relaxation factor. The following modification of the standard value-iteration algorithm can be formulated. The steps 0, 1 and 2 are as before, while step 3 of the standard value-iteration algorithm is modified as follows.

Step 3(a). Determine the states u and v such that

$$V_n(u) - V_{n-1}(u) = m_n \quad\text{and}\quad V_n(v) - V_{n-1}(v) = M_n$$

and compute the relaxation factor

$$\omega = \frac{M_n - m_n}{M_n - m_n + \sum_{j \in I}\{p_{uj}(R_u) - p_{vj}(R_v)\}\{V_n(j) - V_{n-1}(j)\}}. \qquad (3.4.11)$$

Step 3(b). For each $i \in I$, change $V_n(i)$ according to

$$V_n(i) := V_{n-1}(i) + \omega\{V_n(i) - V_{n-1}(i)\}.$$

Step 3(c). $n := n + 1$ and go to step 1.

In the case of a tie when selecting in step 3(a) the state $u(v)$ for which the minimum (maximum) in m_n (M_n) is obtained, the convention is made to choose the minimizing (maximizing) state of the previous iteration when that state is one of the candidates to choose; otherwise choose the first state achieving the minimum (maximum) in m_n (M_n). The ordering of the states may influence the speed of convergence of the modified value-iteration algorithm.

The choice of the dynamic relaxation factor ω is motivated as follows. Suppose that the minimum in the lower bound $m_n = \min_i\{v_n(i) - V_{n-1}(i)\}$ and the maximum in the upper bound $M_n = \max_i\{V_n(i) - V_{n-1}(i)\}$ are achieved by the states u and v. We change the estimate $V_n(i)$ as $V_n(i) = V_{n-1}(i) + \omega\{V_n(i) - V_{n-1}(i)\}$ for all i in order to accomplish at the $(n+1)$th iteration,

$$c_u(R_u) + \sum_{j \in I} p_{uj}(R_u)\overline{V}_n(j) - \overline{V}_n(u) = c_v(R_v) + \sum_{j \in I} p_{vj}(R_v)\overline{V}_n(j) - \overline{V}_n(v),$$

in the implicit hope that the difference between the new upper and lower bounds M_{n+1} and m_{n+1} will decrease more quickly. Using the relation $m_n = V_n(u) - V_{n-1}(u) = c_u(R_u) + \Sigma_j p_{uj}(R_u)V_{n-1}(j) - V_{n-1}(u)$ and the similar relation for M_n, it is a matter of simple algebra to verify from the above condition the expression for ω. We omit the easy proof that $\omega > 0$.

Numerical experiments indicate that using a dynamic relaxation factor in value iteration may greatly enhance the speed of convergence of the algorithm (see also Example 3.5.1 in Section 3.5). The modified value-iteration algorithm is theoretically not guaranteed to converge, but in practice the algorithm will usually work very well. It is important to note that the relaxation factor ω is kept outside the recursion equation (3.4.1) so that the bounds in (3.4.3) are not destroyed. Although the bounds apply, it is no longer true that the sequences $\{m_n\}$ and $\{M_n\}$ are monotonic.

To conclude this section, we apply value iteration to calculate optimal no-claim limits in an insurance problem. The next example shows that periodicities can easily be dealt with when the Markov chains associated with the stationary policies all have the same period. Then no data transformation is needed.

Example 3.4.1 Optimal no-claim limits for vehicle insurance

Consider a vehicle insurance which charges reduced premiums to motorists who do not make claims over one or more years. When an accident occurs the motorist has the option of either making a claim and thereby perhaps losing a reduction in premium, or paying the costs associated with the accident himself. The decision should be based on a no-claim limit which typically depends on the claim history during the current premium year and the time until the next premium payment. We assume a premium structure for which the premium payment is due at the beginning of each year and the payment depends only on the previous payment and the number of claims made in the past year. The premium structure is as shown in Table 3.4.1 where the five possible premiums $\pi(i)$ are increasing in i.

Table 3.4.1   The premium structure

Current premium	Subsequent premium		
	No claim	One claim	Two or more claims
$\pi(1)$	$\pi(2)$	$\pi(1)$	$\pi(1)$
$\pi(2)$	$\pi(3)$	$\pi(1)$	$\pi(1)$
$\pi(3)$	$\pi(4)$	$\pi(1)$	$\pi(1)$
$\pi(4)$	$\pi(5)$	$\pi(2)$	$\pi(1)$
$\pi(5)$	$\pi(5)$	$\pi(3)$	$\pi(1)$

The accidents occur according to a Poisson process at a rate λ per year. The costs associated with any accident have a given probability distribution function F with a positive density f. What are the no-claim limits for which the long-run average cost per year is minimal?

To model this insurance problem as a discrete-time Markov decision process, we assume that the value of λ is such that the probability of more than one accident in a month is negligible. Thus we take a basic unit of one month and take the beginnings of the successive months as the decision epochs. The decisions of whether to claim or not are modelled as follows. Taking decision $a = d$ means that, in the coming month, a claim will be made only when an accident occurs whose damage costs exceed to level d. In view of the premium structure shown in Table 3.4.1, the states of the system are defined as (t, i) with $t = 0, 1, \ldots, 11$ and $i = 1, \ldots, 6$, where

t denotes the number of months until the next premium payment,

$i = 1$ means that the next premium is $\pi(1)$,

$i = 2$ means that the next premium is $\pi(2)$ if no more claims are made until the next premium payment,

$i = 3$ means that the next premium is $\pi(3)$ if no more claims are made until the next premium payment,

$i = 4$ means that the next premium is $\pi(4)$ if no more claims are made until the next premium payment,

$i = 5$ means that the last premium was $\pi(4)$ while the next premium is $\pi(5)$ if no more claims are made until the next premium payment,

$i = 6$ means that the last premium was $\pi(5)$ while the next premium is $\pi(5)$ if no more claims are made until the next premium payment.

Next we specify the one-step expected costs and the one-step transition probabilities. Since the accidents occur according to a Poisson process with an average of λ accidents per year and the probability of more than one accident in a month is assumed to be negligible, we take $\lambda/12$ as the probability of one accident in a month and $1 - \lambda/12$ as the probability of no accident in a month. If an accident with associated costs D occurs in a month and action a is followed in that month, then the accident results in a claim only if $D > a$. This accident involves for the motorist damage costs D if $D \leq a$ and damage costs 0 otherwise. Thus the probability of a claim in a month during which action a is followed equals

$$p(a) = \frac{\lambda}{12}\{1 - F(a)\},$$

while for the motorist the expected damage costs in that month are given by

$$c(a) = \frac{\lambda}{12} \int_0^a sf(s)\,ds.$$

It now follows that for action a in state (t, i) the one-step expected costs are given by $\pi(\min(i, 5)) + c(a)$ for $t = 0$ and by $c(a)$ for $t \neq 0$. Also, for the one-step transition probabilities we distinguish between the states $(0, i)$ and the states (t, i) with $t \neq 0$. In Table 3.4.2 we list the next premium class $\phi(i)$ that results when a claim is made in state (t, i) with $t \neq 0$. If action a is chosen in state (t, i) with $t \neq 0$, then at the beginning of the next month the second state variable equals i with probability $1 - p(a)$ and equals $\phi(i)$ with probability $p(a)$. Noting the effect of the premium payment in state $(0, i)$ and assuming that the action a is chosen in this state, it is readily verified that at the beginning of the next month the second state variable equals $\min(i + 1, 6)$ with probability $1 - p(a)$ and equals $\phi(\min(i + 1, 6))$ with probability $p(a)$. The way in which the first state variable changes each month is obvious. This variable changes from t to $t - 1$ for $t \geq 1$ and from $t = 0$ to $t = 11$.

We have now completely specified the Markov decision model for the insurance problem. The most effective way to compute an average cost optimal policy in this model is to apply the value-iteration algorithm. Define, for $0 \leq t \leq 11$, $1 \leq i \leq 6$ and $n \geq 1$,

$V_{12n+t}(t, i) =$ the minimal expected total cost if the motorist has still an insurance contract for $t + 12n$ months only and the present state is (t, i).

The following recursion relation applies for $n = 1, 2, \ldots$ (verify!):

$$V_{12n+t}(t, i) = \min_a [c(a) + \{1 - p(a)\}V_{12n+t-1}(t - 1, i)$$

$$+ p(a)V_{12n+t-1}(t - 1, \phi(i))] \quad \text{for } 1 \leq t \leq 11, 1 \leq i \leq 6,$$

$$V_{12n}(0, i) = \min_a [\pi(i) + c(a) + \{1 - p(a)\}V_{12(n-1)+11}(11, i + 1)$$

$$+ p(a)V_{12(n-1)+11}(11, \phi(i + 1))] \quad \text{for } 1 \leq i \leq 5,$$

Table 3.4.2 Transition of the premium class

i	1	2	3	4	5	6
$\phi(i)$	1	1	1	1	2	3

$$V_{12n}(0, 6) = \min_a[\pi(5) + c(a) + \{1 - p(a)\}V_{12(n-1)+11}(11, 6)$$

$$+ p(a)V_{12(n-1)+11}(11, \phi(6))],$$

where $V_t(t, i) = 0$ for $0 \le t \le 11$ and $1 \le i \le 6$. We denote by $R_{t,i}(n)$ the action a for which the minimum is assumed in the right side of the recursion relation for $V_{12n+t}(t, i)$. This minimizing action can explicitly be given. To do this, note that the expression between square brackets in the recursion relation for $V_{12n+t}(t, i)$ has the derivative $(\lambda/12)f(a)\{a + V_{12n+t-1}(t-1, i) - V_{12n+t-1}(t-1, \phi(i))\}$ with respect to a. Using $V_{12n+t-1}(t-1, i) \le V_{12n+t-1}(t-1, j)$ for $i \ge j$, it is readily verified that the minimizing action $R_{t,i}(n)$ is given by

$$R_{t,i}(n) = \begin{cases} V_{12n+t-1}(t-1, \phi(i)) - V_{12n+t-1}(t-1, i), & 1 \le t \le 11, \\ & 1 \le i \le 6, \\ V_{12(n-1)+11}(11, \phi(i+1)) - V_{12(n-1)+11}(11, i+1), & t = 0, \\ & 1 \le i \le 5, \\ V_{12(n-1)+11}(11, \phi(6)) - V_{12(n-1)+11}(11, 6), & t = 0, i = 6. \end{cases}$$

The Markov decision model of the insurance problem is indeed periodic, but its special periodicity structure is such that each decision process considered only every twelve months is aperiodic. This observation enables us to define upper and lower bounds that both converge to the minimal average twelve-monthly costs. Defining

$$m_n = \min_{1 \le i \le 6}\{V_{12n}(0, i) - V_{12(n-1)}(0, i)\}, \quad M_n = \max_{1 \le i \le 6}\{V_{12n}(0, i) - V_{12(n-1)}(0, i)\},$$

it can be seen that (3.4.3) holds where the bounds m_n and M_n both converge to g^*, provided we interpret g^* and $g(R(n))$ as the minimal average twelve-monthly costs and the average twelve-monthly costs of policy $R(n)$ (cf. also Su and Deininger 1972). Letting L be the first integer n for which $(M_n - m_n)/m_n \le \varepsilon$, the value-iteration algorithm is stopped with the stationary policy $R(L)$. The minimal average yearly cost g^* is approximated by $(m_L + M_L)/2$. The relative difference percentage of the average yearly cost of policy $R(L)$ and the minimal average yearly cost is at most $100\varepsilon\%$.

As an illustration we give some numerical examples in which

$$\pi(1) = 500, \quad \pi(2) = 375, \quad \pi(3) = 300, \quad \pi(4) = 250,$$

$$\pi(5) = 200, \quad \lambda = 0.5 \quad \text{and} \quad E(D) = 500,$$

where the random variable D represents the costs associated with an accident. The squared coefficient of variation $c_D^2 = \sigma^2(D)/E^2(D)$ is varied as 1 and 4. In order to show that for larger values of c_D^2 the optimal no-claim limits become increasingly sensitive to more than the first two moments of the costs D, we consider for the costs D a lognormal distribution and a gamma distribution each having the same first two moments (cf. also Appendix B). In Table 3.4.3 we give for the various numerical examples the nearly optimal no-claim limits obtained by the value-iteration algorithm starting with the values $V_0(i) = 0$ for all i and having $\varepsilon = 10^{-3}$ as the tolerance number for the stopping criterion. The results in the table reflect the differences in tail behaviour of the various distributions considered. Also, we give in Table 3.4.3 the number L of iterations required by the algorithm and the lower and upper bounds m_L and M_L.

Table 3.4.3 The (nearly) optimal no claim limits

$c_D^2 = 1$			Lognormal: $L = 14$, $m_L = 317.3$, $M_L = 317.6$ $g^* \approx 317.5$								
i \ t	0	1	2	3	4	5	6	7	8	9	10	11
1	190	0	0	0	0	0	0	0	0	0	0	0
2	355	286	276	267	257	248	239	230	221	213	205	197
3	495	499	485	471	457	444	430	417	405	392	380	368
4	418	662	646	630	615	599	584	569	554	539	524	510
5	291	467	464	460	456	452	447	443	438	433	428	423
6	291	255	260	264	269	272	276	279	282	285	287	289

$c_D^2 = 1$			Gamma: $L = 15$, $m_L = 306.5$, $M_L = 306.7$, $g^* \approx 306.6$								
i \ t	0	1	2	3	4	5	6	7	8	9	10	11
1	200	0	0	0	0	0	0	0	0	0	0	0
2	368	289	280	271	262	254	246	238	230	222	215	207
3	509	502	489	476	463	451	438	426	414	402	391	380
4	423	668	653	637	622	608	593	578	564	550	536	522
5	293	470	466	463	459	455	451	446	442	437	433	428
6	293	257	262	266	271	274	278	281	284	287	289	292

$c_D^2 = 4$			Lognormal: $L = 14$, $m_L = 289.1$, $M_L = 289.2$, $g^* \approx 289.1$								
i \ t	0	1	2	3	4	5	6	7	8	9	10	11
1	217	0	0	0	0	0	0	0	0	0	0	0
2	396	300	292	284	276	268	261	253	246	238	231	224
3	533	511	500	489	478	467	457	446	436	426	416	406
4	418	665	652	640	628	615	603	591	579	568	556	544
5	271	448	445	443	441	438	436	433	430	427	424	421
6	271	237	241	245	248	252	255	258	261	264	266	268

$c_D^2 = 4$			Gamma: $L = 14$, $m_L = 259.9$, $M_L = 260.0$, $g^* \approx 259.9$								
i \ t	0	1	2	3	4	5	6	7	8	9	10	11
1	245	0	0	0	0	0	0	0	0	0	0	0
2	427	305	299	293	287	282	276	271	266	260	255	250
3	563	514	505	497	489	481	473	465	457	450	442	434
4	418	665	656	646	636	627	617	608	599	590	581	572
5	263	440	438	436	434	433	431	429	427	424	422	420
6	263	231	234	238	241	244	247	250	253	255	258	260

Appendix. Proof of Theorem 3.4.2

The proof is only given for the special case that the following two assumptions are satisfied:

(i) for each stationary policy R the associated Markov chain $\{X_n\}$ has no two disjoint closed sets;

(ii) $p_{ii}(a) > 0$ for all $i \in I$ and $a \in A(i)$.

Note that assumption (ii) automatically holds when the data transformation (3.4.10) is applied.

We first introduce some notation. Let $R(n)$ be any stationary policy for which the action $R_i(n)$ minimizes the right side of the value-iteration equation (3.4.1) for all $i \in I$. Denote by P_n the stochastic matrix whose (i, j)th element equals $p_{ij}(R_i(n))$ and define the vector V_n by $V_n = (V_n(i),\ i \in I)$. In the last part of the proof of Theorem 3.4.1 it has been shown that

$$V_n - V_{n-1} \le P_{n-1}(V_{n-1} - V_{n-2}) \quad \text{and} \quad V_n - V_{n-1} \ge P_n(V_n - V_{n-1}). \quad (3.4.12)$$

Fix $n \ge 2$. Since $M_n = V_n(i_1) - V_{n-1}(i_1)$ and $m_n = V_n(i_2) - V_{n-1}(i_2)$ for some states i_1 and i_2, we find

$$M_n - m_n \le [P_{n-1}(V_{n-1} - V_{n-2})](i_1) - [P_n(V_{n-1} - V_{n-2})](i_2).$$

Applying repeatedly the inequalities (3.4.12), we find for any $1 \le k < n$

$$M_n - m_n \le [P_{n-1}P_{n-2}\ldots P_{n-k}(V_{n-k} - V_{n-k-1})](i_1)$$
$$- [P_n P_{n-1}\ldots P_{n-k+1}(V_{n-k} - V_{n-k-1})](i_2). \quad (3.4.13)$$

For any stationary policy f denote by $P(f)$ the stochastic matrix whose (i, j)th element is $p_{ij}(f(i))$. Let N denote the number of states of the state space I. We now prove that for any two N-tuples $(f_N, \ldots, f_1)$ and $(g_N, \ldots, g_1)$ of stationary policies and any two states r and s there is a state $j \in I$ such that

$$[P(f_N)\ldots P(f_1)]_{rj} > 0 \quad \text{and} \quad [P(g_N)\ldots P(g_1)]_{sj} > 0. \quad (3.4.14)$$

To prove this, define the sets $S(k)$ and $T(k)$ by

$$S(k) = \{j \in I \,|\, [P(f_k)\ldots P(f_1)]_{rj} > 0\}, \quad k = 1, \ldots, N,$$
$$T(k) = \{j \in I \,|\, [P(g_k)\ldots P(g_1)]_{sj} > 0\}, \quad k = 1, \ldots, N.$$

Since $p_{ii}(a) > 0$ for all $i \in I$ and $a \in A(i)$, we have

$$S(k + 1) \supseteq S(k) \text{ and } T(k + 1) \supseteq T(k) \quad \text{for } k = 1, \ldots, N - 1.$$

Assume now to the contrary that (3.4.14) does not hold. Then $S(N) \cap T(N)$ is empty and so both $S(N) \ne I$ and $T(N) \ne I$. Thus, since the sets $S(k)$ and $T(k)$ are nondecreasing, there are integers v and w with $1 \le v, w < N$ such that $S(v) = S(v + 1)$ and $T(w) = T(w + 1)$. This implies that the set $S(v)$ of states is closed under policy f_{v+1} and the set $T(w)$ of states is closed under policy g_{w+1}. Since $S(N) \cap T(N)$ is empty and the sets $S(k)$ and $T(k)$ are nondecreasing, we must have that $S(v)$ and $T(w)$ are disjoint sets of states. Construct now a stationary policy R with $R(i) = f_{v+1}(i)$ for $i \in S(v)$ and $R(i) = g_{w+1}(i)$ for $i \in T(w)$. Then policy R has the two disjoint closed sets $S(v)$ and $T(w)$. This contradicts assumption (i). Hence the result (3.4.14) must hold. This result implies that there is a number $\rho > 0$ such that

$$\sum_{j \in I} \min \left\{ [P(f_N) \ldots P(f_1)]_{rj}, \ [P(g_N) \ldots P(g_1)]_{sj} \right\} \geq \rho$$

for any two N-tuples $(f_N, \ldots, f_1)$ and $(g_N, \ldots, g_1)$ and for any two states r and s.

The remainder of the proof uses the same ideas as in the proof of Theorem 2.3.2 in Section 2.3. Fix $n > N$ and choose $k = N$ in (3.4.13). This yields

$$M_n - m_n \leq \sum_{j \in I} d_j \left\{ V_{n-N}(j) - V_{n-N-1}(j) \right\},$$

where d_j is a shorthand notation for

$$d_j = [P_{n-1} \ldots P_{n-N}]_{i_1 j} - [P_n \ldots P_{n-N+1}]_{i_2 j}.$$

Write $d^+ = \max(d, 0)$ and $d^- = -\min(d, 0)$. Then $d = d^+ - d^-$ and $d^+, d^- \geq 0$. Thus

$$M_n - m_n \leq \sum_{j \in I} d_j^+ \left\{ V_{n-N}(j) - V_{n-N-1}(j) \right\} - \sum_{j \in I} d_j^- \left\{ V_{n-N}(j) - V_{n-N-1}(j) \right\}$$

$$\leq M_{n-N} \sum_{j \in I} d_j^+ - m_{n-N} \sum_{j \in I} d_j^- = (M_{n-N} - m_{n-N}) \sum_{j \in I} d_j^+,$$

using that $\Sigma_j d_j^+ = \Sigma_j d_j^-$. This identity is a consequence of $\Sigma_j d_j = 0$. Next use the relation $(p - q)^+ = p - \min(p, q)$ to conclude that

$$M_n - m_n \leq (M_{n-N} - m_{n-N}) \left\{ 1 - \sum_{j \in I} \min([P_{n-1} \ldots P_{n-N}]_{i_1 j}, [P_n \ldots P_{n-N+1}]_{i_2 j}) \right\}$$

$$\leq (1 - \rho)(M_{n-N} - m_{n-N}).$$

The sequence $\{M_n - m_n\}$ is nonincreasing; see Theorem 3.4.1. Thus we find that

$$M_n - m_n \leq (1 - \rho)^{[n/N]} (M_0 - m_0) \quad \text{for all } n \geq 1,$$

implying the desired result.

3.5 SEMI-MARKOV DECISION PROCESSES

In the previous sections we have considered a decision model in which the decisions can be made only at fixed epochs $t = 0, 1, \ldots$. However, in many optimization problems the times between consecutive decision epochs are not identical but random. A possible tool for analysing such problems is the semi-Markov decision model. This model concerns a dynamic system which at random points in time is observed and classified into one of a possible number of states. The set of possible states is denoted by I. After observing the state of the system, a decision has to be made, and costs are incurred as a consequence of the decision made. For each state $i \in I$, a set $A(i)$ of possible decisions or actions is available. It is assumed that the state space I and the action sets $A(i)$, $i \in I$, are finite. This controlled dynamic system is called a *semi-Markov decision process* when the following Markovian properties are satisfied. If at a decision epoch the action a is chosen in state i, then the time until, and the state at, the next decision epoch depend only on the present state i and the subsequently chosen action a and are thus independent of the past history of the system. Also, the costs incurred until the next decision epoch depend only on the present state and the action chosen in that state. We note that in specific problems the state occurring at the next transition will often depend on the time until that transition. Also, the costs usually consist of lump costs incurred at discrete points in time and rate costs incurred continuously in time.

The long-run average cost per unit time is taken as the optimality criterion. For this criterion the semi-Markov decision model is in fact determined by the following characteristics:

$p_{ij}(a)$ = the probability that at the next decision epoch the system will be in state j if action a is chosen in the present state i,

$\tau_i(a)$ = the expected time until the next decision epoch if action a is chosen in the present state i,

$c_i(a)$ = the expected costs incurred until the next decision epoch if action a is chosen in the present state i.

In the following it is assumed that $\tau_i(a) > 0$ for all $i \in I$ and $a \in A(i)$. As before, a stationary policy R is a rule which prescribes the same action $R_i \in A(i)$ whenever the system is observed in state i at a decision epoch. Using the finiteness of the state space, it can be shown that under each stationary policy the number of decisions made in a finite time interval is finite with probability 1. We omit the proof of this result. Also, denoting by X_n the state of the system at the nth decision epoch, it follows that under a stationary policy R the embedded stochastic process $\{X_n\}$ is a discrete-time Markov chain with one-step transition probabilities $p_{ij}(R_i)$.

Define the random variable $Z(t)$ by

$$Z(t) = \text{the total costs incurred up to time } t, \quad t \geq 0.$$

Fix now a stationary policy R. Denote by $E_{i,R}$ the expectation operator when the initial state $X_0 = i$ and policy R is used. Then the limit

$$g_i(R) = \lim_{t \to \infty} \frac{1}{t} E_{i,R}[Z(t)]$$

exists for all $i \in I$. This result can be proved using the renewal-reward theorem in Section 1.3. The details are omitted. Just as in the discrete-time decision model, we can give a stronger interpretation for the average cost function $g_i(R)$. If the initial state i is recurrent under policy R, then the long-run *actual* average cost per unit time equals $g_i(R)$ with probability 1. In case the Markov chain $\{X_n\}$ associated with policy R has no two disjoint closed sets, the Markov chain $\{X_n\}$ has a unique equilibrium distribution $\{\pi_j(R), \ j \in I\}$, and $g_i(R) = g(R)$ independently of the initial state $X_0 = i$.

Theorem 3.5.1   *Suppose that the embedded Markov chain $\{X_n\}$ associated with policy R has no two disjoint closed sets. Then*

$$\lim_{t \to \infty} \frac{Z(t)}{t} = g(R) \quad \text{with probability 1}$$

for each initial state $X_0 = i$, where the constant $g(R)$ is given by

$$g(R) = \sum_{j \in I} c_j(R_j)\pi_j(R) \Big/ \sum_{j \in I} \tau_j(r_j)\pi_j(R). \tag{3.5.1}$$

Proof   We give only a sketch of the proof of (3.5.1). The key to the proof of (3.5.1) is that

$$\lim_{t \to \infty} \frac{Z(t)}{t} = \lim_{m \to \infty} \frac{E(\text{costs over the first } m \text{ decision epochs})}{E(\text{time over the first } m \text{ decision epochs})} \tag{3.5.2}$$

with probability 1. To verify this relation, fix a recurrent state r and suppose that $X_0 = r$. Let a cycle be defined as the time elapsed between two consecutive transitions into state r. By the renewal-reward theorem in Section 1.3

$$\lim_{t \to \infty} \frac{Z(t)}{t} = \frac{E(\text{costs incurred during one cycle})}{E(\text{length of one cycle})}$$

with probability 1. By the expected-value version of the renewal-reward theorem,

$$\lim_{m \to \infty} \frac{1}{m} E(\text{costs over the first m decision epochs})$$

$$= \frac{E(\text{costs incurred during one cycle})}{E(\text{number of transitions in one cycle})}$$

and

$$\lim_{m \to \infty} \frac{1}{m} E(\text{time over the first } m \text{ decision epochs})$$

$$= \frac{E(\text{length of one cycle})}{E(\text{number of transitions in one cycle})}$$

Together the above three relations yield (3.5.2). The remainder of the proof is simple. Obviously, we have

$$E(\text{costs over the first } m \text{ decision epochs}) = \sum_{t=0}^{m-1} \sum_{j \in I} c_j(R_j) p_{rj}^{(t)}(R)$$

and

$$E(\text{time over the first } m \text{ decision epochs}) = \sum_{t=0}^{m-1} \sum_{j \in I} \tau_j(R_j) p_{rj}^{(t)}(R).$$

Using these two relations and relation (3.1.5), the desired result now follows easily.

A stationary policy R^* is said to be *average cost optimal* if $g_i(R^*) \leq g_i(R)$ for all $i \in I$ and all stationary policies R. The algorithms for computing an average cost optimal policy in the discrete-time Markov decision model can be generalized to the semi-Markov decision model.

Policy iteration

We first note that the Theorems 3.1.1 and 3.2.1 require only minor modifications for the semi-Markov decision model. These theorems remain valid provided that we replace g by $g\tau_i(R_i)$ in equations (3.1.8) and g by $g\tau_i(\overline{R}_i)$ in the inequality (3.2.1).

The policy-iteration algorithm will be described under the Unichain assumption 3.2.1. This assumption requires that for each stationary policy the embedded Markov chain $\{X_n\}$ has no two disjoint closed sets. The following basic result underlies the algorithm. Suppose that $g(R)$ and $v_i(R)$, $i \in I$, are the average cost and the relative values of a stationary policy R. If a stationary policy $\overline{R}$ is constructed such that, for each state $i \in I$,

$$c_i(\overline{R}_i) - g(R)\tau_i(\overline{R}_i) + \sum_{j \in I} p_{ij}(\overline{R}_i)v_j(R) \leq v_i(R), \qquad (3.5.3)$$

then $g(\overline{R}) \leq g(R)$. Moreover, by the same arguments as used in the proof in the Appendix of Section 3.2, we have that $g(\overline{R}) < g(R)$ if the strict inequality sign holds in (3.5.3) for at least one state i which is recurrent under $\overline{R}$.

Under the Unichain assumption 3.2.1, we can now formulate the following algorithm.

Policy-iteration algorithm

Step 0 (initialization). Choose a stationary policy R.
Step 1 (value-determination step). For the current rule R, compute the average costs $g(R)$ and the relative values $v_i(R)$, $i \in I$, as the unique solution to the linear equations

$$v_i = c_i(R_i) - g\tau_i(R_i) + \sum_{j \in I} p_{ij}(R_i)v_j, \quad i \in I,$$

$$v_s = 0,$$

where s is an arbitrarily chosen state.

Step 2 (policy-improvement step). For each state $i \in I$, determine an action a_i yielding the minimum in

$$\min_{a \in A(i)} \left\{ c_i(a) - g(R)\tau_i(a) + \sum_{j \in I} p_{ij}(a)v_j(R) \right\}.$$

The new stationary policy $\overline{R}$ is obtained by choosing $\overline{R}_i = a_i$ for all $i \in I$ with the convention that $\overline{R}_i$ is chosen equal to the old action R_i when this action minimizes the policy-improvement quantity.

Step 3 (convergence test). If the new policy $\overline{R}$ equals the old policy, the algorithm is stopped with policy R. Otherwise, go to step 1 with R replaced by $\overline{R}$.

In the same way as for the discrete-time Markov decision model, it can be shown that the algorithm converges in a finite number of iterations to an average cost optimal policy. Also, as a consequence of the convergence of the algorithm, there exist numbers g^* and v_i^*, $i \in I$, satisfying

$$v_i^* = \min_{a \in A(i)} \left\{ c_i(a) - g^*\tau_i(a) + \sum_{j \in I} p_{ij}(a)v_j^* \right\}, \quad i \in I. \tag{3.5.4}$$

The constant g^* is uniquely determined as the minimal average cost per unit time. Moreover, each stationary policy' whose actions minimize the right side of (3.5.4) for all $i \in I$ is average cost optimal. The proof of these statements is left as an exercise to the reader.

Value-iteration algorithm

For the semi-Markov decision model the formulation of a value-iteration algorithm is not straightforward. A recursion relation for the minimal expected costs over the first n decision epochs does not take into account the non-identical transition times and thus these costs cannot be related to the minimal average cost per unit time. However, by a data transformation, we can convert the semi-Markov decision model into a discrete-time Markov decision model such that for each stationary policy the average costs per unit time are the same in both models. A value-iteration algorithm for the original semi-Markov decision model is then implied by the value-iteration algorithm for the associated discrete-time Markov decision model. The data transformation to be given below may be considered as an extension of the uniformization technique for continuous-time Markov chains discussed in Section 2.8.

We introduce the following *data transformation*. Choose a number τ with

$$0 < \tau \le \min_{i,a} \tau_i(a).$$

Consider now the discrete-time Markov decision model whose state space, action sets, one-step costs and one-step transition probabilities are given by

$$\overline{I} = I,$$

$$\overline{A}(i) = A(i), \qquad\qquad\qquad i \in \overline{I},$$

$$\overline{c}_i(a) = \frac{c_i(a)}{\tau_i(a)}, \qquad\qquad\qquad i \in \overline{I} \text{ and } a \in A(\overline{i}),$$

$$\overline{p}_{ij}(a) = \begin{cases} \dfrac{\tau}{\tau_i(a)} p_{ij}(a), & j \ne i, i \in \overline{I} \text{ and } a \in \overline{A}(i), \\[2ex] \dfrac{\tau}{\tau_i(a)} p_{ij}(a) + \left[1 - \dfrac{\tau}{\tau_i(a)}\right], & j = i, i \in \overline{I} \text{ and } a \in \overline{A}(i). \end{cases}$$

This discrete-time Markov decision model has the same class of stationary policies as the original semi-Markov decision model. For each stationary policy R there is a one-to-one correspondence between the equilibrium distribution $\{\overline{\pi}_j(R), j \in I\}$ of the Markov chain $\{\overline{X}_n\}$ in the transformed model and the equilibrium distribution $\{\pi_j(R), j \in I\}$ of the Markov chain $\{X_n\}$ in the original model. By direct substitution of the $\overline{p}_{ij}(R_i)$'s in the equilibrium equations (2.3.4) in Section 2.3 it follows that the one-to-one correspondence is given by

$$\overline{\pi}_j(R) = \gamma \tau_j(R_j)\pi_j(R), \quad j \in I,$$

where γ is a normalizing constant. Using this result, it is easily shown that for each stationary policy R the average cost function $\overline{g}_i(R)$ in the transformed model is identical to the average cost function $g_i(R)$ in the original model. The proof is only given for the unichain case. Then, noting that $\gamma = 1/\sum_j \tau_j(R_j)\pi_j(R)$, it follows from the representations (3.1.3) and (3.5.1) that

$$\overline{g}(R) = \sum_{j \in I} \overline{c}_j(R_j)\overline{\pi}_j(R) = \frac{\displaystyle\sum_{j \in I} c_j(R_j)\pi_j(R)}{\displaystyle\sum_{j \in I} \tau_j(R_j)\pi_j(R)} = g(R).$$

Thus the semi-Markov model can be solved by applying the value-iteration algorithm to the transformed discrete-time model. In doing so, it is no restriction to assume in the transformed model that for each stationary policy the associated Markov chain $\{\overline{X}_n\}$ is *aperiodic*. The reason is that by choosing the constant τ less than $\min_{i,a} \tau(i, a)$ we have $\overline{p}_{ii}(a) > 0$ for all i, a and thus the required aperiodicity. Also, in applying the value-iteration algorithm to the transformed model, it is no restriction to assume that each $\overline{c}_i(a) = c_i(a)/\tau_i(a)$ is positive; otherwise, add a sufficiently large positive constant to each $\overline{c}_i(a)$.

An application of the value-iteration algorithm to the transformed discrete-time Markov decision model results in the following algorithm for the original semi-Markov decision model.

Value-iteration algorithm

Step 0. Choose $V_0(i)$ such that $0 \le V_0(i) \le \min_a\{c_i(a)/\tau_i(a)\}$ for all i. Let $n := 1$.

Step 1. Compute the function $V_n(i)$, $i \in I$, from

$$V_n(i) = \min_{a \in A(i)} \left[\frac{c_i(a)}{\tau_i(a)} + \frac{\tau}{\tau_i(a)} \sum_{j \in I} p_{ij}(a)V_{n-1}(j) \right.$$

$$\left. + \left\{ 1 - \frac{\tau}{\tau_i(a)} \right\} V_{n-1}(i) \right], \quad i \in I, \tag{3.5.5}$$

and determine $R(n)$ as a stationary policy whose actions minimize the right side of (3.5.5).

Step 2. Compute the bounds

$$m_n = \min_{j \in I}\{V_n(j) - V_{n-1}(j)\}, \quad M_n = \max_{j \in I}\{V_n(j) - V_{n-1}(j)\}.$$

The algorithm is stopped with policy $R(n)$ when $0 \le (M_n - m_n) \le \varepsilon m_n$, where ε is a prespecified accuracy number. Otherwise, go to step 3.

Step 3. $n := n + 1$ and go to step 1.

Let us assume that the Weak Unichain Assumption 3.3.1 of Section 3.3 is satisfied for the embedded Markov chains $\{X_n\}$ associated with the stationary policies. Then by the aperiodicity property, the algorithm stops after finitely many iterations with a policy $R(n)$ whose average cost function $g_i(R(n))$ satisfies

$$0 \le \frac{g_i(R(n)) - g^*}{g^*} \le \varepsilon, \quad i \in I,$$

where g^* denotes the minimal average cost per unit time. Regarding the choice of τ in the algorithm, it is in general to be recommended taking $\tau = \min_{i,a} \tau_i(a)$ when for this choice the aperiodicity requirement is satisfied; otherwise $\tau = \frac{1}{2} \min_{i,a} \tau_i(a)$ is a reasonable choice. The convergence speed of the above algorithm may be greatly enhanced by using the modified value-iteration algorithm with a dynamic relaxation factor as discussed in Section 3.4.

Linear programming formulation

As the last algorithm for the semi-Markov decision model, we give the linear programming algorithm formulation under the Weak Unichain Assumption 3.3.1 stated in Section 3.3. Using the data transformation for converting the semi-Markov decision model into a discrete-time Markov decision model, and using the change of variable $u_{ia} = x_{ia}/\tau_i(a)$, the reader may easily verify that the linear programming algorithm in Section 3.3 requires only the following modification to the linear program (3.3.1):

$$\text{Minimize} \sum_{i \in I} \sum_{a \in A(i)} c_i(a)u_{ia}$$

subject to

$$\sum_{a \in A(j)} u_{ja} - \sum_{i \in I} \sum_{a \in A(i)} p_{ij}(a)u_{ia} = 0, \quad j \in I,$$

$$\sum_{i \in I} \sum_{a \in A(i)} \tau_i(a)u_{ia} = 1,$$

$$u_{ia} \geq 0.$$

We remark that probabilistic constraints such as the fraction of time that the system is in some subset I_0 of states should not exceed α and the average frequency of taking some action d in some subset I_0 of states should not exceed β can easily be dealt with in the linear programming formulation by adding the constraints $\sum_{i \in I_0, a \in A(i)} \tau_i(a)u_{ia} \leq \alpha$ and $\sum_{i \in I_0} u_{id} \leq \beta$. In the case of probabilistic constraints the average cost optimal policy usually involves randomized decisions (see the discussion in Section 3.3).

We conclude this section with two applications of the semi-Markov decision model. The first application concerns the important problem of optimal sharing of limited sources between competing users in a random environment. In this problem the times between the decision epochs are exponentially distributed. This feature enables us to use the simple trick of introducing *fictitious decision epochs* with a view to getting sparse matrices of one-step transition probabilities. This trick considerably reduces the computational effort of the value-iteration algorithm.

Example 3.5.1 Optimal sharing of memory between processors

In data or computer networks an important problem is the allocation of memory to several types of users. An example of the shared use of memory is the situation in which different types of messages destined for different processors share a common waiting area. Decisions to accept or reject an arriving message are based on the number of waiting messages of each type. The goal is to minimize the average rejection rate of messages or equivalently to maximize the average throughput of messages. Since the different types of messages have different processing times, it may be optimal to reject certain types of messages even when the waiting area is not full.   .

Suppose two processors share a common memory that is able to accommodate a total of M messages. The messages are distinguished by the processor destinations. Messages of the types 1 and 2 destined for the processors 1 and 2 arrive according to independent Poisson processes with respective rates λ_1 and λ_2. When a message arrives a decision to accept or reject that message must be made. A message that is rejected has no further influence on the system. If a message is accepted it stays in the memory until completion of service. The time required to process a message of type i is exponentially distributed with mean $1/\mu_i$. The processor i handles only messages of type i and is able to serve only one message at a time.

The measure of system performance is a weighted sum of the rejection rates of the messages of types 1 and 2, where the respective weights are given by γ_1 and γ_2. Note that for the important case of $\gamma_1 = \gamma_2 = 1$ the minimization of the average rejection rate is equivalent to the maximization of the average throughput.

The sharing problem can be modelled as a semi-Markov decision problem.

Trick of fictitious decision epochs

A straightforward formulation takes the arrival epochs as the only decision epochs. In such a formulation the determination of the one-step transition probabilities is rather complicated but, worse, the vectors $(p_{ij}(a), i \in I)$ of one-step transition probabilities have many nonzero entries. By the nature of the value-iteration algorithm it is computationally burdensome to have many nonzero $p_{ij}(a)$'s. In our specific problem this difficulty can be circumvented by including the service completion epochs as fictitious decision epochs in addition to the real decision epochs, being the arrival epochs of messages. The fictitious decision at the service completion epochs is to leave the system unchanged. Note that the inclusion of these fictitious decision epochs does not change the Markovian nature of the decision processes, since the times between state transitions are exponentially distributed and thus have the memoryless property. It will appear that the inclusion of fictitious decision epochs simplifies not only the formulation of the value-iteration algorithm, but, more importantly, reduces as well the computational effort as compared with a straightforward formulation. The inclusion of the service completion epochs as fictitious decision epochs has as consequence that the state space must be enlarged. We take as state space

$$I = \{(i_1, i_2, k) | i_1, i_2 = 0, 1, \ldots, M; \quad i_1 + i_2 \leq M; \quad k = 0, 1, 2\}.$$

State (i_1, i_2, k) with $k = 1$ or 2 corresponds to the situation in which a message of type k arrives and finds i_1 messages of type 1 and i_2 messages of type 2 being present in the common waiting area. The auxiliary state $(i_1, i_2, 0)$ corresponds to the situation in which the service of a message is completed and i_1 messages of type 1 and i_2 messages of type 2 are left behind in the waiting area. For the states (i_1, i_2, k) with $k = 1$ or 2 the possible actions are denoted by

$$a = \begin{cases} 0, & \text{reject the arriving message,} \\ 1, & \text{accept the arriving message,} \end{cases}$$

with the stipulation that $a = 0$ is the only feasible decision when $i_1 + i_2 = M$. The fictitious decision of leaving the system alone in the state $s = (i_1, i_2, 0)$ is denoted by $a = 0$. Thanks to the fictitious decision epochs, each transition from a given state is to one of the four neighbouring states. In other words, most of the one-step transition probabilities are zero. Further, the nonzero transition probabilities are easy to specify. To find these probabilities, we use the basic properties (B.2) and (B.3) of the exponential distribution in Appendix B. Put for abbreviation

$$\lambda(i_1, i_2) = \lambda_1 + \lambda_2 + \mu_1 \delta(i_1) + \mu_2 \delta(i_2),$$

where the function $\delta(x)$ is defined by $\delta(0) = 0$ and $\delta(x) = 1$ for $x > 0$. Then, for action $a = 0$ in state $s = (i_1, i_2, k)$,

$$\tau(s, a) = \frac{1}{\lambda(i_1, i_2)}$$

and

$$p_{sv}(a) = \begin{cases} \lambda_1/\lambda(i_1, i_2) & \text{for } v = (i_1, i_2, 1), \\ \lambda_2/\lambda(i_1, i_2) & \text{for } v = (i_1, i_2, 2), \\ \mu_1 \delta(i_1)/\lambda(i_1, i_2) & \text{for } v = (i_1 - 1, i_2, 0), \\ \mu_2 \delta(i_2)/\lambda(i_1, i_2) & \text{for } v = (i_1, i_2 - 1, 0). \end{cases}$$

For action $a = 1$ in state $(i_1, i_2, 1)$, we have

$$\tau(s, a) = \frac{1}{\lambda(i_1 + 1, i_2)}$$

and

$$p_{sv}(a) = \begin{cases} \lambda_1/\lambda(i_1 + 1, i_2) & \text{for } v_1 = (i_1 + 1, i_2, 1), \\ \lambda_2/\lambda(i_1 + 1, i_2) & \text{for } v_1 = (i_1 + 1, i_2, 2), \\ \mu_1/\lambda(i_1 + 1, i_2) & \text{for } v_1 = (i_1, i_2, 0), \\ \mu_2 \delta(i_2)/\lambda(i_1 + 1, i_2) & \text{for } v = (i_1 + 1, i_2 - 1, 0). \end{cases}$$

Similarly, for action $a = 1$ in state $(i_1, i_2, 2)$. Obviously

$$c(s, a) = \begin{cases} \gamma_1 & \text{for } s = (i_1, i_2, 1) \text{ and } a = 0, \\ \gamma_2 & \text{for } s = (i_1, i_2, 2) \text{ and } a = 0, \\ 0 & \text{otherwise.} \end{cases}$$

Value-iteration algorithm

Now, having specified the basic elements of the semi-Markov decision model, we are in a position to formulate the value-iteration algorithm for the computation of a (nearly) optimal sharing rule. In the data transformation, we take

$$\tau = \frac{1}{\lambda_1 + \lambda_2 + \mu_1 + \mu_2}.$$

Using the above specifications, the value-iteration scheme becomes quite simple for the specific problem of sharing resources. Note that the expressions for the one-step transition times $\tau_s(a)$ and the one-step transition probabilities $p_{st}(a)$ have a common denominator and so the ratio $p_{st}(a)/\tau_s(a)$ has a very simple form. More importantly, although the above semi-Markov formulation with fictitious decision epochs requires the extra states $(i_1, i_2, 0)$, the number of additions and multiplications per iteration in the value-iteration algorithm is quadratic in the number

M of waiting places rather than of the order M^4 as it would be in the case of a straightforward semi-Markov formulation. Numerical experiments indicate that the respective numbers of iterations for the two value-iteration formulations do not differ greatly so that there is a considerable overall reduction in computational effort when using the formulation with the fictitious decision epochs. This reduction in computational effort becomes larger as M increases.

In specifying the value-iteration scheme (3.5.5), we obviously have to distinguish between the auxiliary states $(i_1, i_2, 0)$ and the other states. For the states $(i_1, i_2, 0)$ we have

$$V_n(i_1, i_2, 0) = \tau\lambda_1 V_{n-1}(i_1, i_2, 1) + \tau\lambda_2 V_{n-1}(i_1, i_2, 2) + \tau\mu_1 V_{n-1}(i_1 - 1, i_2, 0)$$
$$+ \tau\mu_2 V_{n-1}(i_1, i_2 - 1, 0) + \{1 - \tau\lambda(i_1, i_2)\} V_{n-1}(i_1, i_2, 0),$$

with the convention $V_{n-1}(i_1, i_2, 0) = 0$ when i_1 or i_2 is negative. For the states $(i_1, i_2, 1)$ we find

$$V_n(i_1, i_2, 1) = \min[\gamma_1\lambda(i_1, i_2) + \tau\lambda_1 V_{n-1}(i_1, i_2, 1) + \tau\lambda_2 V_{n-1}(i_1, i_2, 2)$$
$$+ \tau\mu_1 V_{n-1}(i_1 - 1, i_2, 0) + \tau\mu_2 V_{n-1}(i_1, i_2 - 1, 0)$$
$$+ \{1 - \tau\lambda(i_1, i_2)\} V_{n-1}(i_1, i_2, 1),$$
$$\tau\lambda_1 V_{n-1}(i_1 + 1, i_2, 1) + \tau\lambda_2 V_{n-1}(i_1 + 1, i_2, 2)$$
$$+ \tau\mu_1 V_{n-1}(i_1, i_2, 0) + \tau\mu_2 V_{n-1}(i_1 + 1, i_2 - 1, 0)$$
$$+ \{1 - \tau\lambda(i_1 + 1, i_2)\} V_{n-1}(i_1, i_2, 1)],$$

provided we put $V_{n-1}(i_1, i_2, 1) = V_{n-1}(i_1, i_2, 2) = \infty$ when $i_1 + i_2 = M + 1$ in order to exclude the unfeasible decision $a = 1$ in the states $(i_1, i_2, 1)$ with $i_1 + i_2 = M$. A similar expression applies to $V_n(i_1, i_2, 2)$. We omit this obvious expression for reasons of space. This completes the specification of the main step of the value-iteration algorithm. The other steps of the algorithm are self-explanatory.

We found by experimentation that the speed of convergence of the above value-iteration algorithm is greatly enhanced by using the relaxation factor (3.4.11). The specification of (3.4.11) follows by noting that for the function $w(s) = V_n(s) - V_{n-1}(s)$,

$$\sum_{t\in I} \overline{p}_{st}(a)w(t) = \tau\lambda_1 w(i_1, i_2, 1) + \tau\lambda_2 w(i_1, i_2, 2) + \tau\mu_1 w(i_1 - 1, i_2, 0)$$

$$+ \tau\mu_2 w(i_1, i_2 - 1, 0) + \{1 - \tau\lambda(i_1, i_2)\} w(i_1, i_2, k)$$

when $s = (i_1, i_2, k)$ and $a = 0$, while

$$\sum_{t\in I} \overline{p}_{st}(a)w(t) = \tau\lambda_1 w(i_1 + 1, i_2, 1) + \tau\lambda_2 w(i_1 + 1, i_2, 2) + \tau\mu_1 w(i_1, i_2, 0)$$

$$+ \tau\mu_2 w(i_1 + 1, i_2 - 1, 0) + \{1 - \tau\lambda(i_1 + 1, i_2)\} w(i_1, i_2, 1)$$

when $s = (i_1, i_2, 1)$ and $a = 1$. The latter equation applies also when $s = (i_1, i_2, 2)$ and $a = 1$ when we interchange the roles of types 1 and 2.

Numerical results

We assume the numerical data

$$\mu_1 = \mu_2 = 1 \quad \text{and} \quad \gamma_1 = \gamma_2 = 1.$$

The arrival rates λ_1 and λ_2 are varied as $(\lambda_1, \lambda_2) = (1.2, 1.0)$, $(1.0, 1.0)$ and $(0.9, 0.7)$, and the number M of waiting places has the two values 10 and 15. We have applied both the conventional value-iteration algorithm and the modified value-iteration algorithm with the dynamic relaxation factor. In all cases the algorithms were initialized with the values $V_0(i) = 0$ for all i. In step 3(a) of the modified value-iteration algorithm the states (i_1, i_2, k) were ordered by letting first the state variable k run from 0 to 2 for i_1 and i_2 fixed, letting next the state variable i_2 run from 0 to $M - i_1$ for i_1 fixed and letting then the state variable i_1 run from 0 to M. We used both tolerance numbers $\varepsilon = 10^{-2}$ and $\varepsilon = 10^{-3}$ in the stopping criterion in order to see the effect on the number of required iterations. We found for the sharing problem that using the modified value-iteration algorithm rather than the conventional value-iteration algorithm reduces the number of iterations on the average by a factor of four. In each numerical example the same nearly optimal sharing rule was found for both $\varepsilon = 10^{-2}$ and $\varepsilon = 10^{-3}$ in either of the algorithms.

The numerical investigations indicate that the optimal sharing rule has the intuitively reasonable property that the acceptance of a message of type 1 [2] in state (i_1, i_2) implies the acceptance of that same message in state $(i_1 - 1, i_2)$ $[(i_1, i_2 - 1)]$. A control rule with this property is conveniently represented by two non-increasing sequences $\{a_k^{(r)}, 0 \le r \le M\}$ of integers for $k = 1, 2$ so that a message of type k finding upon arrival r messages of the other type present is accepted only when less than $a_k^{(r)}$ messages of type k are present and the waiting area is not full. In Table 3.5.1 we display the numerical results. Here $L(\varepsilon)$ denotes the number of iterations required by value iteration when the tolerance number is ε, where the first figures refer to the algorithm with relaxation and the figures within brackets refer to the algorithm without relaxation. The minimal average rejection rate g^* is estimated as $(m_L + M_L)/2$ with $L = L(10^{-3})$.

Extensions

Using the trick of fictitious decision epochs, we can easily extend the value-iteration algorithm to more complex allocation problems. The algorithm needs only a minor adjustment when service completions of messages of type k occur at a general rate of $\mu_k(i_1, i_2)$ whenever i_h messages of type h for $h = 1, 2$ are in the waiting area. The present problem deals with the special case of $\mu_k(i_1, i_2) = \mu_k$ for $i_k \ge 1$ and $\mu_k(i_1, i_2) = 0$ for $i_k = 0$. The value-iteration formulation for the general case follows by an appropriate replacement of μ_k by $\mu_k(i_1, i_2)$. A service completion rate function $\mu_k(i_1, i_2)$ enables us to cover not only the sharing problem with multiple

Table 3.5.1 Numerical results for the sharing problem

	$M = 15$						$M = 10$					
(λ_1, λ_2)	(1.2, 1.0)		(1.0, 1.0)		(0.9, 0.7)		(1.2, 1.0)		(1.0, 1.0)		(0.9, 0.7)	
r	$a_1^{(r)}$	$a_2^{(r)}$	$a_1^{(r)}$	$a_2^{(r)}$	$a_1^{(r)}$	$a_2^{(r)}$	$a_1^{(r)}$	$a_2^{(r)}$	$a_1^{(r)}$	$a_2^{(r)}$	$a_1^{(r)}$	$a_2^{(r)}$
0	11	12	12	12	13	14	7	8	8	8	8	9
1	11	12	12	12	13	13	7	7	7	7	8	8
2	10	11	11	11	12	12	6	7	7	7	7	8
3	10	11	10	10	11	12	6	7	6	6	7	7
4	9	10	10	10	10	11	5	6	6	6	6	6
5	8	10	9	9	10	10	5	5	5	5	5	5
6	8	9	9	9	9	9	4	4	4	4	4	4
7	7	8	8	8	8	8	3	3	3	3	3	3
≥ 8	7	7	7	7	7	7	2	2	2	2	2	2
g^*	0.348		0.204		0.0500		0.436		0.294		0.108	
$L(10^{-2})$	91 (271)		85 (253)		87 (431)		44 (133)		49 (124)		50 (198)	
$L(10^{-3})$	94 (407)		97 (342)		146 (578)		58 (200)		60 (166)		64 (271)	

server groups at each node but also the following related allocation problem. Suppose that a limited number of identical servers are available for different types of competing customers. There are $J \geq 2$ independent customer or user classes and each user class j generates requests for service according to a Poisson process with rate λ_j. The service requirement of a request of type j is exponentially distributed with mean $1/\mu_j$. There are M identical servers who can handle requests of any type with the restriction that a server can handle only one request at a time. No queueing is allowed so that an arriving request gets immediate access to a free server when that request is accepted. The decision to accept or reject a request for service is based on the type of that request and the number of servers occupied by each type of users. The goal is to minimize a weighted sum of the rejection rates of the user types. Problems of this type arise in satellite communication systems, amongst others. The problem of optimal dynamic allocation of servers to different types of competing customers is covered by the above semi-Markov formulation with $\mu_k(i_1, i_2) = \mu_k i_k$ as the service completion rate function. So far we have implicitly assumed that the source populations of users is infinite. However, by making a similar adjustment for the input rates as was done for the service completion rates, we can also handle the situation in which there are only finitely many users of each type where each individual user of type j generates requests for service according to a Poisson process with rate η_j whenever not in service.

In certain applications of Markovian decision processes to inventory and queueing problems, it may happen that the number of states is unbounded. The next example shows that exploiting the structure of the specific application may enable us to cast the problem into a Markovian decision model with a finite state space and thus avoid truncation of the set of original states.

Example 3.5.2 Optimal control of a stochastic service system

A stochastic service system has s identical channels available for providing service, where the number of channels in operation can be controlled by turning channels on or off. For example, the service channels could be checkouts in a supermarket or production machines in a factory. Requests for service are sent to the service facility according to a Poisson process with rate λ. Each arriving request for service is allowed to enter the system and waits in line until an operating channel is provided. The service time of each request is exponentially distributed with mean $1/\mu$. It is assumed that the average arrival rate λ is less than the maximum service rate $s\mu$. A channel that is turned on is only able to handle one request at a time. At any time channels can be turned on or off depending on the number of requests for service in the system. A non-negative switching cost $K(a, b)$ is incurred when adjusting the number of channels on from a to b. For each channel turned on there is an operating cost at a rate of $r > 0$ per unit of time. Also, for each request a holding cost of $h > 0$ is incurred for each unit of time the message is in the system until its service is completed. The objective is to find a rule for controlling the number of channels on such that the long-run average cost per unit time is minimal.

By the lack of memory of the Poisson process and of the exponential distribution, it follows that the state of the system at any time is described by the pair (i, t), where

$i = $ the number of requests for service in the system

$t = $ the number of channels being turned on.

The decision epochs are the epochs at which a new request for service arrives or the service of a request is completed. In this example the number of possible states is unbounded since the state variable i can assume each of the values $0, 1, \ldots$. A brute-force approach would result in a semi-Markov decision formulation in which state the variable i is bounded by a sufficiently large chosen integer U such that the probability of having more than U requests in the system is negligible under any reasonable control rule. This approach would lead to a very large state space when the arrival rate λ is close to the maximum service rate $s\mu$. A more efficient Markovian decision formulation is obtained by restricting the class of control rules rather than truncating the state space. It is intuitively obvious that under each reasonable control rule all of the s channels will be turned on when the number of requests in the system is sufficiently large. In other words, supposing a sufficiently large chosen integer $M \geq c$, it is no real restriction to assume that in the states (i, t) with $i \geq M$ the only feasible action is to turn on all of the s channels. However, this implies that we can restrict the control of the system only to those arrival epochs and service completion epochs at which no more than M requests remain in the system. By doing so, we obtain a semi-Markov decision formulation with as state space

$$I = \{(i, t)|0 \leq i \leq M, \quad 0 \leq t \leq s\},$$

and as action sets

$$A(i, t) = \begin{cases} \{a | a = 0, \ldots, s\}, & 0 \leq i \leq M - 1, \quad 0 \leq t \leq s, \\ \{s\}, & i = M, \quad 0 \leq t \leq s. \end{cases}$$

Here action a in state (i, t) means that the number of channels on is adjusted from t to a. This semi-Markov decision formulation involves the stipulation that, when taking action $a = s$ in state (M, t), the time until the next decision epoch is defined as the time τ_M until the next service completion epoch at which no more than M requests are left behind in the system, given that all of the s channels are always on. Similarly, when taking action $a = s$ in state (M, t), the 'one-step' costs incurred until the next decision epoch are defined as the sum of the switching cost $K(t, s)$ and the holding and operating costs made during the time τ_M. Note that τ_M consists of the time until either an arrival or service completion occurs and, in case an arrival occurs first, the time needed to reduce the number of requests present from $M + 1$ to M. The semi-Markov decision formulation with an embedded state space makes sense only when it is feasible to calculate the one-step expected transition times $\tau_{(M,t)}(s)$ and the one-step expected costs $c_{(M,t)}(s)$. The calculation of these quantities is indeed not difficult when using the fact that service completions occur according to exponentially distributed interoccurrence times with the same means $1/(s\mu)$ as long as all of the s channels are occupied. In other words, whenever M or more requests are in the system, we can equivalently imagine that a single 'superchannel' is servicing messages one at a time at an exponential rate of $s\mu$. This analogy enables us to invoke the relations (1.3.3) and (1.3.4) with $n = 1$ in Section 1.3. Thus we obtain

the expected time needed to reduce the number of requests present from $M + 1$ to M given that all channels are on

$$= \frac{1/(s\mu)}{1 - \lambda/(s\mu)} = \frac{1}{s\mu - \lambda}$$

and

the expected holding and operating costs incurred during the time needed to reduce the number of requests present from $M + 1$ to M given that all channels are on

$$= \frac{hM}{s\mu - \lambda} + \frac{hs\mu}{s\mu - \lambda} \left\{ \frac{1}{s\mu} + \frac{\lambda}{s\mu(s\mu - \lambda)} \right\} + \frac{rs}{s\mu - \lambda}$$

$$= \frac{h(M + 1) + rs}{s\mu - \lambda} + \frac{h\lambda}{(s\mu - \lambda)^2},$$

where the term $hM/(s\mu - \lambda)$ gives the holding costs associated with the M requests being continuously present during the time needed to reduce the number in the system from $M + 1$ to M. Thus we can conclude that

$$\tau_{(M,t)}(s) = \frac{1}{\lambda + s\mu} + \frac{\lambda}{\lambda + s\mu} \left(\frac{1}{s\mu - \lambda} \right)$$

and

$$c_{(M,t)}(s) = K(t,s) + \frac{hM + rs}{\lambda + s\mu} + \frac{\lambda}{\lambda + s\mu}\left\{\frac{h(M+1) + rs}{s\mu - \lambda} + \frac{h\lambda}{(s\mu - \lambda)^2}\right\}.$$

Here we used that $1/(\lambda + s\mu)$ and $\lambda/(\lambda + s\mu)$ give respectively the expected time until the next state transition and the probability that an arrival occurs earlier than a service completion when all of the s channels are busy. We have now done the most delicate part of the specification of the one-step transition times and the one-step costs associated with the semi-Markov decision formulation with the embedded state space I. The other $\tau_{(i,t)}(a)$ and $c_{(i,t)}(a)$ are easily verified to be given by

$$\tau_{(i,t)}(a) = \frac{1}{\lambda + \min(i,a)\mu}, \quad 0 \le i \le M - 1, \quad 0 \le a \le s,$$

and

$$c_{(i,t)}(a) = K(t,a) + \frac{hi + ra}{\lambda + \min(i,a)\mu}, \quad 0 \le i \le M - 1, \quad 0 \le a \le s.$$

It is left to the reader to specify the one-step transition probabilities (verify in particular that $p_{(M,t)(M-1,s)}(s) = s\mu/(\lambda + s\mu)$ and $p_{(M,t)(M,s)}(s) = \lambda/(\lambda + s\mu)$).

From now on it is straightforward to formulate the value-iteration algorithm. Using the data transformation with $\tau = 1/(\lambda + s\mu)$, it is easy to verify that the recurrence relation (3.5.5) becomes:

$$V_n((i,t)) = \min_{0 \le a \le s}\left[\{\lambda + \min(i,a)\mu\}K(t,a) + hi + ra + \frac{\lambda}{\lambda + s\mu}V_{n-1}((i+1,a))\right.$$

$$\left. + \frac{\min(i,a)\mu}{\lambda + s\mu}V_{n-1}((i-1,a)) + \frac{s\mu - \min(i,a)\mu}{\lambda + s\mu}V_{n-1}((i,t))\right]$$

$$\text{for } 0 \le i \le M - 1, \ 0 \le t \le s,$$

$$V_n((M,t)) = \frac{1}{s\mu}(\lambda + s\mu)(s\mu - \lambda)K(t,s) + \frac{h\lambda}{s\mu - \lambda} + hM + rs$$

$$+ \frac{s\mu - \lambda}{\lambda + s\mu}V_{n-1}((M-1,s)) + \frac{\lambda(s\mu - \lambda)}{s\mu(\lambda + s\mu)}V_{n-1}((M,s))$$

$$+ \frac{\lambda}{s\mu}V_{n-1}((M,t)) \qquad \text{for } 0 \le t \le s,$$

with the convention $V_{n-1}((-1,t)) = 0$.

Numerical results

We consider a switching cost function $K(a,b)$ of the form

$$K(a,b) = \kappa|a - b|$$

and assume the numerical data

$$s = 10, \quad \mu = 1, \quad r = 30 \quad \text{and} \quad h = 10.$$

The arrival rate λ is varied as 7 and 8, while the proportionality constant κ for the switching cost has the two values 10 and 25. In each example, we take the bound $M = 20$ for the states (i, t) with $i \geq M$ in which all of the s channels are always turned on. The value-iteration algorithm is started with $V_0((i, t)) = 0$ for all states (i, t) and uses the tolerance number $\varepsilon = 10^{-3}$ for its stopping criterion. Our numerical calculations indicate that for the case of linear switching costs the average cost optimal control rule is characterized by the parameters $s(i)$ and $t(i)$, $i = 0, \ldots, M$, such that in the states (i, t) with $t < s(i)$ the number of channels on is raised up to the level $s(i)$, in the states (i, t) with $s(i) \leq t \leq t(i)$ the number of channels on is left unchanged and in the states (i, t) with $t > t(i)$ the number of channels on is reduced to $t(i)$.

In Table 3.5.2 we give for the various cases the (nearly) optimal values of $s(i)$ and $t(i)$, the number $L(L')$ of iterations required by the value-iteration algorithm with (without) the relaxation factor and the lower and upper bounds m_L and M_L on the minimal average cost per unit time. As in the previous example, the use of a dynamic relaxation factor in the value-iteration algorithm considerably reduces the number of iterations required.

3.6 TAILOR-MADE POLICY-ITERATION ALGORITHMS

The policy-iteration algorithm is a flexible algorithm on account of the freedom in specifying the policy-improvement procedure. In this section we demonstrate how this flexibility can be used to develop tailor-made algorithms for practical applications having a specific structure. In structured applications one typically wishes to consider only policies which have a simple form and thus are easy to implement. The structure of the application considered together with the freedom in specifying the policy-improvement procedure may be exploited to design a tailor-made policy-iteration algorithm which generates a sequence of improving policies within some class of policies having a simple form. Such an algorithm typically has the important advantage of a considerable reduction in the computational effort. In the value-determination step of the algorithm the structure of the simple policy can often be exploited to reduce the system of linear equations for the average cost and the relative values to a considerably smaller system of linear equations on an embedded set of states. An application of this approach will be given to the continuous-review (s, S) inventory system.

Another useful heuristic approach for Markov decision problems is to do only a single iteration in the policy-iteration algorithm starting with a reasonable policy whose relative values are easy to compute. This approach is particularly useful for large-scale Markov decision problems in which the value-determination step is computationally too demanding. The rationale behind this approach is the empirical

Table 3.5.2   Numerical results obtained by value iteration

i	$\lambda = 7, \kappa = 10$		$\lambda = 8, \kappa = 10$		$\lambda = 7, \kappa = 25$		$\lambda = 8, \kappa = 25$	
	$s(i)$	$t(i)$	$s(i)$	$t(i)$	$s(i)$	$t(i)$	$s(i)$	$t(i)$
0	0	3	0	4	0	6	0	6
1	1	4	1	4	1	6	1	7
2	2	4	2	5	2	6	2	7
3	2	5	3	5	3	6	3	7
4	3	6	3	6	3	7	3	8
5	4	6	4	7	4	7	4	8
6	5	7	5	8	5	8	5	8
7	5	8	5	8	5	8	6	9
8	6	9	6	9	6	9	6	9
9	6	9	7	10	6	9	7	10
10	7	10	7	10	7	10	7	10
11	8	10	8	10	7	10	7	10
12	8	10	9	10	7	10	8	10
13	9	10	9	10	8	10	8	10
14	9	10	10	10	8	10	9	10
15	10	10	10	10	8	10	9	10
16	10	10	10	10	9	10	10	10
17	10	10	10	10	9	10	10	10
18	10	10	10	10	9	10	10	10
19	10	10	10	10	10	10	10	10
≥ 20	10	10	10	10	10	10	10	10
$L(L')$	59 (174)		82 (311)		87 (226)		71(250)	
m_L	319.3		367.1		331.5		378.0	
M_L	319.5		367.4		331.8		378.3	

finding that the policy-iteration algorithm typically achieves its largest improvements in costs in the first few iterations. The heuristic approach of applying a single policy-improvement step will be illustrated to the problem of assigning stochastic service requests to one of several different groups of servers.

Before developing a tailor-made algorithm for the continuous-review (s, S) inventory system, we discuss the simple but powerful idea of embedding.

An embedding technique

The general idea of the embedding approach will be described for the semi-Markov decision model which is specified by the basic elements $(I, A(i), p_{ij}(a), \tau_i(a), c_i(a))$. Let R be a given stationary policy for which the embedded Markov chain $\{X_n\}$ describing the state of the system at the decision epochs has no two disjoint closed sets. Let E be an appropriately chosen embedded set of states with $E \neq I$ such that under rule R the set E can be reached from each initial state $i \in I$.

Define now for each $i \in I$ and $j \in E$,

$p_{ij}^E(R)$ = the probability that the first entry state in the set E equals j when the initial state is i and rule R is used,

with the convention that for initial state $i \in E$ the first entry state in set E should be interpreted as the state taken on at the next return to the set E. Also, define, for each $i \in I$,

$\tau_i^E(R)$ = the expected time until the first entry in the set E when the initial state is i and rule R is used,

$c_i^E(R)$ = the expected costs incurred until the first entry in the set E when the initial state is i and rule R is used (excluding the cost incurred at the epoch of the first entry in E).

Let $g(R)$ and $v_i(R)$, $i \in I$, denote the average cost and relative values associated with rule R. It will be shown below that

$$v_i(R) = c_i^E(R) - g(R)\tau_i^E(R) + \sum_{j \in E} p_{ij}^E(R)v_j(R), \quad i \in I. \tag{3.6.1}$$

Thus for rule R the average cost and relative values of the states $i \in E$ can be computed by solving the embedded system of linear equations

$$v_i = c_i^E(R) - g\tau_i^E(R) + \sum_{j \in E} p_{ij}^E(R)v_j, \quad i \in E, \tag{3.6.2}$$

together with a normalization equation $v_s = 0$ for some $s \in E$. This system of linear equations has a unique solution (why?). Once the linear equations (3.6.2) have been solved we can next compute each relative value v_i for $i \notin E$ from (3.6.1) by a single-pass calculation. It will be clear that the above embedding approach may lead to a considerable reduction in computational effort when an appropriate embedded set E can be found which allows for an efficient computation of $p_{ij}^E(R)$, $\tau_i^E(R)$ and $c_i^E(R)$.

The proof of the relation (3.6.1) is not difficult. Choose a state $r \in E$ such that state r can be reached from each initial state $i \in I$ under rule R. Define the functions $T_i(R)$ and $K_i(R)$, $i \in I$, as in (3.1.7). Then, by the semi-Markov version of part (b) of Theorem 3.1.1, we have that the relative values $v_i(R)$, $i \in I$, differ only by some additive constant from the relative values

$$w_i(R) = K_i(R) - g(R)T_i(R), \quad i \in I.$$

Thus it suffices to verify that the particular relative values $w_i(R)$, $i \in I$, satisfy (3.6.1). By conditioning on the first entry state in the set E and using the definitions of the functions $T_i(R)$ and $K_i(R)$, it readily follows that for any $i \in I$,

$$T_i(R) = \tau_i^E(R) + \sum_{\substack{j \in E \\ j \neq r}} p_{ij}^E(R)T_j(R)$$

and

$$K_i(R) = c_i^E(R) + \sum_{\substack{j \in E \\ j \neq r}} p_{ij}^E(R) K_j(R).$$

By subtracting $g(R)$ times the equation for $T_i(R)$ from the equation for $K_i(R)$ and using the definition of $w_i(R)$, we obtain

$$w_i(R) = c_i^E(R) - g(R)\tau_i^E(R) + \sum_{\substack{j \in E \\ j \neq r}} p_{ij}^E(R) w_j(R), \quad i \in I.$$

This proves the desired result (3.6.1) since $w_r(R) = 0$ by definition.

The embedding principle will be used in the next example dealing with a continuous-review (s, S) inventory system with a service level constraint. It is interesting to note that the analysis for this continuous-review system applies with minor modifications equally well to the periodic-review (s, S) inventory system.

Example 3.6.1 A continuous-review (s, S) inventory system

Consider a single-item inventory system in which the demands for the item occur at epochs generated by a Poisson process with rate λ. The demand sizes are independent non-negative random variables having a common discrete probability distribution $\{\phi(j), \ j = 0, 1, \ldots\}$. The demand sizes are independent of the Poisson process. Excess demand is backlogged. The inventory position is continuously reviewed. At any time, a replenishment order of any size can be made. The replenishment lead time is a non-negative constant L. A fixed setup cost of $K > 0$ is incurred for each replenishment order. Also, for each unit kept in stock a holding cost at rate $h > 0$ is incurred. The goal is to minimize the long-run average ordering and holding costs per unit time subject to the service level constraint that in the long run a specified fraction of the demand is satisfied directly from stock on hand. Service level constraints are typically used in practice rather than shortage costs. These costs are usually difficult to specify.

A reasonable control rule is the (s, S) rule with $0 \leq s < S$. Under this rule, the inventory position is ordered up to S if at a demand epoch the inventory position drops to or below s; otherwise no ordering is done. Recall that the inventory position is defined as the net stock plus the total amount on order, see also Section 1.5. Our goal is to develop a tailor-made policy-iteration algorithm for the computation of the best rule within the class of (s, S) rules. To handle the service level constraint, we proceed as follows. We first ignore this constraint and assume a fixed penalty cost of $\pi > 0$ for every requested unit that cannot be satisfied directly from current inventory. Then, by a tailor-made policy-iteration algorithm, we compute an average cost optimal (s, S) rule for the cost structure consisting of ordering, holding and penalty costs. Next we compute the fraction of demand that is satisfied directly from stock on hand. If the service level constraint is not satisfied for the (s, S) rule found, we alter the value of the fixed penalty cost π and compute anew an average cost optimal rule for the adjusted cost structure. We continue in this way until we have found the smallest value of π for which the average cost optimal rule satisfies the service level constraint. In practical applications this Lagrangian approach will usually result in the best (s, S) rule for the goal of minimizing the average ordering and holding costs subject to the service level constraint.

The key element of the above Lagrangian approach is a tailor-made policy-iteration algorithm that computes an average cost optimal (s, S) rule for the cost structure consisting of the ordering costs, the holding costs and a fixed penalty cost $\pi > 0$ for each unit shortage that occurs. For the moment we consider this cost structure and thus ignore temporarily the service level constraint. We shall now formulate a semi-Markov decision model for this particular inventory problem. In order to have a finite state space, we assume that the demand size distribution $\{\phi(j)\}$ has a finite support, that is $\Sigma_{j=0}^{M}\phi(j) = 1$ for some finite $M \geq 1$. Also, we assume upper and lower bounds U and V such that the inventory position is never raised above U, while it is always raised above V when it drops at or below V. For ease of presentation we take $V = 0$. These assumptions are not restrictive for practical applications.

The Markov decision model

The semi-Markov decision model of the inventory problem has as state space

$$I = \{i| - M \leq i \leq U\}.$$

State i corresponds to the situation where a demand has just occurred leaving an inventory position of i units. The demand epochs are the decision epochs. The action taken at a decision epoch is identified with the inventory position just after a replenishment (if any). In other words, action a in state i means the ordering of $a - i$ units. It is always required that $a \geq i$ and $0 < a \leq U$. The one-step transition probabilities $p_{ij}(a)$ and the one-step expected transition times $\tau_i(a)$ are given by

$$p_{ij}(a) = \phi(a - j)$$

with the convention that $\phi(k) = 0$ for $k < 0$, and

$$\tau_i(a) = \frac{1}{\lambda}.$$

The specification of the one-step expected costs $c_i(a)$ is rather subtle. Since excess demand is backlogged and the lead time of each replenishment order is a constant L, the inventory on hand at any time $t + L$ is distributed as the inventory position at time t minus the total demand during a period of length L. Hence, letting $\{T_n\}$ be the sequence of decision epochs, the inventory position just after time T_n unambiguously determines the distribution of the stock on hand at time $T_n + L$. We thus adopt the convention that when choosing action a at the decision epoch T_n, one is charged with the immediate ordering cost (if any) and with an amount $\gamma(a)$

representing the expected holding and penalty costs incurred in $(T_n + L, T_{n+1} + L]$. Clearly, this shift in costs leaves the average cost of any policy unchanged. Thus the one-step expected costs are given by

$$c_i(a) = K\delta(a - i) + \gamma(a),$$

where $\delta(j) = 1$ for $j > 0$ and $\delta(j) = 0$ otherwise. Using the lack of memory of the Poisson process generating the demand epochs, it can be shown that $\gamma(i)$ represents also the expected holding and penalty costs incurred between time L and the first decision epoch following time L in the situation that no stock replenishments are possible and the inventory position at time 0 is i (for a proof see Federgruen and Schechner (1983)). In this situation the net stock at time L equals $i - j$ with probability $r_L(j)$, where the compound Poisson distribution $\{r_L(j), \ j \geq 0\}$ represents the distribution of the total demand during a time interval of length L. Thus we obtain

$$\gamma(a) = \frac{1}{\lambda} h \sum_{j=0}^{a} (a - j) r_L(j) + \pi \left[\sum_{j=0}^{a} r_L(j) \sum_{k \geq a-j} (k - a + j)\phi(k) \right.$$

$$\left. + \left\{ \sum_{j=a+1}^{\infty} r_L(j) \right\} E(D) \right], \qquad a = 0, 1, \ldots, \tag{3.6.3}$$

where $E(D) = \Sigma j \phi(j)$ is the average demand size. It follows easily from (3.6.3) that $\gamma(a)$ satisfies the recurrence relation

$$\gamma(a) = \gamma(a - 1) + \frac{1}{\lambda} \left\{ h \sum_{j=0}^{a-1} r_L(j) - \pi\lambda \sum_{j=0}^{a-1} r_L(j) \sum_{k=0}^{a-1-j} \phi(k) \right\}, \qquad a = 1, 2, \ldots,$$

with $\gamma(0) = \pi E(D)$. This recurrence relation is better suited for computational purposes than the explicit expression (3.6.3). Note that a computational scheme for the $r_L(j)$'s is given in Section 1.2.2.

Tailor-made policy iteration

An intuitively appealing control rule is the (s, S) rule. Under general conditions an (s, S) rule is average cost optimal among all conceivable control rules. The theoretical optimality of an (s, S) rule will not be discussed here. Our goal is to develop a tailor-made policy-iteration algorithm that operates on the class of (s, S) rules. We restrict ourselves to (s, S) rules with a non-negative reorder point s.

First we discuss how to compute the average cost and the relative values for a given control rule R of the (s, S) type. The average cost $g(R)$ and the relative values $v_i(R)$, $i \in I$, are the unique solution of the system of linear equations

$$v_i = \begin{cases} \gamma(i) - g\dfrac{1}{\lambda} + \displaystyle\sum_{j=0}^{M} v_{i-j}\phi(j) & \text{for } s < i \le U, \\[4mm] K + \gamma(S) - g\dfrac{1}{\lambda} + \displaystyle\sum_{j=0}^{M} v_{S-j}\phi(j) & \text{for } -M \le i \le s, \end{cases} \tag{3.6.4}$$

augmented by putting one of the relative values equal to zero, say

$$v_S = 0. \tag{3.6.5}$$

By exploiting the structure of the (s, S) rule, these linear equations can be solved very efficiently. We first observe that the equations (3.6.4) imply that

$$v_i = K + v_S, \quad -M \le i \le s. \tag{3.6.6}$$

This special property shows that for the embedded set of states

$$E = \{i \mid -M \le i \le s\}$$

the embedded system of linear equations (3.6.2) is trivial to solve. To be specific, define for $i > s$,

$\tau_i(s, S) = $ the expected time until the first entry into the set E when the initial state is i and rule $R = (s, S)$ is used,

$c_i(s, S) = $ the expected costs incurred until the first entry into the set E when the initial state is i and rule $R = (s, S)$ is used (excluding the costs incurred at the epoch of the first entry into the set E).

Recalling the convention to assign 'immediate' costs $c_k(a) = K\delta(a - k) + \gamma(a)$ when action a is made in state k, it readily follows by conditioning on the size of the first demand that

$$\tau_i(s, S) = \frac{1}{\lambda} + \sum_{j=0}^{i-s-1} \tau_{i-j}(s, S)\phi(j), \quad i = s+1, s+2, \ldots, \tag{3.6.7}$$

$$c_i(s, S) = \gamma(i) + \sum_{j=0}^{i-s-1} c_{i-j}(s, S)\phi(j), \quad i = s+1, s+2, \ldots. \tag{3.6.8}$$

Hence the quantities $\tau_i(s, S)$ and $c_i(s, S)$, $i > s$, can be computed recursively starting with $\tau_{s+1}(s, S) = \lambda^{-1}/\{1 - \phi(0)\}$ and $c_{s+1}(s, S) = \gamma(s + 1)/\{1 - \phi(0)\}$. Using (3.6.5) and (3.6.6), it follows from (3.6.1) that

$$v_i(R) = c_i(s, S) - g(R)\tau_i(s, S) + K, \quad i > s. \tag{3.6.9}$$

In particular, by taking $i = S$, we obtain

$$g(R) = \frac{K + c_S(s, S)}{\tau_S(s, S)}. \tag{3.6.10}$$

Summarizing, the linear equations (3.6.4) allow for a simple recursive solution. First, the quantities $\tau_i(s, S)$ and $c_i(s, S)$ are computed recursively from (3.6.7) and (3.6.8) for $i = s + 1, \ldots, S$. Next we compute the average cost $g(R)$ from (3.6.10) and obtain the relative values $v_i(R)$ for $s + 1 \le i \le S$ from (3.6.9). Finally, each required $v_i(R)$ for $i > S$ can be obtained from (3.6.9) by a single-pass calculation requiring only the evaluations of $\tau_i(s, S)$ and $c_i(s, S)$ from the recurrence relations (3.6.7) and (3.6.8). Note that the upper bound U is actually not needed when computing the average cost and the relative values. The integer U is only used to bound the state space and may be adjusted during the algorithm without difficulty.

We now turn to the problem of designing a policy-improvement procedure that results in a new rule $\overline{R}$ having the desired (s, S) form. Denoting the policy-improvement test quantity $T_R(i; a)$ by

$$T_R(i; a) = c_i(a) - g(R)\tau_i(a) + \sum_{j \in I} p_{ij}(a)v_j(R),$$

we wish to determine a rule $\overline{R}$ having the (s, S) form such that

$$T_R(i; \overline{R}_i) \le v_i(R) \quad \text{for each } i \in I. \tag{3.6.11}$$

To do this, we first specify $T_R(i; a)$ for the relevant combinations of i and a. We always have for each state i that

$$T_R(i; a) = v_i(R) \quad \text{when } a = R_i. \tag{3.6.12}$$

Using (3.6.5) and (3.6.6), we have

$$T_R(i; a) = \gamma(i) - g(R)\frac{1}{\lambda} + K, \quad i \le s \text{ and } a = i. \tag{3.6.13}$$

For $a > i$ we have

$$T_R(i; a) = K + \gamma(a) - g(R)\frac{1}{\lambda} + \sum_{j=0}^{M} v_{a-j}(R)\phi(j),$$

and so, by (3.6.4),

$$T_R(i; a) = K + v_a(R), \quad i \in I \text{ and } a > \min(i, s). \tag{3.6.14}$$

Using (3.6.6) and (3.6.14), the relation (3.6.11) suggests a search for an integer $\overline{S}$ with $s < \overline{S} \le U$ such that

$$K + v_{\overline{S}}(R) < K + v_S(R) \tag{3.6.15}$$

provided that such an integer exists; otherwise let $\overline{S} = S$. The integer $\overline{S}$ will be the new order-up-to level. The determination of the new reorder point $\overline{s}$ proceeds as follows. We first try to find whether the current value of the reorder point s can be decreased. Since $v_i(R) = K$ for $i \le s$, the relation (3.6.13) suggests a search for an integer s' with $0 \le s' < s$ such that

$$\gamma(i) - g(R)\frac{1}{\lambda} + K < K \quad \text{for } s' < i \le s.$$

If such an integer s' exists, the rule $\overline{R} = (s', \overline{S})$ can easily be shown to satisfy (3.6.11) (verify!). In case the reorder point s cannot be decreased, we try to increase the reorder point by searching for an integer s'' with $s < s'' < \overline{S}$ such that

$$K + v_{\overline{S}}(R) < v_i(R) \quad \text{for } s < i \le s''.$$

If such an integer s'' exists, it is readily verified from (3.6.14) and (3.6.15) that the rule $\overline{R} = (s'', \overline{S})$ satisfies (3.6.11).

We are now in a position to state the algorithm for the inventory problem with penalty costs.

Algorithm (no service level constraint)

Step 0. Choose an upper bound U and an (s, S) rule with $0 \le s < S < U$. An upper bound for the optimal value of S can be shown to be the largest integer U with $\gamma(U) \le K + \gamma_0$ where γ_0 is the minimum value of the unimodal function $\gamma(i)$ (cf. Veinott and Wagner 1965).

Step 1. Let $R = (s, S)$ be the current rule. Compute recursively the quantities $\tau_i(s, S)$ and $c_i(s, S)$ for $i = s + 1, \ldots, S$ from (3.6.7) and (3.6.8). Next compute the average cost $g(R)$ and the relative values $v_i(R)$ for $s < i \le S$ from (3.6.10) and (3.6.9). The other relative values $v_i(R)$, $i > S$, are evaluated from (3.6.9) when required in the next step.

Step 2(a). Determine an integer $\overline{S}$ with $s < \overline{S} < U$ such that $v_{\overline{S}}(R) < 0$ provided such an integer exists; otherwise let $\overline{S} = S$.

Step 2(b). Determine s' as the smallest integer with $0 \le s' < s$ such that $\gamma(i) - g(R)/\lambda < 0$ for $s' < i \le s$. If such an integer s' exists, define $\overline{s} = s'$. Otherwise, let s'' be the largest integer with $s < s'' < \overline{S}$ such that $K + v_{\overline{S}}(R) < v_i(R)$ for $s < i \le s''$. If such an integer s'' exists, define $\overline{s} = s''$ and otherwise let $\overline{s} = s$. This results in a new rule $\overline{R} = (\overline{s}, \overline{S})$.

Step 3. If $\overline{R} = R$, stop; otherwise go to step 1 with R replaced by $\overline{R}$.

We state without proof that this algorithm converges after finitely many iterations to an overall average cost optimal (s, S) rule. The number of iterations required by this algorithm is remarkably small (typically less than 10) and each iteration involves only simple calculations. The above algorithm applies to the inventory problem with a fixed penalty cost π per unit shortage. The algorithm yields as a by-product the service level of the (s, S) rules. Therefore note that the contribution of the shortage cost π to the average cost $g(R)$ is equal to π times the average demand that goes short per unit time. Dividing this contribution by π times the average demand $\lambda E(D)$ per unit time, we find for rule R the fraction of demand that is not satisfied directly from stock on hand. Hence the service level of each (s, S) rule can be computed easily. Note that in the recursive computation of the cost functions $\gamma(i)$ and $c_i(s, S)$ the different contributions of the holding and penalty costs can be dealt with separately.

The goal is to find an (s, S) rule which minimizes the average holding and ordering costs subject to the service level constraint that the fraction of demand to be met directly from stock on hand should be at least a specific value β with $0 < \beta < 1$. To search for such a rule, we apply the above algorithm repeatedly for different values of the fixed penalty cost $\pi > 0$, as in ordinary Lagrangian methods. Since the service levels of the generated (s, S) rules are nondecreasing in π, we continue until we have found the *smallest* value of π for which the associated (s, S) rule resulting from the algorithm still satisfies the service level constraint. In our experience, recovering an average cost optimal (s, S) rule when making (small) changes in the value of π requires very few iterations, provided the algorithm is restarted with the rule that was found to be optimal under the previous value of π. We now state the algorithm for the inventory problem with the service level constraint.

Algorithm (service level constraint)

Step 0. Choose a positive number π.
Step 1. For the current value of the fixed penalty cost π, compute by the preceding algorithm the (s_π, S_π) rule which minimizes the average cost per unit time (initialize this algorithm with the (s, S) rule that was found to be optimal for the previous value of π). Let β_π be the fraction of demand satisfied directly from stock on hand under the (s_π, S_π) rule.
Step 2. Adjust the penalty cost π until the smallest value of $\pi \geq 0$ is found for which $\beta_\pi \geq \beta$ (this procedure can be simplified by using the fact that the nondecreasing function β_π is piecewise constant since there are a finite number of intervals of π values such that within each interval the same optimal (s_π, S_π) rule applies). Go to step 1 each time π is adjusted.

This Lagrangian algorithm will usually find an (s, S) rule that minimizes the average holding and replenishment costs within the class of (s, S) rules satisfying the service level constraint. In the Lagrangian approach it may occasionally occur that the best (s, S) rule is not found, but in those rare cases the average holding and replenishment costs of the obtained (s, S) rule were always very close to the minimal average costs.

Numerical discussion

We consider a number of examples in which the demand distribution $\phi(j)$ is given by the negative binomial distribution

$$\phi(j) = \binom{r + j - 1}{j} p^r (1 - p)^j, \quad j = 0, 1, \ldots,$$

with mean $E(D) = r(1 - p)/p$ and variance $\sigma^2(D) = r(1 - p)/p^2$. The parameters

r and p of the negative binomial distribution are uniquely determined by $E(D)$ and $\sigma^2(D)$ provided $\sigma^2(D)/E(D) > 1$. In the examples we take

$$\lambda = 10, \quad L = 1, \quad E(D) = 5, \quad K = 25 \quad \text{and} \quad h = 1.$$

The squared coefficient of variation $c_D^2 = \sigma^2(D)/E^2(D)$ is varied as $1/2$, 2 and 5, while the required service level β has the values 0.90, 0.95 and 0.99. In Table 3.6.1 we give for the various examples the best (s^*, S^*) rules obtained by the Lagrangian approach and the corresponding average costs $g(s^*, S^*)$ and service levels $\beta(s^*, S^*)$. The exact algorithm can also be used to validate the approximate method discussed in Section 1.5. This approximate method is a sequential approach that first computes the difference $S - s$ according to the economic order quantity formula and next computes the reorder point s. The approximate approach is much easier to apply than the exact approach. The (s_a, S_a) rules in Table 3.6.1 refer to the (s, S) rules that are obtained by the approximate approach using the actual demand densities $\eta(x)$ and $\xi(x)$ in the continuous-review version of equation (1.5.17) in Section 1.5. The exact values of $g(s_a, S_a)$ and $\beta(s_a, S_a)$ are given in the table. The relative error percentage in costs is denoted by

$$\text{Error } \% = \frac{100 \times \{g(s_a, S_a) - g(s^*, S^*)\}}{g(s^*, S^*)} \%.$$

The numerical experiments indicate that in many practical inventory applications the approximate (s_a, S_a) rule performs quite well. Also, the table illustrates the important practical conclusion that the optimal value $S^* - s^*$ of the lot size depends mainly on the holding and ordering costs and only to a slight degree on the required service level when the required grade of customer service is sufficiently high. Further, it may be interesting to note that the optimal lot size shows the tendency to increase when the lead time gets larger.

The policy-iteration algorithm shows not only a remarkably fast convergence, but it also achieves the largest improvements in costs in the first few iterations. These findings underly a heuristic approach for Markov decision problems with a

Table 3.6.1   The best (s^*, S^*) rules and the approximate (s_a, S_a) rules

β	c_D^2	(s^*, S^*)	$\beta(s^*, S^*)$	$g(s^*, S^*)$	(s_a, S_a)	$\beta(s_a, S_a)$	$g(s_a, S_a)$	Error%
0.90	$1/2$	(56, 120)	0.902	59.60	(59, 109)	0.899	60.46	1.4
0.95	$1/2$	(67, 128)	0.952	69.41	(69, 119)	0.950	69.91	0.7
0.99	$1/2$	(87, 145)	0.990	88.51	(88, 138)	0.990	88.57	0.1
0.90	2	(67, 135)	0.902	72.53	(73, 123)	0.904	74.58	2.8
0.95	2	(83, 148)	0.950	86.99	(88, 138)	0.952	88.90	2.2
0.99	2	(115, 178)	0.990	117.99	(119, 169)	0.990	119.45	1.2
0.90	5	(87, 159)	0.901	95.20	(99, 149)	0.910	100.35	5.4
0.95	5	(113, 182)	0.950	119.17	(124, 174)	0.956	124.48	4.5
0.99	5	(168, 235)	0.990	172.81	(176, 226)	0.991	175.90	1.8

multi-dimensional state space. In those decision problems it is usually not feasible to solve the value-determination equations. However, a policy-improvement step offers in general no computational difficulties. This suggests a heuristic approach that determines first a good estimate for the relative values and next applies a single policy-improvement step. By the nature of the policy-iteration algorithm one might expect to obtain a good decision rule by the heuristic approach. How to compute the relative values to be used in the policy-improvement step typically depends on the specific application. The heuristic approach is illustrated in the next example.

Example 3.6.2 Dynamic routing of customers to parallel queues

An important queueing model arising in various practical situations is one in which arriving customers (messages or jobs) have to be assigned to one of several different groups of servers. Problems of this type occur in telecommunication networks and flexible manufacturing. The queueing system consists of n multi-server groups working in parallel, where each group has its own queue. There are s_k servers in group k ($k = 1, \ldots, n$). Customers arrive according to a Poisson process with rate λ. Upon arrival each customer has to be assigned to one of the n server groups. The assignment is irrevocable. The customer waits in the assigned queue until a server becomes available. Each server can handle only one customer at a time.

The problem is to find an assignment rule that would minimize the average sojourn time per customer. This problem will be analysed under the assumption that the service times of the customers are independent and exponentially distributed. The mean service time of a customer assigned to queue k is $1/\mu_k$ ($k = 1, \ldots, n$). It is assumed that $\lambda < \sum_{k=1}^{n} s_k \mu_k$. In what follows we consider the minimization of the overall average number of customers in the system. In view of Little's formula, the minimization of the average sojourn time per customer is equivalent to the minimization of the average number of customers in the system.

Bernoulli-splitting rule

An intuitively appealing control rule is the shortest-queue rule. Under this rule each arriving customer is assigned to the shortest queue. Except for the special case of $s_1 = \ldots = s_n$ and $\mu_1 = \ldots = \mu_n$, this rule is in general not optimal. In particular, the shortest-queue rule may perform quite unsatisfactorily in the situation of a few fast servers and many slow servers.

Another simple rule is the Bernoulli-splitting rule. Under this rule each arrival is assigned with a given probability p_k to queue k ($k = 1, \ldots, n$) irrespective of the queue lengths. This assignment rule produces independent Poisson streams at the various queues, where queue k receives a Poisson stream of rate λp_k. The probabilities p_k must satisfy $\Sigma p_k = 1$ and $\lambda p_k < s_k \mu_k$ for $k = 1, \ldots, n$. The latter condition guarantees that no infinitely long queues can build up. Under the Bernoulli-splitting rule it is easy to give an explicit expression for the overall average number of customers in the system. The separate queues act as independent queues of the $M/M/s$-type. This basic queueing model is discussed in Chapter 4. In the $M/M/s$ queue with arrival rate α and s exponential servers each with service rate μ, the long-run average number of customers in the system equals

$$L(s, \alpha, \mu) = \frac{\rho(s\rho)^s}{s!(1-\rho)^2} \left\{ \sum_{k=0}^{s-1} \frac{(s\rho)^k}{k!} + \frac{(s\rho)^s}{s!(s-\rho)} \right\} + s\rho,$$

where $\rho = \lambda/(s\mu)$. Under the Bernoulli-splitting rule the overall average number of customers in the system equals

$$\sum_{k=1}^{n} L(s_k, \lambda p_k, \mu_k). \tag{3.6.18}$$

The best Bernoulli-splitting rule is found by minimizing this expression with respect to $p_1, \ldots, p_n$ subject to the condition $\Sigma_k p_k = 1$ and $0 \leq \lambda p_k < s_k \mu_k$ for $k = 1, \ldots, n$. This minimization problem must be numerically solved by some search procedure (for the case of $n = 2$, bisection can be used to find the minimum of a unimodal function in a single variable).

Policy-improvement step

The problem of assigning the arrivals to one of the server groups is a Markov decision problem with a multi-dimensional state space. The real decision epochs are the arrival epochs of new customers. The state of the system at a decision epoch is an n-dimensional vector $x = (i_1, \ldots, i_n)$, where i_j denotes the number of customers present in queue j. The action $a = k$ in state x means that the new arrival is assigned to queue k. To deal with the optimality criterion of the long-run average number of customers in the system, we impose the following cost structure on the system. A cost at rate j is incurred whenever there are j customers in the system. Then the long-run average cost per unit time gives the long-run overall average number of customers in the system.

Denote by policy $R^{(0)}$ the best Bernoulli-splitting rule and let $p_k^{(0)}, k = 1, \ldots, n$ be the splitting probabilities associated with policy $R^{(0)}$. It has already been shown that the average cost for rule $R^{(0)}$ is easy to compute. Below it will be shown that the relative values are also easy to obtain for rule $R^{(0)}$. Let us first explain how to derive an improved policy from the Bernoulli-splitting rule $R^{(0)}$. This derivation is based on first principles discussed in Section 3.2. The basic idea of the policy-improvement step is to mimimize for each state x the difference $\Delta(x, a, R^{(0)})$ defined by

$\Delta(x, a, R^{(0)}) =$ the difference in total expected costs over an infinitely long period of time by taking first action a and next using policy $R^{(0)}$ rather than policy $R^{(0)}$ from the beginning onwards when starting in state x.

This difference is well-defined since the Markov chain associated with probability $R^{(0)}$ is aperiodic. Under the Bernoulli-splitting rule the n queues act as independent $M/M/s$ queues. Define for each separate queue j,

$D_j(i) = $ the difference in total expected costs in queue j over
an infinitely long period of time by starting with $i + 1$
customers in queue j rather than with i customers.

Then, for each state $x = (i_1, \ldots, i_n)$ and action $a = k$,

$$\Delta(x, a, R^{(0)}) = \sum_{\substack{j=1 \\ j \neq k}}^{n} p_j^{(0)} [-D_j(i_j) + D_k(i_k)] + p_k^{(0)} \times 0$$

$$= -\sum_{j=1}^{n} p_j^{(0)} D_j(i_j) + D_k(i_k).$$

Since the term $\Sigma_j p_j^{(0)} D_j(i_j)$ does not depend on the action $a = k$, the step of minimizing $\Delta(x, k, R^{(0)})$ over k reduces to the computation of

$$\min_{1 \leq k \leq n} \{D_k(i_k)\}.$$

Hence a remarkably simple expression has to be evaluated in the policy-improvement step applied to the Bernoulli-splitting rule. The suboptimal rule resulting from the single application of the policy-improvement step will be called the *separable rule*. The performance of this rule will be discussed below.

It remains to specify the function $D_k(i)$, $i = 0, 1, \ldots$, for each queue k. To do so, consider an $M/M/s$ queue in isolation, where customers arrive according to a Poisson process with rate α and there are s exponential servers each with service rate μ. Each arrival is admitted to the queue. The state of the system describes the number of customers present. A cost at rate j is incurred when there are j customers present. The long-run average cost per unit time is given by

$$g = L(s, \alpha, \mu).$$

The $M/M/s$ queueing process can be seen as a Markov decision process with a single decision in each state. In the Markov decision formulation it is convenient to consider the state of the system both at the arrival epochs and the service completion epochs. By doing so, we can define in accordance with (3.1.7) the relative costs $w(i)$ by

$$w(i) = \begin{cases} K(i) - gT(i), & i = 1, 2, \ldots, \\ 0, & i = 0, \end{cases} \tag{3.6.19}$$

where

$K(i) = $ the total expected cost incurred until the first return to an empty
system when there are currently i customers present, $i \geq 1$

and

$T(i)$ = the expected time until the first return to an empty system when there
are currently i customers present, $i \geq 1$.

Then, by the economic interpretation of the relative values given in Section 3.1,
we have for any $i = 0, 1, \ldots$ that

$w(i + 1) - w(i)$ = the difference in total expected costs over an infinitely
long period of time by starting in state $i + 1$ rather than in
state i.

The desired function $D_k(i)$ for queue k follows by taking

$$D_k(i) = w_k(i + 1) - w_k(i) \quad \text{with } \alpha = \lambda p_k, s = s_k \text{ and } \mu = \mu_k.$$

The basic functions $K(i)$ and $T(i)$ are easy to compute. By conditioning on the
state after the next transition,

$$T_i = \begin{cases} \dfrac{1}{\alpha + \mu} + \dfrac{\alpha}{\alpha + \mu} T_2 & \text{for } i = 1, \\ \dfrac{1}{\alpha + i\mu} + \dfrac{\alpha}{\alpha + i\mu} T_{i+1} + \dfrac{i\mu}{\alpha + i\mu} T_{i-1} & \text{for } 1 < i \leq s, \end{cases} \tag{3.6.20}$$

$$K_i = \begin{cases} \dfrac{1}{\alpha + \mu} + \dfrac{\alpha}{\alpha + \mu} K_2 & \text{for } i = 1, \\ \dfrac{i}{\alpha + i\mu} + \dfrac{\alpha}{\alpha + i\mu} K_{i+1} + \dfrac{i\mu}{\alpha + i\mu} K_{i-1} & \text{for } 1 < i \leq s. \end{cases} \tag{3.6.21}$$

Further, we have

$$T_i = \frac{i - s}{s\mu - \alpha} + T_s \quad \text{for } i > s,$$

$$K_i = \frac{1}{s\mu - \alpha} \left\{ \frac{1}{2}(i - s)(i - s - 1) + i - s + \frac{\alpha(i - s)}{s\mu - \alpha} \right\} + \frac{s(i - s)}{s\mu - \alpha}$$

$$+ K_s \quad \text{for } i > s.$$

To see the latter relations, note that the time to reach an empty system from state
$i > s$ is the sum of the time to reach state s and the time to reach an empty
system from state s. By the memoryless property of the exponential distribution,
the multi-server $M/M/s$ queue operates in fact as a single-server $M/M/1$ queue
with service rate $s\mu$ when s or more customers are present. Next, by applying the
general formulas (1.3.3) and (1.3.4) in Section 1.3, we find the formulas for T_i and
K_i when $i > s$.

Substituting the expressions for T_{s+1} and K_{s+1} in (3.6.20) and (3.6.21) with
$i = s$, we get two systems of linear equations for T_i, $1 \leq i \leq s$ and K_i, $1 \leq i \leq s$.
Once these systems of linear equations have been solved, we can next compute T_i
and K_i for any desired $i > s$.

Summarizing, the heuristic algorithm proceeds as follows.

Heuristic algorithm

Step 1. Compute the best values $p_k^{(0)}$, $k = 1, \ldots, n$, of the Bernoulli-splitting probabilities by minimizing the expression (3.6.18) subject to $\Sigma_{k=1}^{n} p_k = 1$ and $0 \leq \lambda p_k < s_k \mu_k$ for $k = 1, \ldots, n$.

Step 2. For each queue $k = 1, \ldots, n$, solve the systems of linear equations (3.6.20) and (3.6.21) with $\alpha = \lambda p_k^{(0)}$, $s = s_k$ and $\mu = \mu_k$. Next compute for each queue k the function $w_k(i)$ from (3.6.19) with $\alpha = \lambda p_k^{(0)}$, $s = s_k$ and $\mu = \mu_k$.

Step 3. For each state $x = (i_1, \ldots, i_n)$, determine an index k_0 achieving the minimum in

$$\min_{1 \leq k \leq n} \{w_k(i_k + 1) - w_k(i_k)\}.$$

The separable decision rule assigns a new arrival in state $x = (i_1, \ldots, i_n)$ to queue k_0.

Numerical results

Let us consider the numerical data

$$s_1 = 10, \quad s_2 = 1, \quad \mu_1 = 1 \quad \text{and} \quad \mu_2 = 9.$$

The traffic load ρ defined by

$$\rho = \lambda/(s_1 \mu_1 + s_2 \mu_2)$$

is varied as $\rho = 0.2$, 0.5, 0.7, 0.8 and 0.9. Table 3.6.2 gives both the long-run average sojourn time per customer for the heuristic separable rule and the minimal average sojourn time per customer. In addition the table gives the long-run average sojourn time per customer under the shortest-expected-delay (SED) rule. Under this rule an arriving customer is assigned to the queue in which his expected individual delay is smallest. The results in the table show that this intuitively appealing control policy performs very unsatisfactorily for the case of heterogeneous services, particularly when ρ increases. However, the heuristic separable rule shows an excellent performance for all values of ρ.

Table 3.6.2 The average sojourn times

ρ	SED	Separable	Optimal
0.2	0.192	0.191	0.191
0.5	0.647	0.453	0.436
0.7	0.894	0.578	0.575
0.8	1.029	0.674	0.671
0.9	1.481	0.931	0.930

EXERCISES

3.1 Consider a periodic review production/inventory problem where the demands for a single product in the successive weeks are independent random variables with a common discrete probability distribution $\{\phi(j), j = 0, \ldots, N\}$. Any demand in excess of on-hand inventory is lost. At the beginning of each week it has to be decided whether to start a production run or not. The lot size of each production run consists of a fixed number of Q units. The production lead time is one week so that a batch delivery of the entire lot occurs at the beginning of the next week. Due to capacity restrictions on the inventory, a production run is never started when the on-hand inventory is greater than M. The following costs are involved. A fixed setup cost of $K > 0$ is incurred for a new production run started after the production facility has been idle for some time. The holding costs incurred during a week are proportional to the on-hand inventory at the end of that week, where $h > 0$ is the proportionality constant. A fixed lost-sales cost of $\pi_0 > 0$ is incurred for each unit of excess demand. Formulate the problem of finding an average cost optimal production rule as a Markov decision problem.

3.2 Consider Exercise 2.2 in Chapter 2. Verify by policy iteration or linear programming that the control rule given in this exercise is average cost optimal.

3.3 A stamping machine produces six-cornered plates of the following form:

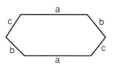

The machine has three pairs of adjustable knives. In the picture these pairs are denoted by a, b and c. Each pair of knives can fall from the correct position during the stamping of a plate. The following five situations can occur: (1) all the three pairs have the correct position, (2) only the pairs b and c have the correct position, (3) only the pair b has the correct position, (4) only the pair c has the correct position and (5) no pair has the correct position. The probabilities q_{ij} that during a stamping a change from situation i to situation j occurs are given by

$$(q_{ij}) = \begin{pmatrix} \frac{3}{4} & \frac{1}{4} & 0 & 0 & 0 \\ 0 & \frac{1}{2} & \frac{1}{4} & \frac{1}{4} & 0 \\ 0 & 0 & \frac{3}{4} & 0 & \frac{1}{4} \\ 0 & 0 & 0 & \frac{1}{2} & \frac{1}{2} \\ 0 & 0 & 0 & 0 & 1 \end{pmatrix}.$$

After each stamping it is possible to adjust the machine such that all pairs of knives have again the correct position. The following costs are involved. The cost of bringing all pairs of knives into the correct position is 10. Each plate produced when j pairs of knives have the wrong position involves an adjustment cost of $4j$. Compute by policy iteration or linear programming a maintenance rule which minimizes the average cost per stamping. (*Answer*: always adjust the machine when not all of the pairs of knives have the correct position; the minimal average cost per stamping is 3.5.)

3.4 An electronic equipment having two identical devices is inspected every day. Redundancy has been built into the system so that the system is still operating if only one device works. The system goes down when both devices are no longer working. The failure rate of a device depends both on its age and condition of the other device. A device working successfully for m days will be failed the next day with probabilities $r_1(m)$ and $r_2(m)$ when, respectively, the other device is currently being overhauled or is working. It is assumed that both $r_1(m)$ and $r_2(m)$ are equal to 1 when m is sufficiently large. A device that is found in the failure state upon inspection has to be overhauled. An overhaul of a failed device takes T_0 days. Also a preventive overhaul of a working device is possible. Such an overhaul takes T_1 days. It is assumed that $1 \leq T_1 \leq T_0$. At each inspection it has to be decided to overhaul one or both of the devices, or let them continue working for the next day. The goal is to minimize the long-run fraction of time the system is down. Formulate a value-iteration algorithm for the computation of an optimal maintenance rule. (*Hint*: define the states (i, j), $(i, -k)$ and $(-h, -k)$. The first state means that both devices are working for i and j days respectively, the second state means that one device is working for i days and the other is being overhauled with a remaining overhaul time of k days, and the third state means that both devices are being overhauled with remaining overhaul times of h and k days.)

3.5 Two furnaces in a steel works are used to produce pig-iron for working-up elsewhere in the factory. Each furnace needs overhauling from time to time because of failure during operation or to prevent such a failure. Assuming an appropriately chosen time unit, an overhaul of a furnace always takes a fixed number of L periods. The overhaul facility is capable of overhauling both furnaces simultaneously. A furnace just overhauled will operate successfully during i periods with probability q_i, $i = 1, \ldots, M$. If a furnace has failed it must be overhauled, otherwise there is an option of either a preventive overhaul or letting the furnace operate for the next period. Since other parts of the steel works are affected when not all furnaces are in action, a loss of revenue of $c(j)$ is incurred for each period during which j furnaces are out of action, $j = 1, 2$. No cost is incurred if both furnaces are working. Formulate a value-iteration algorithm for the computation of an average cost optimal overhauling policy. (This example is based on Stengos and Thomas (1980).)

3.6 An electricity plant has two generators $j = 1$ and 2 for generating electricity. The required amount of electricity fluctuates during the day. The 24 hours of a day are divided into six consecutive periods of 4 hours each. The amount of electricity required in period k is d_k kWh for $k = 1, \ldots, 6$. Also, the generator j has a capacity of generating c_j kWh of electricity per period of 4 hours for $j = 1, 2$. An excess of electricity produced during a period cannot be used for a next period. At the beginning of each period k it has to be decided which generators to use for that period. The following costs are involved. An operating cost of r_j is incurred for each period in which generator j is used. Also, a setup cost of S_j is incurred each time generator j is turned on after having been idle for some time. Develop a policy-iteration algorithm that exploits the fact that the state transitions are deterministic. Solve for the numerical data $d_1 = 20$, $d_2 = 40$, $d_3 = 60$, $d_4 = 90$, $d_5 = 70$, $d_6 = 30$, $c_1 = 40$, $c_2 = 60$, $r_1 = 1000$, $r_2 = 1100$, $S_1 = 500$ and $S_2 = 300$.

3.7 Every week a repairman server travels to customers in five towns on the successive working days of the week. The repairmen visits Amsterdam (town 1) on Monday, Rotterdam (town 2) on Tuesday, Brussels (town 3) on Wednesday, Aachen (town 4) on Thursday and Arnhem (town 5) on Friday. In the different towns it may be necessary to replace a certain crucial piece of an electronic equipment rented by customers. The probability distribution of the number of replacements required at a visit to town j is given by $\{p_j(k), k \geq 0\}$ for $j = 1, \ldots, 5$. The numbers of required replacements on the successive days are independent of each other. The repairman is able to carry at most M spare parts. In case the number of spare parts the repairman carries is not enough to satisfy the demand in a town, another repairman has to be sent the next day to that town to complete the remaining replacements. The cost of such a special mission to town j is K_j. At the end of each day the repairman may decide to send for a replenishment of the spare parts to the town where the repairman is. The cost of sending such a replenishment to town j is a_j. Develop a value-iteration algorithm for the computation of an average cost optimal policy and indicate how to formulate converging lower and upper bounds on the minimal costs. Write a computer program and solve for the numerical data $M = 5$, $K_j = 200$ for all j, $a_1 = 60$, $a_2 = 30$, $a_3 = 50$, $a_4 = 25$, $a_5 = 100$, where the probabilities $p_j(k)$ are given as follows:

k \ j	1	2	3	4	5
0	0.5	0.25	0.375	0.3	0.5
1	0.3	0.5	0.375	0.5	0.25
2	0.2	0.25	0.25	0.2	0.25

(*Answer*: in town i the number of spare parts is raised up to the level 5 when the stock is at or below s_i and otherwise no replenishment is made, where $s_1 = s_2 = s_3 = 1$, $s_4 = 3$ and $s_5 = 0$ with average weekly costs of 39.07.)

3.8 Suppose now in Example 3.1.1 of Section 3.1 that the repair times are stochastic. A preventive repair takes either 0 or 2 days each with probability $1/2$, whereas a repair upon failure takes either 1, 2 or 3 days each with probability $1/3$. Compute by policy iteration or linear programming an average cost optimal policy.

3.9 A cargo liner operates between the five harbours $A_1, \ldots, A_5$. A cargo shipment from harbour A_i to harbour A_j ($j \neq i$) takes a random number τ_{ij} of days (including load and discharge) and yields a random payoff of ξ_{ij}. The shipment times τ_{ij} and the payoffs ξ_{ij} are normally distributed with respective means $\mu(\tau_{ij})$ and $\mu(\xi_{ij})$ and respective standard deviations $\sigma(\tau_{ij})$ and $\sigma(\xi_{ij})$. We assume the following numerical data:

The values $\mu(\tau_{ij})\{\sigma(\tau_{ij})\}$

i \ j	1	2	3	4	5
1	—	$3[1/2]$	$6[1]$	$3[1/2]$	$2[1/2]$
2	$4[1]$	—	$1[1/4]$	$7[1]$	$5[1]$
3	$5[1]$	$1[1/4]$	—	$6[1]$	$8[1]$
4	$3[1/2]$	$8[1]$	$5[1]$	—	$2[1/2]$
5	$2[1/2]$	$5[1]$	$9[1]$	$2[1/2]$	—

The values $\mu(\xi_{ij})\{\sigma(\xi_{ij})\}$

i \ j	1	2	3	4	5
1	—	$8[1]$	$12[2]$	$6[1]$	$6[1]$
2	$20[3]$	—	$2[1/2]$	$14[3]$	$16[2]$
3	$16[3]$	$2[1/2]$	—	$18[3]$	$16[1]$
4	$6[1]$	$10[2]$	$20[2]$	—	$6[1/2]$
5	$8[2]$	$16[3]$	$20[2]$	$8[1]$	—

Compute by policy iteration or linear programming a sailing route for which the long-run average reward per day is maximal (*Answer*: $A_1 \rightarrow A_5 \rightarrow A_4 \rightarrow A_3 \rightarrow A_2 \rightarrow A_1$ with a maximal average reward of 4 per day.)

3.10 Consider Exercise 1.16 of Chapter 1 and assume that the project types $j = 1, \ldots, N$ are (re)numbered according to $E(\xi_j)/E(\tau_j) \geq E(\xi_{j+1})/E(\tau_{j+1})$ for all j. Use the policy-improvement procedure to show that the optimal acceptance set A equals $\{1, \ldots, v\}$, where v is the smallest integer such that

$$\frac{\displaystyle\sum_{j=1}^{v} \lambda_j E(\xi_j)}{1 + \displaystyle\sum_{j=1}^{v} \lambda_j E(\tau_j)} > \frac{E(\xi_{v+1})}{E(\tau_{v+1})},$$

where $E(\xi_{N+1})/E(\tau_{N+1}) = 0$ by convention.

3.11 Consider a production facility that operates only intermittently to manufacture a single product. The production will be stopped if the inventory is sufficiently high while the production will be restarted when the inventory has dropped sufficiently low. Customers asking for this product arrive according to a Poisson process with rate λ, and the demand of each customer equals one unit. The demand which cannot be satisfied directly from stock on hand is lost. Also, a finite-capacity C for the inventory is assumed. In a production run, any desired lot size can be produced. The production time of a lot size of Q units is a random variable T_Q having a probability density $f_Q(t)$. The lot size is added to the inventory at the end of the production run. After the completion of a production run, a new production run is started or the facility is closed down. At each point of time the production can be restarted. The production costs for a lot size of $Q \geq 1$ units consist of a fixed setup cost $K > 0$ and a variable cost c per unit produced. Also, there is a holding cost of $h > 0$ per unit kept in stock per unit time, and a lost-sales cost of $\pi > 0$ is incurred for each lost demand. The goal is to minimize the long-run average cost per unit time. Formulate a semi-Markov decision model.

3.12 Consider a flexible manufacturing facility producing parts, one at a time, for two assembly lines. The time needed to produce one part for assembly line k is exponentially distributed with mean $1/\mu_k$, $k = 1, 2$. Each part produced for line k is put into the buffer for line k. This buffer for line k has space for only N_k parts including the part (if any) in assembly. Each line takes parts one at a time from its buffer as long as the buffer is not empty. At line k, the assembly time for one part is exponentially distributed with mean $1/\lambda_k$, $k = 1, 2$. The production times at the flexible manufacturing facility and the assembly times at the lines are mutually independent. A real-time control for the flexible manufacturing facility is exercised. After each production at this central facility, it must be decided what type of part is to be produced next. The system cannot produce for a line whose buffer is full. Also, the system cannot remain idle if not all the buffers are full. The control is based on the full knowledge of the buffer status at both lines. The system incurs a lost-opportunity cost at a rate of γ_k per unit time when line k is idle. The goal is to control the production at the flexible manufacturing facility in such a way that the long-run average cost per unit time is minimal. Formulate a semi-Markov decision model. (This problem is based on Seidman and Schweitzer (1984.))

3.13 Consider a two-item continuous-review inventory system with the possibility of coordinating replenishment orders for the items. The demand processes for items 1 and 2 are independent Poisson processes with respective demand rates λ_1 and λ_2. At each time a replenishment order for either one or both items can be placed. The lead time of any replenishment order is zero. No backlogging of excess demand is allowed so that a replenishment order for an item must be placed when the stock of that item becomes zero by a demand transaction. There is a maximum capacity C_i for the inventory of item i, $i = 1, 2$. The fixed cost of a replenishment order consisting only of item i is $K_i > 0$ ($i = 1, 2$), while the fixed cost of a replenishment order including both items is $K > 0$ with $K < K_1 + K_2$. For each unit of item i a holding cost of $h_i > 0$ is incurred per unit of time the unit is kept in stock. Formulate a value-iteration algorithm for the computation of an average cost optimal policy (be aware of possible periodicities!). Write a computer program and solve for the numerical data $\lambda_1 = 2$, $\lambda_2 = 3$, $C_1 = C_2 = 10$, $K_1 = K_2 = 10$, $K = 15$ and $h_1 = h_2 = 1$. (*Answer*: the minimal average cost is 1.733.)

3.14 Consider the control problem formulated at the end of Example 2.7.3 in Section 2.7 of Chapter 2. Develop an effective value-iteration algorithm and write a computer program to verify that the optimal control rule is indeed as stated.

3.15 Consider a communication system with two processors for transmitting messages. Each of the two processors has its own waiting area, where the area for processor i consists of N_i waiting places including one for the message (if any) being transmitted. No jockeying is allowed between the queues. Each processor is only able to transmit one message at a time, and the time needed for transmission of a message at processor i is exponentially distributed with mean $1/\mu_i$. The messages arrive at the system according to a renewal process with interarrival-time density $a(x)$. A message finding upon arrival that both waiting areas are full is lost; otherwise the message is assigned to one of the processors having unoccupied waiting places. The goal is to find an assignment rule which minimizes the long-run fraction of messages lost. Develop a value-iteration algorithm for computing an optimal assignment rule. For the special case in which the messages arrive according to a Poisson process, formulate an alternative value-iteration algorithm with sparse transition matrices.

3.16 Consider two independent queues that are attended by a single server. Units requesting service arrive at the ith queue according to a Poisson process with rate λ_i, and their service requirements have a general distribution with probability density $b_i(x)$. The ith queue has a finite waiting room of capacity N_i (excluding any unit in service). An arrival at queue i, finding that all of the waiting places there are occupied, is lost. The server is only able to handle one queue at a time and provides non-preemptive service. Once the server has finished service of a unit at a particular queue, it has to be decided which queue to service next. It takes no time to switch from one queue to the other. The goal is to find a service rule minimizing the long-run fraction of arrivals lost. Develop a value-iteration algorithm for the computation of such a rule. For the special case in which the service requirements have an exponential distribution, formulate an alternative value-iteration algorithm with sparse transition matrices.

3.17 In Example 3.5.2 of Section 3.5, suppose now that the requests for service are generated by a finite number of sources with the same average idle time $1/\eta$. That is, each source generates new requests for service according to a Poisson process with rate η whenever that source has no requests waiting or being served at the service facility. Develop a value-iteration algorithm for the computation of an average cost optimal control rule. Write a computer program and solve for the numerical data $N = 50$, $\eta = 0.16$, $s = 10$, $\mu = 1$, $r = 30$, $h = 10$ and $K = 10$ when $K(a, b) = K|a - b|$. (*Answer*: the minimal average cost is 299.98.)

3.18 Develop an effective value-iteration algorithm for the computation of an optimal acceptance/rejection rule for the control problem in Exercise 2.30 of Chapter 2 when a rejection cost of γ_i is incurred each time a customer of type i is turned away. Write a computer program and solve for the numerical data $c = 50$, $b_1 = 3$, $b_2 = 2$, $\lambda_1 = 10$, $\lambda_2 = 5$, $1/\mu_1 = 1$, $1/\mu_2 = 1$, $\gamma_1 = 3$ and $\gamma_2 = 1$. (*Answer*: the minimal average costs per unit time are 2.694.)

3.19 Consider the control problem stated in Exercise 2.34 of Chapter 2. Suppose now that we wish to minimize the average time spent in the system by a message subject to the probabilistic constraint that the fraction of time during which the slower line 2 is used does not exceed a prespecified value α. Give a linear programming formulation for solving this problem.

3.20 Messages arrive at one of the outgoing communication lines in a message-switching centre according to a Poisson process with rate λ. The message length is exponentially distributed with mean $1/\mu$. An arriving message finding the communication line idle is provided with service immediately; otherwise the message is stored in a buffer until access to the line can be given. The communication line is only able to transmit one message at a time, but has available two possible transmission rates σ_1 and σ_2 with $0 < \sigma_1 < \sigma_2$. It is assumed that the faster transmission rate σ_2 is larger than the offered load λ/μ. At any time the line may switch from one transmission rate to the other. A fixed cost of $K > 0$ is incurred when switching from rate σ_1 to the faster rate σ_2. An operating cost at rate $r_i > 0$ is incurred when the line is transmitting a message using rate σ_i, while an operating cost at rate $r_0 \geq 0$ is incurred when the line is idle. For each message a holding cost of $h > 0$ is incurred for each unit of time the message is in the system until its transmission is completed. Under the two-critical numbers rule (i_1, i_2) the communication line switches from transmission rate σ_1 to σ_2 when the number of messages in the system reaches the level i_1, and switches from rate σ_2 to σ_1 again when the number of messages in the system reduces to the level i_2. Develop a tailor-made policy-iteration algorithm that operates on the class of (i_1, i_2) policies.

3.21 Consider Example 3.6.2 in Section 3.6. Suppose now that there are $n = 2$ service groups each consisting of a single server, where the service time of a customer assigned to server i has the general probability distribution function $B_i(t)$. Derive a heuristic rule for assigning the arrivals to one of the two servers. (*Hint*: approximate the remaining service time of a service in progress at an arrival epoch by the equilibrium excess distribution and use the results in Example 1.3.4 in Section 1.3 to obtain the value function for the Bernouilli splitting rule.)

BIBLIOGRAPHIC NOTES

The policy-iteration method for the discrete-time Markov decision model was developed in Howard (1960) and was extended to the semi-Markov decision model by De Cani (1964), Howard (1964), Jewell (1963) and Schweitzer (1965). A theoretical foundation to Howard's policy-iteration method was given in Blackwell (1962); see also Denardo and Fox (1968) and Veinott (1966). Linear programming formulations for the Markov decision model were first given by De Ghellinck (1960) and Manne (1960) and streamlined later by Denardo and Fox (1968), Derman (1970) and Hordijk and Kallenberg (1979, 1984). The computational usefulness of the value-iteration algorithm was greatly enlarged by Odoni (1969) and Hastings (1971) who introduced lower and upper bounds on the minimal average costs and on the average costs of the policies generated by the algorithm. These authors extended the original value iteration bounds of MacQueen (1966) for the discounted cost case to the average cost case. The first proof of the geometric convergence of the undiscounted value-iteration algorithm was given by White

(1963) under a very strong recurrence condition. The proof in Section 3.4 is along the same lines as the proof given in Van der Wal (1980). General proofs for the geometric convergence of value iteration can be found in Bather (1973) and Schweitzer and Federgruen (1979). The paper of Zijm (1987) deals with the exponential convergence of the value function in continuous-time Markov decision chains. The modified value-iteration algorithm with a dynamic relaxation factor comes from Popyack *et al* (1979). The data-transformation technique converting a semi-Markov decision model into an equivalent discrete-time Markov decision model was introduced in Schweitzer (1971); see also the related work of Lippman (1975) for continuous-time Markov decision chains. The embedding technique discussed in Section 3.6 is adapted from De Leve *et al* (1977), see also Tijms (1980).

A survey of applications of Markov decision models is given in White (1985). Markov decision theory has many applications to replacement and maintenance problems, see e.g. Derman and Lieberman (1967), Golabi *et al* (1982), Kawai (1983), Stengos and Thomas (1980) and Tijms and Van der Duyn Schouten (1985). The insurance application of Example 3.4.1 is taken from Norman and Shearn (1980); see also Kolderman and Volgenant (1985). The resource allocation problem in Example 3.5.1 was motivated by Foschini and Gopinath (1983) and its Markov decision solution was discussed in Tijms and Eikeboom (1986). The treatment of the (s, S) inventory control model in Example 3.6.1 is adapted from Federgruen *et al* (1984). The heuristic solution for the customer assignment problem in Example 3.6.2 comes from Krishnan and Ott (1987). The heuristic approach of attacking a multi-dimensional Markov decision problem through decomposition and a single policy-improvement step goes back to Norman (1970) and has been successfully applied in Krishnan and Ott (1986, 1987) and Wijngaard (1979). Another useful heuristic approach to handle large-scale Markov decision problems is discussed in Schweitzer and Seidman (1985).

REFERENCES

Bather, J. (1973). 'Optimal decision procedures for finite Markov chains', Adv. Appl. Prob., **5**, 521–540.

Bellman, R. (1957). *Dynamic Programming*, Princeton University Press, Princeton.

Blackwell, D. (1962). 'Discrete dynamic programming', *Ann. Math. Statist.*, **33**, 719–726.

De Cani, J.S. (1964). 'A dynamic programming algorithm for embedded Markov chains when the planning horizon is at infinity', *Management Sci.*, **10**, 716–733.

De Ghellinck, G. (1960). 'Les problèmes de decisions sequentielles', *Cahiers Centre Etudes Recherche Opér.*, **2**, 161–179.

De Leve, G., Federgruen, A. and Tijms, H.C. (1977). 'A general Markov decision method I: model and method, II: applications', *Adv. Appl. Prob.*, **9**, 296–315, 316–335.

Denardo, E.V. (1982). Dynamic Programming, Prentice-Hall, Englewood Cliffs, NJ.

Denardo E.V., and Fox, B.L. (1968). 'Multichain Markov renewal programs', *SIAM J. Appl. Math.*, **16**, 468–487.

Derman, C. (1970). *Finite State Markovian Decision Processes*, Academic Press, New York.

Derman, C., and Lieberman, G.J. (1967). 'A Markovian decision model for a joint replacement and stocking problem', *Management Sci.*, **13**, 609–617.

Federgruen , A., Groenevelt, H., and Tijms, H.C. (1984). 'Coordinated replenishments in a multi-item inventory system with compound Poisson demands', *Management Sci.*, **30**, 344–357.

Federgruen, A., and Schechner, Z. (1983). 'Cost formulas for continuous review inventory models with fixed delivery lags', *Operat. Res.*, **31**, 957–965.

Foschini, G.J., and Gopinath, B. (1983). 'Sharing memory optimally', *IEEE Trans. Commun.*, **31**, 352–359.

Golabi, K., Kulkarni, R.B., and Way, C.B. (1982). 'A statewide pavement management system', *Interfaces*, **12**, no. 6, 5–21.

Hastings, N.A.J. (1971). 'Bounds on the gain of a Markov decision process', *Operat. Res.*, **19**, 240–244.

Hordijk, A., and Kallenberg, L.C.M. (1979). 'Linear programming and Markov decision chains', *Management Sci.*, **25**, 352–362.

Hordijk, A., and Kallenberg, L.C.M. (1984). 'Constrained undiscounted stochastic dynamic programming' *Math. Operat. Res.*, **9**, 276–289.

Howard, R.A. (1960). *Dynamic Programming and Markov Processes*, Wiley, New York.

Howard, R.A. (1964). 'Research in semi-Markovian decision structures', *J. Operat. Res. Soc. Japan*, **6**, 163–199.

Jewell, W.S. (1963), 'Markov renewal programming: I and II', *Operat. Res.*, **11**, 938–971.

Kawai, H. (1983). 'An optimal ordering and replacement policy of a Markovian degradation system under complete observation, Part I', *J. Operat. Res. Soc. Japan*, **26**, 279–290.

Kolderman, J., and Volgenant, A. (1985). 'Optimal claiming in an automobile insurance system with bonus-malus structure', *J. Operat. Res. Soc.*, **36**, 239–247.

Krishnan, K.R., and Ott, T.J. (1986). 'State-dependent routing for telephone traffic: theory and results', in: *Proceedings of 25th IEEE Conference on Decision and Control (Athens, Greece)*, pp. 2124–2128.

Krishnan, K.R., and Ott, T.J. (1987). 'Joining the right queue: a Markov decision rule', in: *Proceedings of 26th IEEE Conference on Decision and Control (Los Angeles, CA)*, IEEE, New York, pp. 1863–1868.

Lippman, S.A. (1975). 'Applying a new device in the optimization of exponential queueing systems', *Operat. Res.*, **23**, 687–710.

MacQueen, J. (1966). 'A modified dynamic programming method for Markovian decision problems', *J. Math. Appl. Math.*, **14**, 38–43.

Manne, A. (1960). 'Linear programming and sequential decisions', *Management Sci.*, **6**, 259–267.

Norman, J.M. (1972). *Heuristic Procedures in Dynamic Programming*, Manchester University Press, Manchester.

Norman, J.M. and Shearn, D.C.S. (1980). 'Optimal claiming on vehicle insurance revisited', *J. Operat. Res. Soc.*, **31**, 181–186.

Odoni, A. (1969). 'On finding the maximal gain for Markov decision processes', *Operat. Res.*, **17**, 857–860.

Popyack, J.L., Brown, R.L. and White, III, C.C. (1979). 'Discrete versions of an algorithm due to Varaiya', *IEEE Trans. Automat. Contr.*, **24**, 503–504.

Schweitzer, P.J. (1965). *Perturbation Theory and Markovian Decision Processes*, Ph.D. dissertation, Massachusetts Institute of Technology.

Schweitzer, P.J. (1971). 'Iterative solution of the functional equations of undiscounted Markov renewal programming', *J. Math. Anal. Appl.*, **34**, 495–501.

Schweitzer, P.J., and Federgruen, A. (1979). 'Geometric convergence of value iteration in multichain Markov decision problems', *Adv. Appl. Prob.*, **11**, 188–217.

Schweitzer, P.J., and Seidman, A. (1985). 'Generalized polynomial approximations in Markovian decision processes', *J. Math. Anal. Appl.*, **110**, 568-582.

Seidman, A., and Schweitzer, P.J. (1984). 'Part selection policy for a flexible manufacturing cell feeding several production lines', *AIEE Trans.*, **16**, 355-362.

Stengos, D., and Thomas, L.C. (1980). 'The blast furnaces problem', *European J. Operat. Res.*, **4**, 330-336.

Su, Y., and Deininger, R. (1972). 'Generalization of White's method of successive approximations to periodic Markovian decision processes', *Operat. Res.*, **20**, 318-326.

Tijms, H.C. (1980). 'An algorithm for average cost denumerable state semi-Markov decision problems with applications to controlled production and queueing systems', in: *Recent Developments in Markov Decision Processes*, eds. R. Hartley, L.C. Thomas and D.J. White, pp. 143-179, Academic Press, New York.

Tijms, H.C., and Eikeboom, A.M. (1986). 'A simple technique in Markovian control with applications to resource allocation in communication networks', *Operat. Res. Letters*, **5**, 25-32.

Tijms, H.C., and Van der Duyn Schouten, F.A. (1985). 'A Markov decision algorithm for optimal inspections and revisions in a maintenance system with partial information', *European J. Operat. Res.*, **21**, 245-253.

Van der Wal, J. (1980). 'The method of value oriented successive approximations for the average reward Markov decision process', *OR Spektrum*, **1**, 233-242.

Veinott, A.F. Jr., (1966). 'On finding optimal policies in discrete dynamic programming with no discounting', *Ann. Math. Statist.*, **37**, 1284-1294.

Veinott, A.F. Jr., and Wagner, H.M. (1965). 'Computing optimal (s, S) policies', *Management Sci.*, **11**, 522-555.

Wagner, H.M. (1975). *Principles of Operations Research*, 2nd edn, Prentice-Hall, Englewood Cliffs, NJ.

White, D.J. (1963). 'Dynamic programming, Markov chains and the method of successive approximations', *J. Math. Anal. Appl.*, **6**, 373-376.

White, D.J. (1985). 'Real applications of Markov decision processes', *Interfaces*, **15**, no. 6, 73-78.

Wijngaard, J. (1979). 'Decomposition for dynamic programming in production and inventory control', *Engineering and Process Econom.*, **4**, 385-388.

Zijm, W.H.M. (1987). 'Exponential convergence in undiscounted continuous-time Markov decision chains', *Math. Operat. Res.*, **12**, 700-717.

CHAPTER 4

Algorithmic Analysis
of Queueing Models

4.0   INTRODUCTION

Queueing models have their origin in the study of design problems of automatic telephone exchanges and were first analysed by the queueing pioneer A. K. Erlang in the early 1900s. In planning telephone systems to meet given performance criteria, questions were asked such as 'how many lines are required in order to give a certain grade of service?' or 'what is the probability that a delayed customer has to wait more than a certain time before getting a connection?' Similar questions arise in the design of many other systems: 'how many terminals are needed in a computer system in order to keep the probability of wait of a user below a prespecified value?', 'what will be the effect on the average waiting time of customers when changing the size of a maintenance staff to service leased equipment?', 'how much storage space is needed in buffers at work stations in an assembly line in order to keep the probability of blocking below a specified acceptable level?'.

These design problems and many others concern, in fact, facilities serving a community of users, where both the times at which the users ask for service and the lengths of the times that the requests for service will occupy facilities are stochastic, so that inevitably congestion occurs and queues may build up. In the first stage of design, the system engineer usually needs quick answers to a variety of questions as posed above. Queueing theory constitutes a basic tool for making first-approximation estimates of queue sizes and probabilities of delays. Such a simple tool should in general be preferred to simulation, especially when a large number of different configurations in the design problem is possible.

In this chapter we discuss a number of basic queueing models that have proved to be useful in analysing a wide variety of stochastic service systems. The emphasis will be on algorithms and approximations rather than on mathematical aspects. We feel that there is a need for such a treatment in view of the increased use of queueing models in modern technology. Actually, the application of queueing theory in the performance analysis of computer and communication systems has stimulated much practically oriented research on computational aspects of queueing models. It is to these aspects that the present chapter is addressed. Here considerable attention is paid to robustness results. While it was seen in Chapter 2 that many loss systems (no access of arrivals finding all servers busy) are exactly or nearly insensitive to the distributional form of the service time except for its first moment, it will be demonstrated in this chapter that many delay systems (full access of arrivals) and many delay-loss systems (limited access of arrivals) allow for two-moment approximations. The approximate methods for complex queueing models are usually based on exact results for simpler related models and on asymptotic expansions. The usefulness of asymptotic expansions can be hardly overestimated.

Algorithmic analysis of queueing systems is more than getting numerical answers. The essence of algorithmic probability is to find probabilistic ideas which make the computations transparent and natural. However, once an algorithm has been developed according to these guidelines, one should always verify that it works in practice. The algorithms presented in this chapter have all been thoroughly tested. The cornerstones used in this book for the algorithmic analysis of queues are:

- The embedded Markov chain method.
- The continuous-time Markov chain approach together with the phase method.
- Renewal-theoretic methods.
- Asymptotic expansions.

These methods are of course not the only computational tools for queueing models. Other techniques such as the root-finding method and the matrix-geometric method are not discussed here, but can be found in the books of Chaudry and Templeton (1983) and Neuts (1981).

This chapter is organized as follows. Section 4.1 discusses some basic concepts including phase-type distributions and Little's formula. In Section 4.2 we derive algorithms for computing the state probabilities and the waiting-time probabilities in the single-server queue with Poisson input and general service times. These results are extended in Section 4.3 to the single-server queue with batch Poisson input. The single-server queue with general interarrival times and service times is the subject of Section 4.4. Section 4.5 deals with multi-server queues with Poisson input including both the case of single arrivals and the case of batch arrivals. Tractable exact results are only obtained for the special case of deterministic services and exponential services. For the case of general service times we derive several approximations. These approximations include two-moment approximations that are based on exact results for simpler models and use a linear interpolation with respect to the squared coefficient of variation of the service time. In Section 4.6 the multi-server queue with renewal input is discussed. In particular, attention is paid to the tractable models with exponential services and deterministic services. Queues with finite-source input are dealt with in Section 4.7. In the final Section 4.8 we consider finite-capacity queueing systems with limited access of arrivals. In particular, attention is paid to approximations for the overflow probability. Throughout this chapter many numerical results are given in order to provide insight into the performance of the solution methods.

4.1   BASIC CONCEPTS FOR QUEUEING SYSTEMS

In this section we discuss a number of basic concepts for queueing systems. The discussion is restricted to queueing systems with only one service node. However, the fundamental results below are also useful for networks of queues.

Let us start with giving Kendall's notation for a number of standard queueing models in which the source of population of potential customers is assumed to be infinite. The customers arrive singly and are served singly. A queueing system having waiting room for an unlimited number of customers can be described by a three-part code $a/b/c$. The first symbol a specifies the interarrival-time distribution, the second symbol b specifies the service-time distribution and the third symbol c specifies the number of servers. Some examples of Kendall's shorthand notation are:

1. $M/G/1$ – Poisson (Markovian) input, General service-time distribution, 1 server.

2. $M/D/c$ – Poisson input, Deterministic service times, c servers.

3. $GI/M/c$ – General, Independently distributed interarrival times, exponential (Markovian) service times, c servers.

4. $GI/G/c$ – General, Independently distributed interarrival times, General service-time distribution, c servers.

The above notation can be extended to cover other queueing systems. For example, queueing systems having waiting room only for M customers (including those in service) are often abbreviated by a four-part code $a/b/c/M$. The notation $GI^X/G/c$ is used for infinite-capacity queueing systems in which customers arrive in batches and the batch size is distributed according to the random variable X.

Phase-type distributions

In queueing applications it is often convenient to approximate the interarrival time and/or the service time by distributions that are built out of a finite sum or a finite mixture of *exponentially* distributed components, or a combination of both. These distributions are called *phase-type distributions*. For practical purposes it usually suffices to use finite mixtures of Erlangian distributions with the *same* scale parameters and Coxian-2 distributions. These distributions are discussed in detail in Appendix B. The class of Coxian-2 distributions contains the hyperexponential distribution of order two as special case. The hyperexponential distribution has always a coefficient of variation greater than or equal to 1. This distribution is particularly suited to model irregular interarrival (or service) times which have the feature that most outcomes tend to be small and large outcomes occur only occasionally. The class of mixtures of Erlangian distributions with the same scale parameters is much more versatile than the class of Coxian-2 distributions and allows us to cover any positive value of the coefficient of variation. In particular, a mixture of E_{k-1} and E_k distributions with the same scale parameters is convenient to represent regular interarrival (or service) times which have a coefficient of variation smaller than or equal to 1. The theoretical basis for the use of mixtures of Erlangian distributions with the same scale parameters is provided by Theorem 2.9.1 in Section 2.9 of Chapter 2. This theorem states that each non-negative random variable can be approximated arbitrarily closely by a random sum of exponentially distributed phases with the same means. This explains why finite mixtures of Erlangian distributions with the same scale parameters are widely used for queueing calculations.

Performance measures

It is convenient to use the *GI/G/c/c+N* queue as a vehicle to introduce some basic notation. Thus we assume a multi-server queue with c identical servers and a waiting room of capacity $N(\leq \infty)$ for customers awaiting to be served. A customer who finds upon arrival $c + N$ other customers present is rejected and has no further influence on the system. Otherwise, the arriving customer is admitted to the system and waits in queue until a server becomes available. The customers arrive according to a renewal process. In other words, the interarrival times are positive independent random variables having a common probability distribution function $A(t)$. The service times of the customers are independent random variables with a common probability distribution function $B(x)$ and are also independent of the arrival process. The queue discipline specifying which customer is to be served next is first-come-first-served unless stated otherwise. A server cannot be idle when customers are waiting in queue and a busy server works at unity rate. A customer leaves the system upon service completion. Denote by

$$\lambda = \text{the long-run average arrival rate of customers,}$$

$$E(S) = \text{the mean service time of a customer.}$$

The generic variable S denotes the service time of a customer. Note that $\lambda = 1/E(A)$, where the generic variable A denotes the interarrival time. An important quantity is the offered load which is defined as $\lambda E(S)$. This dimensionless quantity indicates the average amount of work that is offered to the system per unit time. In infinite-capacity queueing systems ($N = \infty$) the offered load should be less than the maximum load the system can handle in order to avoid that infinitely long queues ultimately build up. Therefore the following assumption is made for the infinite-capacity case

Assumption 4.1.1   *For the GI/G/c queue the server utilisation ρ defined by*

$$\rho = \frac{\lambda E(S)}{c}$$

is less than 1.

It will be seen below that in the *GI/G/c* queue the quantity ρ can be interpreted as the long-run fraction of time a given server is busy. This explains the name of server utilisation. In many single-server systems it is desirable to have a server utilisation of no more than 0.8 (say), since otherwise a small increase in the offered load may lead to considerable increases in the average queue size and average delay per customer. However, multi-server queueing systems allow for higher values of the server utilisation ρ so that a small increase in the offered load does not cause a dramatic degradation in system performance; see also the discussion in Whitt (1992).

In addition to Assumption 4.1.1, we make the following technical assumption.

Assumption 4.1.2   *(a) The interarrival-time distribution $A(t)$ or the service-time distribution $B(t)$ has a positive density on some interval.*
(b) The probability that the interarrival time A is larger than the service time S is positive.

Define a cycle as the time elapsed between two consecutive arrivals who find the system empty. Then, under the Assumptions 4.1.1 and 4.1.2, it can be shown that the expected value of the cycle length is always finite. The proof of this result is quite deep and is not given here; see Wolff (1989).

Let us now define the following random variables:

$L(t)$ = the number of customers in the system at time t (including those in service),

$L_q(t)$ = the number of customers in the queue at time t (excluding those in service),

D_n = the amount of time spent by the nth accepted customer in the queue (excluding service time),

R_n = the amount of time spent by the nth accepted customer in the system (including service time).

The continuous-time stochastic processes $\{L(t),\ t \geq 0\}$ and $\{L_q(t),\ t \geq 0\}$ and the discrete-time stochastic processes $\{D_n,\ n \geq 0\}$ and $\{R_n,\ n \geq 0\}$ are all regenerative. The regeneration epochs are the epochs at which an arriving customer finds the system empty. The regeneration cycles have finite means. Thus the following long-run averages exist:

$$L = \lim_{t \to \infty} \frac{1}{t} \int_0^t L(u)\,du \qquad \text{(the long-run average number in system)}$$

$$L_q = \lim_{t \to \infty} \frac{1}{t} \int_0^t L_q(u)\,du \qquad \text{(the long-run average number in queue)}$$

$$W_q = \lim_{n \to \infty} \frac{1}{n} \sum_{k=1}^n D_k \qquad \text{(the long-run average delay in queue)}$$

$$W = \lim_{n \to \infty} \frac{1}{n} \sum_{k=1}^n R_k \qquad \text{(the long-run average wait in system)}.$$

These long-run averages are constants with probability 1. From Theorem 1.3.4 in

Section 1.3, it follows that the long-run averages L_q and W_q can be computed from

$$L_q = \sum_{j=c}^{c+N} (j - c) p_j \quad \text{and} \quad W_q = \int_0^\infty \{1 - W_q(x)\}\, dx,$$

where p_j and $W_q(x)$ are defined by

$$p_j = \lim_{t \to \infty} P\{L(t) = j\}, \quad j = 0, 1, \ldots$$

and

$$W_q(x) = \lim_{n \to \infty} P\{D_n \le x\}, \quad x \ge 0.$$

These limits exist and represent proper probability distributions which are independent of the initial condition of the system; see Theorem 1.3.3 in Section 1.3. The probabilities p_j and $W_q(x)$ can be interpreted respectively as the long-run fraction of time that j customers are in the system and as the long-run fraction of accepted customers whose delay in queue is at most x.

It is important to note that the distribution of the number of customers in the system is invariant to the order of service when the queue discipline is *service-time independent* and *work-conserving*. Here service-time independent means that the rule for selecting a next customer to be served does not depend on the service time of a customer, while work-conserving means that the work or service requirement of a customer is not affected by the queue discipline. Queue disciplines having these properties include first-come-first-served, last-come-first-served and service in random order. The waiting-time distribution will obviously depend on the order of service.

Finally, letting the random variable $I_n = 1$ if the nth arrival is rejected and letting $I_n = 0$ otherwise, we define the constant P_{rej} by

$$P_{\text{rej}} = \lim_{n \to \infty} \frac{1}{n} \sum_{k=1}^n I_k.$$

In other words, P_{rej} is the long-run fraction of customers rejected.

Little's formula

The most basic result for queueing systems is Little's formula. This formula relates certain averages like the average number of customers in the queue and the average delay in queue per customer. The formula of Little is valid for almost any queueing system. In particular, for the $GI/G/c/c+N$ queue, we have the following fundamental relations,

$$L_q = \lambda \left(1 - P_{\text{rej}}\right) W_q \quad \text{and} \quad L = \lambda (1 - P_{\text{rej}}) W. \qquad (4.1.1)$$

Note that $P_{\text{rej}} = 0$ when $N = \infty$. To motivate these relations, we argue as follows. Assume that for each *accepted* customer the system incurs a cost according to a specific rule. Next we apply the general principle:

> the long-run average cost incurred by the system per unit time
> = (the long-run average arrival rate of accepted customers)
> × (the long-run average cost incurred per accepted customer). (4.1.2)

A proof of this principle for regenerative systems can be found in Section 1.6. To see the first relation in (4.1.1), assume that for each accepted customer the system incurs a cost at a rate of 1 while the customer is in queue. Then the long-run average cost incurred per accepted customer equals W_q. On the other hand, since the system incurs a cost at a rate of j while j customers are waiting in queue, the long-run average cost incurred by the system per unit time equals L_q. The first relation in (4.1.1) now follows by noting that the long-run average arrival rate of accepted customers is $\lambda(1 - P_{\text{rej}})$. The second relation in (4.1.1) is obtained by assuming that for each accepted customer the system incurs a cost at a rate of 1 while the customer is in the system.

As another example of Little's formula, we have:

$$\text{the long-run average number of busy servers} = \lambda(1 - P_{\text{rej}})E(S). \qquad (4.1.3)$$

To see this relation, assume that each accepted customer pays at a rate of 1 while in service. Then, the reward earned for an accepted customer is equal to the service time of the customer. On the other hand, the system earns a reward at a rate of j when j servers are busy. Hence the average reward earned by the system per unit time is equal to the average number of busy servers. The relation (4.1.3) now follows.

The result (4.1.3) has two interesting implications. First, since each of the c servers carries the same load on the average, we have

$$\text{the long-run fraction of time a given server is busy} = \frac{1}{c}\lambda(1 - P_{\text{rej}})E(S).$$

In particular, the long-run fraction of time a given server is busy equals ρ in the *GI/G/c* queue. Secondly, since p_j represents the long-run fraction of time that j customers are present, the long-run average number of busy servers is also given by the expression $\Sigma_{j=0}^{c-1} jp_j + c\Sigma_{j\geq c}p_j$. Thus we obtain the useful identity

$$\sum_{j=1}^{c-1} jp_j + c\left(1 - \sum_{j=0}^{c-1} p_j\right) = \lambda(1 - P_{\text{rej}})E(S). \qquad (4.1.4)$$

In particular, we find the general relation $p_0 = 1 - \lambda E(S)$ for the *GI/G/1* queue. The above relations can be easily extended to queueing systems with batch arrivals.

4.2 THE *M/G/*1 QUEUE

In the *M/G/1* queueing system customers arrive according to a Poisson process with rate λ and the service time S of a customer has a general probability distribution function $B(x)$ with $B(0) = 0$. It is assumed that the server utilisation $\rho = \lambda E(S)$ is smaller than 1.

In Section 4.2.1 we derive a recursive algorithm for the computation of the state probabilities. For the $M/G/1$ queue the customer-average state probabilities are identical to the time-average state probabilities by the property Poisson arrivals see time averages. The state probabilities can be obtained by analysing a Markov chain embedded at the service completion epochs and using that the long-run fraction of service completions leaving j customers behind equals the long-run fraction of arrivals finding j other customers present. However, we prefer to use a regenerative approach rather than the embedded Markov chain approach. The regenerative approach leads directly to a numerically stable recursion scheme for the state probabilities and allows in a natural way for generalizations to more complex queueing models. Using the technique of generating functions, we also derive an asymptotic expansion for the state probabilities. Since an explicit expression is available for the generating function of the state probabilities, the Fast Fourier Transform method provides an alternative method to compute the state probabilities. In Section 4.2.2 we discuss the computation of the waiting-time probabilities. Also attention is paid to an approximation of the waiting-time distribution by the sum of two exponential functions. This approximation is based on the asymptotic expansion of the tail of the waiting-time distribution. Further, we discuss a simple but generally useful two-moment approximation for the waiting-time percentiles.

4.2.1 The state probabilities

The time-average probabilities p_j, $j = 0, 1, \ldots$, can be interpreted as the long-run fraction of time that j customers are in the system. Using a basic result from the theory of regenerative processes and a simple up- and downcrossings argument, we derive a numerically stable recursion scheme for the state probabilities p_j.

Theorem 4.2.1 *The state probabilities p_j satisfy the recursion*

$$p_j = \lambda a_{j-1} p_0 + \lambda \sum_{k=1}^{j} a_{j-k} p_k, \quad j = 1, 2, \ldots, \tag{4.2.1}$$

where the constants a_n are given by

$$a_n = \int_0^\infty e^{-\lambda t} \frac{(\lambda t)^n}{n!} \{1 - B(t)\} \, dt, \quad n = 0, 1, \ldots .$$

Proof The stochastic process $\{L(t), t \geq 0\}$ describing the number of customers in the system is regenerative. The process regenerates itself each time an arriving customer finds the system empty. Denoting by a cycle the time elapsed between two consecutive arrivals who find the system empty, we define the random variables

$T = $ the length of one cycle,

T_j = the amount of time that j customers are present during one cycle.

The expected length of one cycle is finite (cf. also Example 1.3.4 in Section 1.3). By Theorem 1.3.2 in Section 1.3, we have

$$p_j = \frac{E(T_j)}{E(T)}, \quad j = 0, 1, \dots . \tag{4.2.2}$$

In particular, using that $E(T_0) = 1/\lambda$ by the lack of memory of the Poisson arrival process,

$$p_0 = \frac{1}{\lambda E(T)}. \tag{4.2.3}$$

The following simple idea is crucial for the derivation of a recurrence relation for the probabilities p_j. Divide a cycle into a random number of disjoint intervals separated by the service completion epochs and calculate $E(T_j)$ as the sum of the contributions from the disjoint intervals to the expected sojourn time in state j. Thus we define the random variable N_k by

N_k = the number of service completion epochs in one cycle at which k customers are left behind, $k = 0, 1, \dots .$

Moreover, using the lack of memory of the Poisson arrival process, we define the quantity A_{kj} by

A_{kj} = the expected amount of time that j customers are present during a service time starting when k customers are present.

Then, noting that the first service in a cycle starts with one customer present, it follows that

$$E(T_j) = A_{1j} + \sum_{k=1}^{j} E(N_k)A_{kj}, \quad j = 1, 2, \dots . \tag{4.2.4}$$

It should be pointed out that Wald's equation is used to justify that $E(N_k)A_{kj}$ is the contribution to $E(T_j)$ of those service intervals starting with k customers present.

To find another relation between $E(T_j)$ and $E(N_k)$, we use the simple observation that for each $k = 0, 1, \dots,$

the number of downcrossings from state $k + 1$ to state k in one cycle = the number of upcrossings from state k to state $k + 1$ in one cycle.

The expected number of downcrossings of the $\{L(t)\}$ process from state $k + 1$ to state k in one cycle is by definition equal to $E(N_k)$. On the other hand, since the arrival process is a Poisson process, we have by Corollary 1.7.2 in Section 1.7 that the expected number of upcrossings from state k to state $k + 1$ in one cycle is equal to $\lambda E(T_k)$. Thus we find

$$E(N_k) = \lambda E(T_k), \quad k = 0, 1, \dots . \tag{4.2.5}$$

Together the relations (4.2.2)–(4.2.5) imply that

$$p_j = \lambda p_0 A_{1j} + \sum_{k=1}^{j} \lambda p_k A_{kj}, \quad j = 1, 2, \dots \,. \tag{4.2.6}$$

To specify the constants A_{kj}, suppose that at epoch 0 a service starts when k customers are present. Define the random variable $I_j(t) = 1$ if at time t the service is still in progress and j customers are present and let $I_j(t) = 0$ otherwise. Then, for $j \geq k$,

$$A_{kj} = E\left[\int_0^\infty I_j(t)\,dt\right] = \int_0^\infty E[I_j(t)]\,dt$$

$$= \int_0^\infty P\{I_j(t) = 1\}\,dt = \int_0^\infty \{1 - B(t)\}e^{-\lambda t}\frac{(\lambda t)^{j-k}}{(j-k)!}\,dt. \tag{4.2.7}$$

Together (4.2.6) and (4.2.7) yield the desired result.

The recursion (4.2.1) enables us to compute recursively $p_1, p_2, \dots$ starting with $p_0 = 1 - \rho$. Note that (4.1.4) implies that $p_0 = 1 - \rho$. The recursion scheme is numerically stable since the calculations involve only additions with positive numbers and thus cannot cause a loss of significant digits. For most service-time distributions of practical interest numerical integration can be avoided for the computation of the constants a_n. Using the identity (1.2.3) in Section 1.2, explicit expressions for the a_n's can be given for the cases of deterministic services, generalized Erlangian services and Coxian-2 services. The details are left to the reader.

The computational effort of the recursion scheme can often be considerably reduced by using an asymptotic expansion for the state probabilities. Before giving this asymptotic expansion, we derive the Pollaczek–Khintchine formula for the average queue size. To do so, we need the generating function $P(z)$ defined by

$$P(z) = \sum_{j=0}^{\infty} p_j z^j, \quad |z| \leq 1.$$

Multiplying both sides of (4.2.1) by z^j and summing over j, it is a matter of simple algebra to derive that

$$P(z) - p_0 = \lambda p_0 z \sum_{n=0}^{\infty} a_n z^n + \lambda\{P(z) - p_0\} \sum_{n=0}^{\infty} a_n z^n.$$

Using that $p_0 = 1 - \rho$, we thus obtain

$$P(z) = (1 - \rho)\frac{1 - \lambda(1 - z)A(z)}{1 - \lambda A(z)}, \tag{4.2.8}$$

where $A(z) = \sum_{n=0}^{\infty} a_n z^n$ is given by

$$A(z) = \int_0^\infty \{1 - B(t)\} e^{-\lambda(1-z)t}\, dt.$$

Pollaczek–Khintchine formula

A formula for the long-run average number of customers in queue follows from the representation $L_q = \sum_{j=1}^{\infty}(j-1)p_j$ and the relation $P'(1) = \sum_{j=1}^{\infty} j p_j$. Using the explicit expression (4.2.8) and L'Hospital's rule, we find after some algebra that

$$L_q = \frac{\lambda^2 E(S^2)}{2(1-\rho)}. \tag{4.2.9}$$

This is the famous Pollaczek–Khintchine formula. Using Little's formula $L_q = \lambda W_q$, it follows that the long-run average delay in queue per customer is given by

$$W_q = \frac{\lambda E(S^2)}{2(1-\rho)}. \tag{4.2.10}$$

This formula was already derived by a different approach in Section 2.9.

Asymptotic expansion for the state probabilities

The representation (4.2.8) shows that the generating function $P(z)$ is the ratio of two functions $N(z)$ and $D(z)$. These functions allow for an analytic continuation outside the unit circle when the following assumption is made.

Assumption 4.2.1   (a)

$$\int_0^\infty e^{st}\{1 - B(t)\}\, dt < \infty$$

for some $s > 0$.
   (b)

$$\lim_{s \to B} \int_0^\infty e^{st}\{1 - B(t)\}\, dt = \infty,$$

where

$$B = \sup\left[s \left| \int_0^\infty e^{st}\{1 - B(t)\}\, dt < \infty \right. \right].$$

The assumption requires that the service-time distribution has no extremely long tail as is the case in most situations of practical interest. Using Assumption 4.2.1, it follows from relation (C.12) in Appendix C that

$$p_j \approx \sigma \tau^{-j} \quad \text{for } j \text{ large enough}, \tag{4.2.11}$$

where τ is the unique solution of the equation

$$\int_0^\infty e^{-\lambda(1-\tau)t}\{1 - B(t)\}\,dt = \frac{1}{\lambda} \qquad (4.2.12)$$

on the interval $(1, 1 + B/\lambda)$ and the constant σ is given by

$$\sigma = \frac{(1-\rho)}{\lambda^2}\left[\int_0^\infty te^{-\lambda(1-\tau)t}\{1 - B(t)\}\,dt\right]^{-1}. \qquad (4.2.13)$$

It is empirically found that the asymptotic expansion (4.2.11) applies already for relatively small values of j. It turns out that the closer the traffic intensity ρ is to 1, the earlier the asymptotic expansion applies.

The asymptotic expansion can be used to reduce the computational effort of the recursion scheme (4.2.1). Since $p_{j-1}/p_j \approx \tau$ for j large enough, the recursive calculations can be halted as soon as the ratio p_{j-1}/p_j has sufficiently converged to the constant τ.

The Fast Fourier Transform (FFT) method provides an alternative method for the computation of the state probabilities using the explicit expression for the generating function $P(z)$ of the state probabilities. This method, which is discussed in Section 1.2.2, can be further speeded up by using the asymptotic expansion (4.2.11). In a straightforward application of the FFT method we must choose an integer $M = 2^m$ such that $\Sigma_{j \geq M} p_j$ is negligibly small. This integer M tends to become quite large when the traffic intensity ρ gets close to 1. However, for the case of non-light traffic we can do much better by using (4.2.11). Let $L(= 2^r)$ be an appropriately chosen integer such that the asymptotic expansion (4.2.11) holds with sufficient accuracy for $j \geq L$. Then, by (4.2.8),

$$\sum_{j=0}^{L-1} p_j z^j \approx (1-\rho)\frac{1 - \lambda(1-z)A(z)}{1 - \lambda A(z)} - \frac{\sigma(z/\tau)^L}{1 - z/\tau}. \qquad (4.2.14)$$

Using this explicit expression for $\Sigma_{j=0}^{L-1} p_j z^j$ we next apply the FFT method to obtain the numerical values of $p_0, \ldots, p_{L-1}$. The other probabilities follow from $p_j = (1/\tau)^{j-L+1} p_{L-1}$ for $j \geq L$. In most cases of practical interest the integer L will be much smaller than a truncation integer M for which $\Sigma_{j \geq M} p_j$ is negligibly small. Thus a combination of the FFT method and the asymptotic expansion usually leads to a computational improvement over a straightforward application of the FFT method.

4.2.2 The waiting-time probabilities

In this subsection it is assumed that customers are served in order of arrival. Under this assumption it was shown in Section 1.8 that the complementary waiting-time distribution function $W_q^c(x) = 1 - W_q(x)$ is the unique solution to the integral equation

$$W_q^c(x) = \lambda \int_x^\infty \{1 - B(y)\}\,dy + \lambda \int_0^x W_q^c(x - y)\{1 - B(y)\}\,dy, \quad x \geq 0.$$

In particular, the long-run fraction of customers who have to wait in queue is given by

$$P_{\text{delay}} = \rho. \tag{4.2.15}$$

The integral equation can numerically be solved by using the discretization algorithm discussed in Section 1.1.3. However, when a high accuracy is required, this approach is computationally rather demanding even when it is combined with the asymptotic expansion for $W_q^c(x)$.

Using the continuous-time Markov chain approach a computationally tractable algorithm for $W_q(x)$ can be given when the service-time distribution is a mixture of Erlangian distributions with the same scale parameters. This algorithm is discussed in detail in Section 2.9.

An explicit expression for $W_q(x)$ can be given when the service time has a Coxian-2 distribution. In Appendix C it is shown that for Coxian-2 services

$$W_q^c(x) = \alpha e^{-\beta x} + \gamma e^{-\delta x} \quad \text{for all } x \geq 0. \tag{4.2.16}$$

It will be shown below how to compute the constants α, β, γ and δ.

A simple approximation to the waiting-time probabilities

Let us assume that Assumption 4.2.1 is satisfied. Then, as was shown in Section 1.8,

$$W_q^c(x) \approx \gamma e^{-\delta x} \quad \text{for } x \text{ large enough}, \tag{4.2.17}$$

with

$$\delta = \lambda(\tau - 1) \quad \text{and} \quad \gamma = \frac{\sigma}{\tau - 1},$$

where the constants τ and σ are given by (4.2.12) and (4.2.13).

It is empirically found that the asymptotic expansion for $W_q^c(x)$ is accurate enough for practical purposes already for relatively small values of x. Why not improve this first-order estimate by adding a second exponential term? This suggests the following approximation to $W_q^c(x)$:

$$W_{\text{app}}^c(x) = \alpha e^{-\beta x} + \gamma e^{-\delta x}, \quad x \geq 0. \tag{4.2.18}$$

The constants α and β are found by matching the behaviour of $W_q(x)$ at $x = 0$ and the first moment of $W_q(x)$. Using that $1 - W_q(0) = P_{\text{delay}}$ and $W_q = \int_0^\infty \{1 - W_q(x)\}\,dx$, it follows that

$$\alpha = P_{\text{delay}} - \gamma \quad \text{and} \quad \beta = \alpha \left\{ W_q - \gamma/\delta \right\}^{-1}, \tag{4.2.19}$$

where explicit expressions for P_{delay} and W_q are given by (4.2.15) and (4.2.10). It should be pointed out that the approximation (4.2.18) can be applied only when $\beta > \delta$ since otherwise $W^c_{\text{app}}(x)$ is not in accordance with the asymptotic expansion (4.2.17) as x gets large. Numerical experiments indicate that $\beta > \delta$ holds for a wide class of service-time distributions of practical interest. Further support to the approximation (4.2.18) is provided by the fact that the approximation is exact for Coxian-2 services. It is remarked that for the particular case of deterministic services the approximation can be improved by a piecewise exponential function. The latter approximation is discussed in Section 4.6.2 in the more general context of the $GI/D/c$ queue.

Numerical investigations show that the approximation (4.2.18) performs quite satisfactorily for all values of x. In Table 4.2.1 we give the exact values of $W^c_q(x)$, the approximate values in (4.2.18) and the asymptotic values in (4.2.17) for the E_{10} and E_3 distributions. The server utilisation ρ is varied as $\rho = 0.2, 0.5, 0.8$. In all examples the normalization $E(S) = 1$ is used. The numerical results in the table confirm that the larger the traffic load ρ, the earlier the asymptotic expansion (4.2.17) applies.

A two-moment approximation for the waiting-time percentiles

In applications it often happens that only the first two moments of the service time are available. In these situations two-moment approximations may be very helpful. However, such approximations should not be used blindly. Numerical experiments indicate that the waiting-time probabilities are rather insensitive to more than the

Table 4.2.1   The waiting-time probabilities

		Erlang-10			Erlang-3		
	x	exact	approx	asympt	exact	approx	asympt
$\rho = 0.2$	0.10	0.1838	0.1960	0.3090	0.1839	0.1859	0.2654
	0.25	0.1590	0.1682	0.2222	0.1594	0.1615	0.2106
	0.50	0.1162	0.1125	0.1282	0.1209	0.1212	0.1432
	0.75	0.0755	0.0694	0.0739	0.0882	0.0875	0.0974
	1.00	0.0443	0.0413	0.0427	0.0626	0.0618	0.0663
$\rho = 0.5$	0.10	0.4744	0.4862	0.5659	0.4744	0.4764	0.5332
	0.25	0.4334	0.4425	0.4801	0.4342	0.4361	0.4700
	0.50	0.3586	0.3543	0.3651	0.3664	0.3665	0.3810
	0.75	0.2808	0.2745	0.2776	0.3033	0.3026	0.3088
	1.00	0.2127	0.2102	0.2111	0.2484	0.2476	0.2502
$\rho = 0.8$	0.10	0.7833	0.7890	0.8219	0.7834	0.7844	0.8076
	0.25	0.7557	0.7601	0.7756	0.7562	0.7571	0.7708
	0.50	0.7020	0.6998	0.7042	0.7074	0.7074	0.7131
	0.75	0.6413	0.6381	0.6394	0.6577	0.6573	0.6597
	1.00	0.5812	0.5801	0.5805	0.6097	0.6093	0.6103

first two moments of the service time S provided that the squared coefficient of variation c_S^2 is not too large (say, $0 \leq c_S^2 \leq 2$) and the service-time density satisfies a reasonable shape constraint. The sensitivity becomes less and less manifest as the traffic intensity ρ gets closer to 1.

The motivation for the two-moment approximation is provided by the Pollaczek–Khintchine formula for the average delay in queue. The expression (4.2.10) for W_q can be written as

$$W_q = \frac{1}{2}\left(1 + c_S^2\right)\frac{E(S)}{1 - \rho}, \qquad (4.2.20)$$

where $c_S^2 = \sigma^2(S)/E^2(S)$. Denote now by $W_q(\exp)$ and $W_q(\det)$ the average delay in queue for the special cases of exponential services ($c_S^2 = 1$) and deterministic services ($c_S^2 = 0$). Then the formula (4.2.20) is equivalent to the representations

$$W_q = \frac{1}{2}\left(1 + c_S^2\right)W_q(\exp) \qquad (4.2.21)$$

and

$$W_q = (1 - c_S^2)W_q(\det) + c_S^2 W_q(\exp). \qquad (4.2.22)$$

A natural question is whether the representations (4.2.21) and (4.2.22) can be used as a basis for approximations to the waiting-time probabilities. Numerical investigations reveal that the waiting-time probabilities themselves do not allow for two-moment approximations of the forms (4.2.21) and (4.2.22), but the waiting-time percentiles do allow for such two-moment approximations.

The pth percentile $\xi(p)$ of the waiting-time distribution function $W_q(x)$ is defined by

$$W_q(\xi(p)) = p.$$

Since $W_q(0) = 1 \div \rho$, the percentile $\xi(p)$ is only defined for $1 - \rho < p < 1$. Denote by $\xi_{\exp}(p)$ and $\xi_{\det}(p)$ the percentile $\xi(p)$ for the special cases of exponential services and deterministic services with the same means $E(S)$. The representation (4.2.21) suggests the first-order approximation

$$\xi_{\text{app1}}(p) = \frac{1}{2}\left(1 + c_S^2\right)\xi_{\exp}(p), \qquad (4.2.23)$$

while the representation (4.2.22) suggests the second-order approximation

$$\xi_{\text{app2}}(p) = (1 - c_S^2)\xi_{\det}(p) + c_S^2\xi_{\exp}(p). \qquad (4.2.24)$$

In Section 2.9 it was shown that $W_q^c(x) = \rho\exp[-\mu(1 - \rho)x]$ for all $x \geq 0$ when the service time has an exponential distribution with mean $1/\mu = E(S)$. Hence $\xi_{\exp}(p)$ is trivially computed from

$$\xi_{\exp}(p) = \frac{E(S)}{1 - \rho}\ln\left(\frac{\rho}{1 - p}\right).$$

Table 4.2.2   The waiting-time percentiles $\eta(p)$

		$c_S^2 = 0.5$					$c_S^2 = 2$				
p		0.2	0.5	0.9	0.99	0.999	0.2	0.5	0.9	0.99	0.999
$\rho = 0.2$	exact	0.252	0.70	2.06	3.90	5.73	0.32	1.20	4.53	9.30	14.1
	app1	0.209	0.65	2.16	4.32	6.48	0.42	1.30	4.32	8.63	13.0
	app2	0.261	0.73	1.98	3.87	5.76	0.31	1.14	4.67	9.52	14.4
$\rho = 0.5$	exact	0.390	1.09	3.34	6.54	9.75	0.54	2.00	7.12	14.46	21.8
	app1	0.335	1.04	3.45	6.91	10.37	0.67	2.08	6.91	13.82	20.7
	app2	0.405	1.10	3.33	6.55	9.77	0.53	1.96	7.16	14.53	21.9
$\rho = 0.8$	exact	0.907	2.64	8.52	16.95	25.37	1.53	5.14	17.49	35.16	52.8
	app1	0.837	2.60	8.63	17.27	25.91	1.67	5.20	17.27	34.55	51.8
	app2	0.925	2.63	8.52	16.95	25.38	1.50	5.14	17.49	35.19	52.9

A relatively simple algorithm for the computation of $\xi_{\text{det}}(p)$ is given in Section 4.5.2 in the more general context of the $M/D/c$ queue. For higher values of p(say, $p \geq 1 - \frac{1}{2}\rho$) the percentile $\xi_{\text{det}}(p)$ can be simply computed from the asymptotic expansion of $W_q^c(x)$ for deterministic services.

In Table 4.2.2 we give some numerical results. In the table we work with the conditional waiting-time percentiles $\eta(p)$ rather than with the percentiles $\xi(p)$. The percentiles $\eta(p)$ are defined by

$$P\{D_\infty \leq \eta(p)|D_\infty > 0\} = p, \qquad (4.2.25)$$

where the generic variable D_∞ has $W_q(x)$ as probability distribution function. The conditional percentiles $\eta(p)$ are defined for all $0 < p < 1$. Note the relationship $\eta(p_1) = \xi(p_0)$ where $p_0 = 1 - (1 - p_1)\rho$. The approximate relationships (4.2.23) and (4.2.24) apply also to the conditional percentiles $\eta(p)$. Table 4.2.2 gives the exact and approximate values of $\eta(p)$ for E_2 services ($c_S^2 = 0.5$) and H_2 services ($c_S^2 = 2$) with the gamma normalization. The numerical results show an excellent performance of the second-order approximation for all values of ρ and p. The first-order approximation $\frac{1}{2}(1 + c_S^2)\eta_{\exp}(p)$ is only useful for quick engineering calculations when ρ is not too small (say, $\rho > 0.5$) and p is sufficiently close to 1 (say, $p > 1 - \rho$).

4.3   THE $M^X/G/1$ QUEUE WITH BATCH INPUT

Queueing systems with customers arriving in batches rather than singly have many applications in practice, for example in telecommunication. A useful model is the single-server $M^X/G/1$ queue where batches of customers arrive according to a Poisson process with rate λ and the batch size X has a discrete probability distribution $\{\beta_j, \ j = 1, 2, \ldots\}$ with finite mean β. The customers are served individually by a single server. The service times of the customers are independent random variables with a common probability distribution function $B(t)$. Denoting by the generic variable S the service time of a customer, it is assumed that the server utilisation ρ defined by

$$\rho = \lambda \beta E(S)$$

is smaller than 1. The analysis given for the $M/G/1$ queue can be generalized to the $M^X/G/1$ queue. In Section 4.3.1 we give an algorithm for the state probabilities. The computation of the waiting-time probabilities is discussed in Section 4.3.2.

4.3.1 The state probabilities

The stochastic process $\{L(t), \ t \geq 0\}$ describing the number of customers in the system is regenerative. The process regenerates itself each time an arriving batch finds the system empty. The cycle length has a continuous distribution with finite mean. Thus the process $\{L(t)\}$ has a limiting distribution $\{p_j\}$. The probability p_j can be interpreted as the long-run fraction of time that j customers are in the system. The probability p_0 allows for the explicit expression

$$p_0 = 1 - \rho. \tag{4.3.1}$$

To see this, we apply the general principle (4.1.2) underlying Little's formula. Assume that the system earns a reward at rate 1 whenever a customer is in service. Then the long-run average reward per unit time represents the long-run fraction of time the server is busy which in turn is equal to $1 - p_0$. The long-run average reward earned per customer is equal to E(S), while the long-run average arrival rate of customers is $\lambda\beta$. This shows that $1 - p_0 = \lambda\beta E(S)$.

A recursion scheme for the other p_j 's is given in the following generalization of Theorem 4.2.1.

Theorem 4.3.1 *The state probabilities p_j satisfy the recursion*

$$p_j = \lambda p_0 \sum_{s=1}^{j} \beta_s a_{j-s} + \lambda \sum_{k=1}^{j} \left(\sum_{i=0}^{k} p_i \sum_{s>k-i} \beta_s \right) a_{j-k}, \quad j = 1, 2, \ldots, \tag{4.3.2}$$

where

$$a_n = \int_0^\infty r_n(t)\{1 - B(t)\} \, dt, \quad n = 0, 1, \ldots$$

with $r_n(t)$ denoting the probability that exactly n customers arrive during a time interval of length t.

Proof The proof is along the same lines as the proof of Theorem 4.2.1. The only modification is with respect to the up- and downcrossings relation (4.2.5). We now use the up- and downcrossings argument that,

the number of downcrossings from a state in the set $\{k + 1, k + 2, \ldots\}$ to a state outside this set during one cycle = the number of upcrossings from a state outside the set $\{k+1, k+2, \ldots\}$ to a state in this set during one cycle.

Thus relation (4.2.5) generalizes to

$$E(N_k) = \sum_{i=0}^{k} E(T_i)\lambda \sum_{s>k-i} \beta_s, \quad k = 0, 1, \ldots .$$

The remainder of the proof is analogous to the proof of Theorem 4.2.1.

The recursion scheme (4.3.2) is not as easy to apply as the recursion scheme (4.2.1). The reason is that the computation of the constants a_n is quite burdensome. In general numerical integration must be used, where each function evaluation in the integration procedure requires an application of Adelson's recursion scheme for the computation of the compound Poisson probabilities $r_n(t)$, $n \geq 0$, cf. Section 1.2.2.

The best general-purpose approach for the computation of the state probabilities is the Fast Fourier Transform method. An explicit expression for the generating function

$$P(z) = \sum_{j=0}^{\infty} p_j z^j, \quad |z| \leq 1$$

can be given. It is a matter of tedious algebra to derive from (4.3.2) that

$$P(z) = (1-\rho)\frac{1 - \lambda A(z)\{1 - \beta(z)\}}{1 - \lambda A(z)\{1 - \beta(z)\}/(1-z)}, \tag{4.3.3}$$

where

$$\beta(z) = \sum_{j=1}^{\infty} \beta_j z^j \quad \text{and} \quad A(z) = \int_0^{\infty} e^{-\lambda\{1-\beta(z)\}t}(1 - B(t))\, dt.$$

The derivation uses that $e^{-\lambda\{1-\beta(z)\}t}$ is the generating function of the compound Poisson probabilities $r_n(t)$; see relation (1.2.12) in Section 1.2.2. Moreover, the derivation uses that the generating function of the convolution of two discrete probability distributions is the product of the generating functions of the two probability distributions. The other details of the derivation of (4.3.3) are left to the reader.

For most service-time distributions of practical interest no numerical integration is required to evaluate the function $A(z)$ in the FFT method. Similarly to the representation (4.2.14) for the $M/G/1$ queue, the computational effort of the FFT method can be reduced by using the asymptotic expansion of p_j for j large enough. This asymptotic expansion exists when it is assumed that the batch-size distribution and the service-time distribution have no extremely long tails. Let us make the following assumption.

Assumption 4.3.1   *(a) The convergence radius R of $\beta(z) = \Sigma_{j=1}^{\infty} \beta_j z^j$ is larger than 1. Moreover, $\int_0^{\infty} e^{st}\{1 - B(t)\}\,dt < \infty$ for some s > 0.*
(b) $\lim_{s \to B} \int_0^{\infty} e^{st}\{1-B(t)\}dt = \infty$, where $B = \sup[s| \int_0^{\infty} e^{st}\{1 - B(t)\}dt] < \infty$.
(c) $\lim_{x \to R_0} \beta(x) = 1 + B/\lambda$ for some number R_0 with $1 < R_0 \le R$.

Under this assumption we obtain from the general result (C.8) in Appendix C that

$$p_j \approx \sigma \tau^{-j} \quad \text{for } j \text{ large enough,} \tag{4.3.4}$$

where τ is the unique solution to the equation

$$\lambda A(\tau)\{1 - \beta(\tau)\} = 1 - \tau$$

on $(1, R_0)$ and the constant σ is given by

$$\sigma = (1 - \rho)(1 - \tau)\left[\lambda A'(\tau)\{1 - \beta(\tau)\} - \frac{(1 - \tau)\beta'(\tau)}{1 - \beta(\tau)} + 1\right]^{-1}.$$

A formula for the average queue size

The long-run average number of customers in queue is $L_q = \Sigma_{j=1}^{\infty}(j - 1)p_j$. Using the relation $P'(1) = \Sigma_{j=1}^{\infty} jp_j$, we obtain after some algebra from (4.3.3) that

$$L_q = \frac{1}{2}(1 + c_S^2)\frac{\rho^2}{1 - \rho} + \frac{\rho}{2(1 - \rho)}\left[\frac{E(X^2)}{E(X)} - 1\right]. \tag{4.3.5}$$

Note that the first part of the expression for L_q gives the average queue size in the standard M/G/1 queue, while the second part reflects the additional effect of the batch size.

The formula (4.3.5) implies directly an expression for the long-run average delay in queue per customer. By Little's formula,

$$L_q = \lambda \beta W_q. \tag{4.3.6}$$

4.3.2 The waiting-time probabilities

In this section it is assumed that customers from different batches are served in order of arrival of the batches and customers from a same batch are served in order of their position in the batch. Denote by D_n the delay in queue of the customer who receives the nth service. The waiting-time distribution function $W_q(x)$ is defined by

$$W_q(x) = \lim_{n \to \infty} \frac{1}{n} \sum_{k=1}^{n} P\{D_k \le x\}, \quad x \ge 0.$$

In other words, $W_q(x)$ represents the long-run fraction of customers whose delay in queue is no more than x. It is pointed out that $\lim_{n \to \infty} P\{D_n \le x\}$ need not exist in the batch-arrival queue (to see this, consider the case of a constant batch size of 2; then $P\{D_n > 0\} = 1$ for n even and $P\{D_n > 0\} < 1$ for n odd).

We first give an exact algorithm for the computation of $W_q(x)$ when the service-time distribution is a mixture of Erlangian distributions. Next we discuss an extension of the approximations given for the $M/G/1$ queue.

Exact algorithm

It is assumed that the service-time distribution has the density

$$b(t) = \sum_{i=1}^{m} q_i \mu^i \frac{t^{i-1}}{(i-1)!} e^{-\mu t}, \quad t \geq 0, \tag{4.3.7}$$

where $q_i \geq 0 (i = 1, \ldots, m)$ and $\sum_{i=1}^{m} q_i = 1$. Any service-time density can be arbitrarily closely approximated by a density of this form; cf. Section 2.9 in Chapter 2. For the service-time density (4.3.7) the service time of a customer can be seen as a random sum of independent phases, where the phases must be sequentially processed and each phase has an exponentially distributed length with mean $1/\mu$. The probability that the service of a customer requires i phases is q_i. In view of this interpretation, we can apply the continuous-time Markov chain approach. To do so, define the random variable $X(t)$

$X(t) =$ the number of uncompleted service phases present at time t.

Denote by $\{f_j, \ j = 0, 1, \ldots\}$ the equilibrium distribution of the continuous-time Markov chain $\{X(t)\}$. In other words, f_j is the long-run fraction of time that j uncompleted phases are present. By the property Poisson arrivals see time averages the probability f_j can also be interpreted as the long-run fraction of batches finding upon arrival j uncompleted service phases present. Closely related to the probabilities f_j are the probabilities z_j defined by

$z_j =$ the long-run fraction of customers who have j uncompleted phases in front of them just after arrival.

Suppose for the moment that the z_j 's have been computed. Then it becomes simple to calculate $W_q(x)$. To do so, consider a customer who finds upon arrival j uncompleted phases in front of him. The conditional waiting time of this customer is the sum of j independent exponentials with common mean $1/\mu$ and is therefore Erlang-j distributed. Thus, by conditioning, we find

$$W_q(x) = \sum_{j=0}^{\infty} z_j \left(1 - \sum_{k=0}^{j-1} e^{-\mu x} \frac{(\mu x)^k}{k!} \right), \quad x \geq 0.$$

This series representation can be written as

$$W_q(x) = 1 - \sum_{k=0}^{\infty} e^{-\mu x} \frac{(\mu x)^k}{k!} \left(1 - \sum_{j=0}^{k} z_j \right), \quad x \geq 0 \tag{4.3.8}$$

using an interchange of the order of summation. The latter series representation is better suited for numerical calculations.

It remains to compute the z_j 's. Let

η_k = the long-run fraction of customers taking the kth position in their batch.

An expression for η_k will be given below. Using that f_i represents the long-run fraction of batches who find upon arrival i uncompleted phases present, it follows that

$$z_j = \sum_{i=0}^{j} f_i \sum_{k=1}^{j-i+1} \eta_k q_{j-i}^{(k-1)}, \quad j = 0, 1, \ldots, \tag{4.3.9}$$

where $q_j^{(n)}$ denotes the probability that n customers represent a total of j service phases. In general the probabilities $q_s^{(n)}$ must be recursively computed from $q_j^{(n)} = \sum_{m=1}^{j-1} q_m q_{j-m}^{(n-1)}$ starting with $q_j^{(0)} = 1$ for $j = 0$ and $q_j^{(0)} = 0$ for $j \geq 1$. To find η_k, we use the Renewal-Reward Theorem. Fix k. Suppose that a reward of 1 is earned for each customer taking the kth position in its batch. Then the long-run average reward per customer equals η_k. On the other hand, this long-run average equals the expected reward earned for one batch divided by the expected batch size. Thus

$$\eta_k = \frac{1}{\beta} \sum_{j \geq k} \beta_j, \quad k = 1, 2, \ldots . \tag{4.3.10}$$

The formula (4.3.9) requires the f_i 's. These probabilities are found by equating the rate at which the continuous-time Markov chain $\{X(t)\}$ leaves the set of states $\{i, i+1, \ldots\}$ to the rate at which the process enters this set. This gives

$$\mu f_i = \sum_{k=0}^{i-1} f_k \lambda \sum_{s \geq i-k} \nu_s, \quad i = 1, 2, \ldots, \tag{4.3.11}$$

where ν_s denotes the probability that a batch represents a total of s service phases. Obviously,

$$\nu_s = \sum_{n=1}^{s} \beta_n q_s^{(n)}, \quad s = 1, 2, \ldots .$$

Together the relations (4.3.9) and (4.3.11) provide an algorithm for the z_j's. The recursion scheme (4.3.11) is started with $f_0 = 1 - \rho$.

The computational effort of the recursion scheme (4.3.11) can be considerably reduced by using that

$$\frac{f_j}{f_{j+1}} \approx \tau \quad \text{for } j \text{ large enough}$$

for some constant $\tau > 1$. This asymptotic expansion requires that the batch-size distribution has no extremely long tail. Using the generating function of the f_j 's (see also Section 2.9) and (C.8) in Appendix C, an analytical expression for τ can be given. This expression is not given here. Another approach is simply to estimate τ during the recursive calculations.

Approximations for the waiting-time probabilities

Suppose that Assumption 4.3.1 is satisfied and let $b(t)$ denote the density of the service-time distribution function $B(t)$. Then the following asymptotic expansion applies

$$1 - W_q(x) \approx \gamma e^{-\delta x} \qquad \text{for } x \text{ large enough,}$$

where δ is the smallest positive solution to

$$\sum_{j=1}^{\infty} \beta_j \left\{ \int_0^{\infty} e^{\delta t} b(t) \, dt \right\}^j = 1 + \frac{\delta}{\lambda}$$

and the constant γ is given by

$$\gamma = \frac{(1 - \rho)\delta}{\lambda \beta} \left[1 - \lambda \int_0^{\infty} t e^{\delta t} b(t) \, dt \sum_{j=1}^{\infty} j \beta_j \left\{ \int_0^{\infty} e^{\delta t} b(t) \, dt \right\}^{j-1} \right]^{-1}$$

$$\times \left[1 - \int_0^{\infty} e^{\delta t} b(t) \, dt \right]^{-1}.$$

The proof of this result is quite deep and will not be given here; see Van Ommeren (1989).

The waiting-time probabilities can be calculated through the asymptotic expansion already for relatively small values of x. Just as in the case of the $M/G/1$ queue, the complementary waiting-time distribution $W_q^c(x) = 1 - W_q(x)$ can be approximated by

$$W_{\text{app}}^c(x) = \alpha e^{-\beta x} + \gamma e^{-\delta x} \qquad \text{for all } x \geq 0.$$

The constants α and β are the same as in (4.2.19), where W_q is now given (4.3.6) and P_{delay} is given by

$$P_{\text{delay}} = 1 - \frac{1 - \rho}{\beta}.$$

The explicit expression for P_{delay} is easy to see. By the property Poisson arrivals see time averages, the long-run fraction of batches that find upon arrival the server idle equals $p_0 = 1 - \rho$. Thus the long-run fraction of customers who directly enter service is equal to $(1 - \rho)/\beta$.

Two-moment approximation for the waiting-time percentiles

Just as in the $M/G/1$ queue the expression for W_q allows for the representation

$$W_q = (1 - c_S^2)W_q(\text{det}) + c_S^2 W_q(\text{exp}),$$

where $W_q(\text{det})$ and $W_q(\text{exp})$ denote the average delay in queue for the special cases of deterministic services and exponential services with the same means $E(S)$. Unlike in the $M/G/1$ queue, the average delay W_q cannot be written as $\frac{1}{2}(1 + c_S^2)W_q(\text{exp})$ because of the second term in the right side of (4.3.5).

In analogy with the approximation (4.2.24), it is suggested to approximate the percentiles of $W_q(x)$ by

$$\xi_{\text{app}}(p) = (1 - c_S^2)\xi_{\text{det}}(p) + c_S^2 \xi_{\text{exp}}(p) \quad \text{for } P_{\text{delay}} < p < 1.$$

The performance of this approximation will be discussed in Section 4.5.5 in the more general context of the $M^X/G/c$ queue.

4.4 THE *GI/G/*1 QUEUE

This section deals with the $GI/G/1$ queue in which the interarrival times and the service times both have general probability distributions. The server utilisation ρ is assumed to be smaller than 1. Computationally tractable results can be obtained only for special cases. However, the exact results for simpler models may be used as a basis for approximations to the complex $GI/G/1$ model, see also the discussion in Section 4.6 The discussion will concentrate on the computation of the waiting-time probabilities for the cases of generalized Erlangian services and Coxian-2 services. The case of deterministic services is dealt with in the context of the $GI/D/c$ queue in Section 4.6. The discussion below assumes that service is in order of arrival.

Generalized Erlangian services

Suppose that the service-time density $b(t)$ is given by (4.3.7). In other words, with probability $q_i (i = 1, \ldots, m)$ the service time of a customer is the sum of i independent phases each having an exponential distribution with mean $1/\mu$. Thus we can define the embedded Markov chain $\{X_n\}$ by

$X_n =$ the number of uncompleted service phases present just before the arrival of the nth customer.

Denoting by $\{\pi_j, \ j = 0, 1, \ldots\}$ the equilibrium distribution of this Markov chain, we find by the same arguments as used to derive (4.3.8) that

$$W_q(x) = 1 - \sum_{k=0}^{\infty} e^{-\mu x} \frac{(\mu x)^k}{k!} \left(1 - \sum_{j=0}^{k} \pi_j\right), \quad x \geq 0. \qquad (4.4.1)$$

Thus we have a computationally useful algorithm for the waiting-time distribution when the probabilities π_j can be efficiently computed. These probabilities are the unique solution to the equilibrium equations

$$\pi_j = \sum_{k=0}^{\infty} \pi_k p_{kj}, \quad j = 0, 1, \ldots \tag{4.4.2}$$

together with the normalizing equation $\Sigma_{j=0}^{\infty}\pi_j = 1$, where the p_{ij}'s are the one-step transition probabilities of the Markov chain $\{X_n\}$. The p_{ij}'s are easily found. Using that the service completions of phases occur according to a Poisson process with rate μ as long as the server is busy, it is readily seen that for any $i \geq 0$

$$p_{ij} = \sum_{k=\max(j-i,1)}^{m} q_k \int_0^{\infty} e^{-\mu t} \frac{(\mu t)^{i+k-j}}{(i+k-j)!} a(t)\, dt \quad \text{for } 1 \leq j \leq i+m,$$

where $a(t)$ denotes the probability density of the interarrival time. The infinite system of linear equations can be effectively solved by using the geometric tail approach discussed in Section 2.3.3. It will be shown below that a constant $0 < \eta < 1$ can be computed such that

$$\frac{\pi_{j+1}}{\pi_j} \approx \eta \quad \text{for } j \text{ large enough.} \tag{4.4.3}$$

Then, for an appropriately chosen integer M, the probability π_j can be replaced by $\pi_M \eta^{j-M}$ for $j \geq M$. In this way the infinite system (4.4.2) is reduced to a finite system of linear equations. For practical purposes a relatively small value of M usually suffices. The constant $\eta \in (0, 1)$ is the unique solution of the equation

$$\sum_{i=1}^{m} q_i \eta^{m-i} \int_0^{\infty} e^{-\mu(1-\eta)t} a(t)\, dt - \eta^m = 0. \tag{4.4.4}$$

The derivation of the results (4.4.3) and (4.4.4) is interesting, but may be skipped without loss of continuity. Define the generating function $\Pi(z)$ by $\Pi(z) = \Sigma_{j=0}^{\infty}\pi_j z^j$, $|z| \leq 1$. It seems not easy to derive an expression for $\Pi(z)$ directly from the balance equations (4.4.2). An indirect derivation uses the relation

$$\Pi(z) = \lim_{n \to \infty} E(z^{X_n}).$$

The key idea for the evaluation of this limit is to imagine that actual and fictitious service completions of phases occur according to a Poisson process with rate μ independently of the state of the system. By doing so and using the notation $x^+ = \max(x, 0)$, we can write X_{n+1} as

$$X_{n+1} = (X_n + Q_n - C_n)^+,$$

where Q_n is the number of service phases represented by the nth arrival and C_n is the number of Poisson events occurring in the interarrival time between the customers n and $n + 1$. Thus we find

$$E(z^{X_{n+1}}) = E\left[z^{(X_n + Q_n - C_n)^+}\right] = \sum_{i=1}^{m} q_i E\left[z^{(X_n + i - C_n)^+}\right]$$

$$= \sum_{i=1}^{m} q_i \left[E(z^{X_n + i - C_n}|X_n - i + C_n > 0)P\{X_n + i - C_n > 0\}\right.$$

$$\left. + E(z^0|X_n - i + C_n \le 0)P\{X_n + i - C_n \le 0\}\right].$$

Further, we have

$$E(z^{X_n + i - C_n}) = E(z^{X_n + i - C_n}|X_n + i - C_n > 0)P\{X_n + i - C_n > 0\}$$

$$+ E(z^{X_n + i - C_n}|X_n + i - C_n \le 0)P\{X_n + i - C_n \le 0\}.$$

Combining the two relations yields

$$E(z^{X_{n+1}}) = \sum_{i=1}^{m} q_i \left[E(z^{X_n + i - C_n})\right.$$

$$\left. + \left\{1 - E(z^{X_n + i - C_n}|X_n + i - C_n \le 0)\right\} P\{X_n + i - C_n \le 0\}\right].$$

Letting $n \to \infty$, we find

$$\Pi(z) = \sum_{i=1}^{m} q_i z^i \Pi(z) C(z^{-1}) + \sum_{i=1}^{m} q_i \sum_{m=0}^{\infty} b_{mi}(1 - z^{-m}),$$

where $C(z) = E(z^{C_n})$ and $b_{mi} = \lim_{n \to \infty} P\{X_n + i - C_n = -m\}$. The limits b_{mi} exist since the Markov chain $\{X_n\}$ is aperiodic. The function $C(z)$ is given by

$$C(z) = \sum_{j=0}^{\infty} \left\{\int_0^{\infty} e^{-\mu t} \frac{(\mu t)^j}{j!} a(t)\, dt\right\} z^j = \int_0^{\infty} e^{-\mu(1-z)t} a(t)\, dt.$$

Thus we find

$$\Pi(z) = \frac{\displaystyle\sum_{i=1}^{m} q_i \sum_{m=0}^{\infty} b_{mi}(1 - z^{-m})}{1 - C(z^{-1}) \displaystyle\sum_{i=1}^{m} q_i z^i}.$$

The generating function $\Pi(z)$ is the ratio of two functions that allow an analytic continuation outside the unit circle. Using the general result (C.8) in Appendix C, we next find the desired results (4.4.3) and (4.4.4).

Coxian-2 services

Suppose that the service time S of a customer has a Coxian-2 distribution with parameters (b, μ_1, μ_2). That is, S is distributed as U_1 with probability $1 - b$ and

S is distributed as $U_1 + U_2$ with probability b, where U_1 and U_2 are independent exponentials with respective means $1/\mu_1$ and $1/\mu_2$. Then the waiting-time distribution function $W_q(x)$ allows for the explicit expression

$$1 - W_q(x) = a_1 e^{-\eta_1 x} + a_2 e^{-\eta_2 x}, \quad x \geq 0, \tag{4.4.5}$$

where η_1 and η_2 with $0 < \eta_1 < \min(\mu_1, \mu_2) \leq \eta_2$ are the roots of

$$x^2 - (\mu_1 + \mu_2)x + \mu_1\mu_2 - \{\mu_1\mu_2 - (1-b)\mu_1 x\} \int_0^\infty e^{-xt} a(t)\, dt = 0.$$

The function $a(t)$ denotes the interarrival-time density. The constants a_1 and a_2 are given by

$$a_1 = [-\eta_1^2 \eta_2 + \eta_1\eta_2(\mu_1 + \mu_2) - \eta_2\mu_1\mu_2]/[\mu_1\mu_2(\eta_1 - \eta_2)]$$

$$a_2 = [\eta_1\eta_2^2 - \eta_1\eta_2(\mu_1 + \mu_2) + \eta_1\mu_1\mu_2]/[\mu_1\mu_2(\eta_1 - \eta_2)].$$

A derivation of this explicit result can be found in Cohen, 1982. In particular P_{delay} and W_q are given by

$$P_{\text{delay}} = 1 - \frac{\eta_1\eta_2}{\mu_1\mu_2} \quad \text{and} \quad W_q = -\frac{(\mu_1 + \mu_2)}{\mu_1\mu_2} + \frac{1}{\eta_1} + \frac{1}{\eta_2}.$$

Since the computation of the roots of a function of a single variable is standard fare in numerical analysis, the above results are very easy to use for practical purposes. Bisection is a safe and fast method to compute the roots.

General service times

The general $GI/G/1$ queue is very difficult to analyse. In general one has to resort to approximations. There are several approaches to obtain approximate numerical results:

(a) Approximate the service-time distribution by a mixture of Erlangian distributions or a Coxian-2 distribution.
(b) Approximate the continuous-time model by a discrete-time model and use Fast Fourier Transform methods.
(c) Use two-moment approximations based on the exact solutions of the $GI/D/1$ queue and the $GI/M/1$ queue.

The approaches (a) and (b) lead to approximate numerical results for the waiting-time probabilities. Moreover, the approach (a) requires that the squared coefficient of variation of the service time is not too large (say, $0 \leq c_S^2 \leq 2$). The approach (c) can only be applied to certain performance measures such as the average queue size, the average waiting time per customer and the (conditional) waiting-time percentiles. The latter approach is discussed in more detail in Section 4.6 dealing with the $GI/G/c$ queue.

Let us now briefly discuss the powerful approach (b) for the *GI/G/*1 queue. This approach is based on Lindley's integral equation. Define the random variables

D_n = the delay in queue of the nth customer,

S_n = the service time of the nth customer,

A_{n+1} = the interarrival time between the nth and $(n + 1)$th customer.

Let us for ease assume that the service times and interarrival times have probability densities $b(t)$ and $a(t)$. In the same way as in Section 1.8 of Chapter 1, we obtain

$$D_{n+1} = \max(0, D_n + U_n), \quad n = 1, 2, \ldots,$$

where $U_n = S_n - A_{n+1}$. Using this recurrence equation, it is not difficult to show that the waiting-time distribution function $W_q(x)$ satisfies the so-called Lindley's integral equation

$$W_q(x) = \int_{-\infty}^{x} W_q(x - t)c(t)\, \mathrm{d}t, \quad x \geq 0,$$

where $c(t)$ is the probability density of the U_n's. Note that $c(t)$ is the convolution of $a(-t)$ and $b(t)$. A discretized version of Lindley's integral equation can be solved effectively by using the Fast Fourier Transform method. The details will not be given here, but can be found in Ackroyd (1980) and Tran-Gia (1986).

KLB-approximation

Using a hybrid combination of basic queueing results and experimental analysis, the following two-moment approximations for the delay probability and the average delay in queue per customer were obtained by Krämer and Langenbach-Belz (1976):

$$P_{\text{delay}}^{\text{KLB}} = \rho + (c_A^2 - 1)\rho(1 - \rho) \times \begin{cases} \dfrac{1 + c_A^2 + \rho c_S^2}{1 + \rho(c_S^2 - 1) + \rho^2(4c_A^2 + c_S^2)} & \text{if } c_A^2 \leq 1, \\[2ex] \dfrac{4\rho}{c_A^2 + \rho^2(4c_A^2 + c_S^2)} & \text{if } c_A^2 > 1, \end{cases}$$

$$W_q^{\text{KLB}} = \frac{\rho E(S)}{2(1 - \rho)}(c_A^2 + c_S^2) \times \begin{cases} \exp\left\{ \dfrac{-2(1 - \rho)(1 - c_A^2)^2}{3\rho(c_A^2 + c_S^2)} \right\} & \text{if } c_A^2 \leq 1, \\[2ex] \exp\left\{ \dfrac{-(1 - \rho)(c_A^2 - 1)}{c_A^2 + 4c_S^2} \right\} & \text{if } c_A^2 > 1. \end{cases}$$

These approximations are only useful for rough estimates in practical engineering provided that the traffic load on the system is not small and c_A^2 is not large. In fact, one should be very careful in using the KLB-approximation when c_A^2 is larger than 1. A reason for this is that performance measures in queueing systems are usually much more sensitive to the shape of the interarrival-time density than to the

Table 4.4.1   Some numerical results for the $GI/G/1$ queue

		$\rho = 0.2$		$\rho = 0.5$		$\rho = 0.8$	
		P_{delay}	W_q	P_{delay}	W_q	P_{delay}	W_q
$D/E_4/1$	exact	0.000	0.000	0.047	0.017	0.446	0.319
	KLB	0.005	0.000	0.091	0.009	0.457	0.257
$D/E_2/1$	exact	0.001	0.000	0.116	0.078	0.548	0.757
	KLB	0.009	0.000	0.143	0.066	0.557	0.717
$E_4/D/1$	exact	0.009	0.002	0.163	0.050	0.578	0.386
	KLB	0.021	0.000	0.188	0.028	0.621	0.344
$E_2/D/1$	exact	0.064	0.024	0.323	0.177	0.702	0.903
	KLB	0.064	0.016	0.313	0.179	0.719	0.920
$E_2/H_2/1$	exact	0.110	0.203	0.405	1.095	0.752	4.825
	KLB	0.088	0.239	0.375	1.169	0.743	4.917
$H_2/E_2/1$	exact	0.336	0.387	0.650	1.445	0.870	5.281
	KLB	0.255	0.256	0.621	1.103	0.869	4.756

shape of the service-time density, particularly when the traffic load on the system is light. To illustrate the KLB approximation, we give in Table 4.4.1 some numerical results. The H_2 distributions in the table refer to a hyperexponential distribution with the gamma normalization and a squared coefficient of variation of 2.

4.5   MULTI-SERVER QUEUES WITH POISSON INPUT

Multi-server queues are notoriously difficult and a simple algorithmic analysis is possible only for special cases. In principle any practical queueing process could be modelled as a Markov process by incorporating sufficient information in the state description, but the dimensionality of the state space would grow quickly beyond any practical bound and thus obstructing an exact solution. In many situations, however, one resorts to approximation methods for calculating measures of system performance. Useful approximations for complex queueing systems are often obtained through exact results for simpler related queueing systems.

In this section we discuss both exact and approximate solution methods for the state probabilities and the waiting-time probabilities in multi-server queues. We first present exact methods for the tractable models of the $M/M/c$ queue and the $M/D/c$ queue. This is done in the Sections 4.5.1 and 4.5.2. In Section 4.5.3 we consider the $M/G/c$ queue with general service times and give several approximations including two-moment approximations based on exact results for the $M/M/c$ queue and the $M/D/c$ queue. Special attention is paid to the $M/G/\infty$ queue in Section 4.5.4. This is a very useful model for practical applications with a very large number of servers. In Section 4.5.5 we consider the $M^X/G/c$ queue with batch arrivals. In particular, the $M^X/M/c$ queue and the $M^X/D/c$ queue are dealt with.

4.5.1 The $M/M/c$ queue

This queueing system assumes a Poisson arrival process with rate λ, exponentially distributed service times with mean $E(S) = 1/\mu$ and c identical servers. It is supposed that the server utilisation ρ defined by

$$\rho = \frac{\lambda E(S)}{c}$$

is smaller than 1. Since both the Poisson process and the exponential distribution are memoryless, the process describing the number of customers in the system is a continuous-time Markov chain. Using the familiar technique of equating the rate at which the process leaves the set of states of having at least j customers present to the rate at which the system enters that set, we obtain for the time-average probabilities p_j the recursion relation

$$\min(j, c)\mu p_j = \lambda p_{j-1}, \quad j = 1, 2, \ldots . \tag{4.5.1}$$

An explicit solution for the p_j's can be given, but this explicit solution is of little use for computational purposes. A simple computational scheme can be based on the recursion relation (4.5.1). To do so, note that

$$p_j = \rho^{j-c+1} p_{c-1}, \quad j \geq c.$$

Algorithm
Step 0. Initialize $\overline{p}_0 := 1$.
Step 1. For $j = 1, \ldots, c - 1$, let $\overline{p}_j := \lambda \overline{p}_{j-1}/(j\mu)$.
Step 2. Calculate the normalizing constant γ from

$$\gamma = \sum_{j=0}^{c-1} \overline{p}_j + \frac{\rho \overline{p}_{c-1}}{1 - \rho}.$$

Normalize the $\overline{p}_j$'s according to

$$p_j := \frac{1}{\gamma} \overline{p}_j \quad \text{for } j = 0, 1, \ldots, c - 1.$$

Step 3. For any $j \geq c$, $p_j := \rho^{j-c+1} p_{c-1}$.

By the property Poisson arrivals see time averages, the time-average probabilities p_j give also the long-run fraction of customers who find upon arrival j other customers present. Since $p_j = \rho^{j-c+1} p_{c-1}$ for $j \geq c$, we thus find that the long-run fraction of customers who are delayed is given by

$$P_{\text{delay}} = \frac{\rho}{1 - \rho} p_{c-1}. \tag{4.5.2}$$

The delay probability for the $M/M/c$ queue is often called *Erlang's delay probability*.

Using the representation $L_q = \sum_{j=c}^{\infty}(j-c)p_j$, we find that the long-run average number of customers in queue is given by

$$L_q = \frac{\rho^2}{(1-\rho)^2}p_{c-1}. \tag{4.5.3}$$

The waiting-time distribution function $W_q(x)$ can also be given explicitly when service is in order of arrival. Therefore note that the conditional waiting time of a customer finding upon arrival $j \geq c$ other customers present is distributed as the sum of $j - c + 1$ independent exponentials with common mean $1/(c\mu)$ and thus has an E_{j-c+1} distribution. Using that the long-run fraction of customers who find upon arrival j other customers present is equal to p_j, it follows by conditioning that

$$1 - W_q(x) = \sum_{j=c}^{\infty} p_j \sum_{k=0}^{j-c} e^{-c\mu x}\frac{(c\mu x)^k}{k!}, \quad x \geq 0.$$

Since $p_j = \rho^{j-c+1}p_{c-1}$ for $j \geq c$, it follows that

$$1 - W_q(x) = \frac{\rho}{1-\rho}p_{c-1}e^{-c\mu(1-\rho)x}, \quad x \geq 0. \tag{4.5.4}$$

In particular, the average delay in queue of a customer equals .

$$W_q = \frac{\rho}{c\mu(1-\rho)^2}p_{c-1}, \tag{4.5.5}$$

in agreement with (4.5.3) and Little's formula $L_q = \lambda W_q$.

4.5.2 The $M/D/c$ queue

In this model the arrival process of customers is a Poisson process with rate λ, the service time of a customer is a constant D and c identical servers are available. It is assumed that the traffic intensity $\rho = \lambda D/c$ is smaller than 1.

An exact algorithm analysis of the $M/D/c$ queue goes already back to Crommelin (1932) and is based on the following observation. Since the service times are equal to the constant D, any customer in service at time t will have left the system at time $t + D$, while the customers present at time $t + D$ are exactly those customers either waiting in queue at time t or having arrived in $(t, t + D)$. Hence, letting $p_j(s)$ be the probability of having j customers in the system at time s, it follows by conditioning on the number of customers present at time t that

$$p_j(t+D) = \sum_{k=0}^{c} p_k(t)e^{-\lambda D}\frac{(\lambda D)^j}{j!} + \sum_{k=c+1}^{c+j} p_k(t)e^{-\lambda D}\frac{(\lambda D)^{j-k+c}}{(j-k+c)!}$$

for $j = 0, 1, \ldots$, using that the number of arrivals in a time D is Poisson distributed with mean λD. Next, by letting $t \to \infty$ in these equations, we find that the time-average probabilities p_j satisfy the linear equations

$$p_j = e^{-\lambda D} \frac{(\lambda D)^j}{j!} \sum_{k=0}^{c} p_k + \sum_{k=c+1}^{c+j} p_k e^{-\lambda D} \frac{(\lambda D)^{j-k+c}}{(j-k+c)!}, \quad j \geq 0. \qquad (4.5.6)$$

Also, we have the normalizing equation $\Sigma_{j=0}^{\infty} p_j = 1$. This infinite system of linear equations can be reduced to a finite system of linear equations by using the geometric tail approach discussed in Section 2.3.3. It will be shown below that the state probabilities p_j exhibit the geometric tail behaviour

$$p_j \approx \sigma \tau^{-j} \quad \text{for } j \text{ large enough}, \qquad (4.5.7)$$

where τ is the unique solution of the equation

$$e^{\lambda D(1-\tau)} \tau^c = 1 \qquad (4.5.8)$$

on the interval $(1, \infty)$ and the constant σ is given by

$$\sigma = (c - \lambda D\tau)^{-1} \sum_{k=0}^{c-1} p_k (\tau^k - \tau^c). \qquad (4.5.9)$$

To solve the equation for τ, it is recommended to use logarithms. Thus τ should be solved from $\lambda D(1 - \tau) + c \ln (\tau) = 0$. For an appropriately chosen integer M the infinite system of linear equations for the p_j's is reduced to a finite system by replacing p_j by $p_M(1/\tau)^{j-M}$ for $j > M$. This reduction approach leads to a relatively small system of linear equations that can usually be solved by a standard Gaussian elimination method. Thus we can conclude that the computation of the state probabilities in the $M/D/c$ queue is a routine matter, see also Remark 4.5.1 below.

To verify (4.5.7), we need the generating function $P(z) = \Sigma_{j=0}^{\infty} p_j z^j, |z| \leq 1$. Multiplying both sides of (4.5.6) by z^j and summing over j, we find that

$$P(z) = e^{\lambda D(z-1)} \sum_{k=0}^{c} p_k + \sum_{j=1}^{\infty} z^j \sum_{i=1}^{j} p_{c+i} e^{-\lambda D} \frac{(\lambda D)^{j-i}}{(j-i)!}$$

$$= e^{\lambda D(z-1)} \sum_{k=0}^{c} p_k + \sum_{i=1}^{\infty} p_{c+i} z^i \sum_{j=i}^{\infty} e^{-\lambda D} \frac{(\lambda D)^{j-i}}{(j-i)!} z^{j-i}$$

$$= e^{\lambda D(z-1)} \sum_{k=0}^{c} p_k + e^{\lambda D(z-1)} \sum_{i=1}^{\infty} p_{c+i} z^i.$$

Next it is readily verified that

$$
P(z) = \frac{\displaystyle\sum_{k=0}^{c-1} p_k(z^k - z^c)}{1 - e^{\lambda D(1-z)} z^c}. \tag{4.5.10}
$$

The generating function $P(z)$ is the ratio of two functions that allow for an analytic continuation outside the unit circle. The desired result (4.5.7) next follows from the general result (C.8) in Appendix C.

Using the representation $L_q = \Sigma_{j=c}^{\infty}(j - c)p_j$ and the general relation (4.1.4), we obtain after considerable algebra from (4.5.10) that

$$
L_q = \frac{1}{2c(1 - \rho)} \left[(c\rho)^2 - c(c - 1) + \sum_{j=0}^{c-1} \{c(c - 1) - j(j - 1)\}p_j \right]. \tag{4.5.11}
$$

Next an expression for the average delay in queue per customer follows by using Little's formula $L_q = \lambda W_q$.

Remark 4.5.1   An alternative approach to computing the state probabilities is to use the Fast Fourier Transform method. We cannot directly apply this method, since the right side of (4.5.10) involves the unknowns $p_0, \ldots, p_{c-1}$. An explicit expression for the generating function $P(z)$ could have been available if the values of $p_0, \ldots, p_{c-1}$ had been known. However, this difficulty can be circumvented as follows. Using an initial guess for $p_0, \ldots, p_{c-1}$, the application of the FFT method to (4.5.10) yields estimates for the p_j's including new estimates for $p_0, \ldots, p_{c-1}$. The FFT method is now repeatedly applied until the estimates for $p_0, \ldots, p_{c-1}$ have been sufficiently converged. It turns out from empirical studies that this iterative procedure converges very quickly provided that after each iteration the estimates for the p_j's are normalized in order to satisfy the equation

$$
\sum_{j=0}^{c-1} jp_j + c \sum_{j=c}^{\infty} p_j = c\rho.
$$

This equation expresses that the average number of busy servers is $c\rho$.

Waiting-time probabilities

We next show how to compute the waiting-time distribution function $W_q(x)$ when service is in order of arrival. An explicit expression in the form of an infinite series can be derived for $W_q(x)$. This infinite series consists of terms that alternate in sign and is thus of little use for computational purposes. Therefore the explicit expression is not further discussed. A computationally tractable algorithm is obtained by combining a recursion scheme and an asymptotic expansion. We first derive a recursive relation for $W_q(x)$ using an ingenious argument due to Crommelin, 1932.

Fix $x > 0$ and define

$$m = \left[\frac{x}{D}\right] \quad \text{and} \quad u = x - mD,$$

where $[a]$ is the largest integer contained in a. Consider now a marked customer. The delay in queue of the marked customer is no more than x if and only if at most $mc + c - 1$ customers among those present just before the arrival of the marked customer remain in the system a time u later. This is directly seen by noting that exactly mc customers will be served during a time mD when all c servers are continuously busy. Thus

the long-run fraction of customers whose delay in queue is no more than x = the long-run fraction of arrivals for which at most $mc + c - 1$ customers among those present just before the occurrence of the arrival remain in the system a time u later.

Let the random variable $I(t) = 1$ if at most $mc + c - 1$ customers among those present at time t remain in the system at time $t + u$ and let $I(t) = 0$ otherwise. Also, let the random variable $I_n = 1$ if at most $mc + c - 1$ customers among those present just before the arrival epoch of the nth customer remain in the system a time u later and let $I_n = 0$ otherwise. Then, by the property Poisson arrivals see time averages,

$$\lim_{n \to \infty} \frac{1}{n} \sum_{k=1}^{n} I_k = \lim_{t \to \infty} \frac{1}{t} \int_0^t I(y) \, dy,$$

see Theorem 1.7.1. As argued above, $\lim_{n \to \infty}(1/n)\Sigma_{k=1}^n I_k$ equals $W_q(x)$ with probability 1. Next we develop the other limit. For any $v \geq 0$, define

$$b_v(u) = \lim_{t \to \infty} P\{ \text{at most } v \text{ customers among those present at time } t \text{ remain in the system at time } t + u\}.$$

Then $\lim_{t \to \infty}(1/t) \int_0^t I(y) \, dy = b_{mc+c-1}(u)$ with probability 1 and so

$$W_q(x) = b_{mc+c-1}(u). \tag{4.5.12}$$

To find the probabilities $b_v(u)$, use the fact that $0 \leq u < D$ to conclude that

$P\{\text{at most } j \text{ customers are in the system at time } t + u\}$

$$= \sum_{k=0}^{j} P\{k \text{ arrivals occur in } (t, t+u] \text{ and at most } j - k \text{ customers among those present at time } t \text{ remain in the system at time } t + u\}.$$

Letting $t \to \infty$ in this equation and using the lack of memory of the Poisson process, we find

$$\sum_{i=0}^{j} p_i = \sum_{k=0}^{j} e^{-\lambda u} \frac{(\lambda u)^k}{k!} b_{j-k}(u), \quad j = 0, 1, \ldots . \tag{4.5.13}$$

By subtracting the equations in (4.5.13) corresponding to $j = n$ and $j = n - 1$, we obtain the recursion scheme

$$b_n(u) = e^{-\lambda u} p_n - \sum_{k=0}^{n-1} b_k(u) \frac{(\lambda u)^{n-1-k}}{(n-1-k)!} \left(\frac{\lambda u}{n-k} - 1 \right), \quad n = 0, 1, \ldots .$$

$$(4.5.14)$$

Summarizing, the waiting-time probabilities can be computed as follows.

Algorithm for $W_q(x)$
Step 0. Let $m := [x/D]$ and $u := x - mD$.
Step 1. Apply the recursion scheme (4.5.14) until $b_{mc+c-1}(u)$ has been obtained.
Step 2. $W_q(x) := b_{mc+c+1}(u)$.

The following comment on this procedure should be made. Since the recursion scheme (4.5.14) involves the taking of differences, it will ultimately be hampered by roundoff errors for large values of λu, particularly when c and ρ increase. Therefore the algorithm should be combined with the asymptotic expansion for $W_q(x)$. In situations of practical interest the asymptotic expansion typically works long before the recursion scheme offers numerical difficulties. The asymptotic expansion for $W_q(x)$ is given by

$$1 - W_q(x) \approx \gamma e^{-\delta x} \quad \text{for } x \text{ large enough}, \tag{4.5.15}$$

where

$$\delta = \lambda(\tau - 1) \quad \text{and} \quad \gamma = \frac{\sigma}{(\tau - 1)\tau^{c-1}}$$

with τ and σ as in (4.5.8) and (4.5.9). To prove the result (4.5.15), define the generating function $\overline{B}_u(z)$ by

$$\overline{B}_u(z) = \sum_{j=0}^{\infty} \{1 - b_j(u)\} z^j.$$

Using that the generating function of the convolution of two probability distributions is the product of the generating functions of the two distributions, it readily follows from (4.5.13) and (4.5.10) that

$$\overline{B}_u(z) = \frac{\left[1 - e^{\lambda D(z-1)} z^c - e^{\lambda u(1-z)} \sum_{k=0}^{c} p_k(z^k - z^c) \right] \Big/ (1 - z)}{1 - e^{\lambda D(z-1)} z^c}.$$

Next the desired result (4.5.15) is obtained by using the general result (C.8) in Appendix C.

4.5.3 The $M/G/c$ queue

In this multi-server model with c servers the arrival process of customers is a Poisson process with rate λ and the service time S of a customer has a general probability distribution function $B(t)$. It is assumed that $\rho = \lambda E(S)/c$ is smaller than 1.

The $M/G/c$ queue with general service times permits no simple analytical solution, not even for the average waiting time. Useful approximations can be obtained by the regenerative approach discussed in Section 4.2.1. In applying this approach to the multi-server queue we encounter the difficulty that the number of customers left behind at a service completion epoch does not provide sufficient information to describe the future behaviour of the system. In fact we need the additional information of the elapsed service times of the other services (if any) still in progress. A full inclusion of this information in the state description would lead to an intractable analysis. However, as an approximation, we will aggregate the information of the elapsed service times in such a way that the resulting approximate model enables us to carry through the regenerative analysis. A closer look at the regenerative approach reveals that we need only a suitable approximation to the time until the next service completion epoch. We now make the following approximation assumption with regard to the behaviour of the process at the service completion epochs.

Approximation Assumption 4.5.1   (a) *If at a service completion epoch k customers are left behind in the system with $1 \leq k < c$, then the time until the next service completion epoch is distributed as $\min(S_1^e, \ldots, S_k^e)$, where $S_1^e, \ldots, S_k^e$ are independent random variables having each the residual life distribution function*

$$B_e(t) = \frac{1}{E(S)} \int_0^t \{1 - B(x)\}\, dx, \quad t \geq 0,$$

as probability distribution function.

   (b) *If at a service completion epoch k customers are left behind in the system with $k \geq c$, then the time until the next service completion is distributed as S/c, where S denotes the original service time of a customer.*

This approximation assumption can be motivated as follows. First, if not all c servers are busy, the $M/G/c$ queueing system may be treated as an $M/G/\infty$ queueing system in which a free server is immediately provided to each arriving customer. For the $M/G/\infty$ queue in statistical equilibrium it was shown by Tákacs (1962) that the remaining service time of any busy server is distributed as the residual life in a renewal process with the service times as the interoccurrence times. The equilibrium excess distribution of the service time is given by $B_e(t)$; see also Theorem 1.1.11 in Section 1.1. Second, if all of the c servers are busy, then the $M/G/c$ queue may be approximated by an $M/G/1$ queue in which the single server works c times as fast as each of the c servers in the original multi-server system. It is pointed out that the Approximation Assumption holds exactly for both the case of $c = 1$ server and the case of exponentially distributed service times.

Approximations to the state probabilities

Under the Approximation Assumption the recursion scheme derived in Section 4.2.1 for the $M/G/1$ queue can be extended to the $M/G/c$ queue to yield approximations p_j^{app} to the state probabilities p_j. These approximations are given in the next theorem whose lengthy proof may be skipped at first reading.

Theorem 4.5.1   *Under the Approximation Assumption,*

$$p_j^{\mathrm{app}} = \frac{(c\rho)^j}{j!} p_0^{\mathrm{app}}, \quad j = 0, 1, \ldots, c - 1, \tag{4.5.16}$$

$$p_j^{\mathrm{app}} = \lambda a_{j-c} p_{c-1}^{\mathrm{app}} + \lambda \sum_{k=c}^{j} b_{j-k} p_k^{\mathrm{app}}, \quad j = c, c + 1, \ldots, \tag{4.5.17}$$

where the constants a_n and b_n are given by

$$a_n = \int_0^\infty \{1 - B_e(t)\}^{c-1}\{1 - B(t)\}e^{-\lambda t} \frac{(\lambda t)^n}{n!} \, dt, \quad n = 0, 1, \ldots,$$

$$b_n = \int_0^\infty \{1 - B(ct)\}e^{-\lambda t} \frac{(\lambda t)^n}{n!} \, dt, \quad n = 0, 1, \ldots .$$

Proof   In the same way as in the proof of Theorem 4.2.1 in Section 4.2 we find

$$p_j^{\mathrm{app}} = \lambda p_0^{\mathrm{app}} A_{0j} + \sum_{k=1}^{j} \lambda p_k^{\mathrm{app}} A_{kj}, \quad j = 1, 2, \ldots, \tag{4.5.18}$$

where the constant A_{kj} is defined by

   A_{kj} = the expected amount of time that j customers are present during the time until the next service completion epoch when a service has just been completed with k customers left behind in the system.

By the same argument as used to derive (4.2.7), we find under the Approximation Assumption that

$$A_{kj} = \int_0^\infty \{1 - B(ct)\}e^{-\lambda t} \frac{(\lambda t)^{j-k}}{(j-k)!} \, dt \quad \text{for } k \geq c \text{ and } j \geq k.$$

However, the problem is to find a tractable expression for A_{kj} when $0 \leq k \leq c-1$. An explicit expression for A_{kj} involves a multi-dimensional integral when $0 \leq k \leq c - 1$. Fortunately, this difficulty can be circumvented by defining, for any $1 \leq k \leq c$ and $j \geq k$, the probability $M_{kj}(t)$ by

   $M_{kj}(t) = P\{\, j - k$ customers arrive during the next t time units and the service of none of these customers is completed in the next t time units when only $c - k$ servers are available for the new arrivals$\}$.

Then, using the Approximation Assumption,

$$A_{kj} = \int_0^\infty \{1 - B_e(t)\}^k M_{kj}(t)\, dt, \quad 1 \le k \le c - 1, \ j \ge k. \qquad (4.5.19)$$

Further, we have

$$A_{0j} = \int_0^\infty \{1 - B(t)\} M_{1j}(t)\, dt, \quad j \ge 1.$$

The definition of $M_{kj}(t)$ implies that

$$M_{kk}(t) = e^{-\lambda t}, \quad k \ge 1$$

and $\qquad\qquad\qquad\qquad\qquad\qquad\qquad\qquad\qquad\qquad\qquad (4.5.20)$

$$M_{cj}(t) = e^{-\lambda t} \frac{(\lambda t)^{j-c}}{(j - c)!}, \quad j \ge c.$$

Next we derive a differential equation for $M_{kj}(t)$ when $j > k$. By conditioning on what may happen in the first Δt time units, we find for any $1 \le k \le c - 1$ and $j > k$ that

$$M_{kj}(t + \Delta t) = (1 - \lambda \Delta t) M_{kj}(t) + \lambda \Delta t \{1 - B(t)\} M_{k+1,j}(t) + o(\Delta t), \quad t > 0.$$

Hence, for any $1 \le k \le c - 1$ and $j > k$,

$$M'_{kj}(t) = -\lambda M_{kj}(t) + \lambda \{1 - B(t)\} M_{k+1,j}(t), \quad t > 0.$$

Multiplying both sides of this differential equation by $\{1 - B_e(t)\}^k$, integrating over t and using (4.5.19), we find after partial integration that

$$A_{kj} = B_{k+1,j} - \frac{k}{\lambda E(S)} B_{kj}, \quad 1 \le k \le c - 1, \ j > k, \qquad (4.5.21)$$

where B_{kj} is a shorthand notation for

$$B_{kj} = \int_0^\infty \{1 - B_e(t)\}^{k-1} \{1 - B(t)\} M_{kj}(t)\, dt.$$

Next it is easy to establish the recursion scheme for p_j^{app}. To verify (4.5.16), we use induction. Obviously, (4.5.16) holds for $j = 0$. Suppose now that (4.5.16) holds for $j = 0, \ldots, n - 1$ for some $1 \le n \le c - 1$. Then, by (4.5.18) and (4.5.21),

$$p_n^{\text{app}} (1 - \lambda A_{nn}) = \lambda p_0^{\text{app}} A_{0n} + \sum_{k=1}^{n-1} \lambda p_k^{\text{app}} \left\{ B_{k+1,n} - \frac{k}{\lambda E(S)} B_{kn} \right\}$$

$$= \sum_{k=0}^{n-1} \lambda p_k^{\text{app}} B_{k+1,n} - \sum_{k=1}^{n-1} \lambda p_{k-1}^{\text{app}} B_{kn} = \lambda p_{n-1}^{\text{app}} B_{nn} \qquad (4.5.22)$$

where the second equality uses $A_{0n} = B_{1n}$ and the induction assumption that $p_k^{app} = c\rho p_{k-1}^{app}/k$ for $1 \le k \le n-1$. Using partial integration it is readily verified that $B_{nn} = (1-\lambda A_{nn})E(S)/n$. Hence we obtain from (4.5.22) that $p_n^{app} = c\rho p_{n-1}^{app}/n$ which completes the induction step. To verify (4.5.17), we first note that

$$\lambda p_0^{app} A_{0j} + \sum_{k=1}^{c-1} \lambda p_k^{app} A_{kj} = \lambda p_{c-1}^{app} B_{cj} \quad \text{for } j \ge c. \qquad (4.5.23)$$

The derivation of this relation is similar to that of (4.5.22). Inserting (4.5.23) into (4.5.18) and using (4.5.20), the desired relation (4.5.17) follows.

The recursion scheme for p_j^{app} is easy to apply in practice. In general the constants a_n and b_n have to be evaluated by numerical integration. However, this is no serious problem for service-time distributions of practical interest. The computational effort of the approximation algorithm depends only to a slight degree on c (=the number of servers), as opposed to exact methods for which the computing times quickly increase when c gets larger. The computational burden of the recursion scheme can be reduced by using an asymptotic expansion for p_j^{app}. This asymptotic expansion will be discussed later in this section. It will then be seen that p_j^{app}/p_{j-1}^{app} is asymptotically exact.

Further support to the quality of the approximation is provided by the result

$$P_{delay} = P_{delay}^{exp}, \qquad (4.5.24)$$

where P_{delay}^{exp} denotes the delay probability in the $M/M/c$ queue. It has long been known that Erlang's delay probability gives a good approximation to the delay probability in the general $M/G/c$ queue. To prove (4.5.24), sum both sides of (4.5.17) over $j \ge c$. This yields

$$\sum_{j=c}^{\infty} p_j^{app} = \frac{\rho}{1-\rho} p_{c-1}^{app}. \qquad (4.5.25)$$

By the relation (4.5.1) and (4.5.16), we have

$$p_j^{exp} = \frac{c\rho}{j} p_{j-1}^{exp} \quad \text{and} \quad p_j^{app} = \frac{c\rho}{j} p_{j-1}^{app} \quad \text{for } 1 \le j \le c-1,$$

where p_j^{exp} denotes the state probability p_j in the $M/M/c$ queue. Hence, for some constant γ, $p_j^{app} = \gamma p_j^{exp}$ for $0 \le j \le c-1$. Rewriting $\Sigma_{j=c}^{\infty} p_j$ as $1 - \Sigma_{j=0}^{c-1} p_j$, it next follows from (4.5.25) that

$$\frac{\gamma\rho}{1-\rho} p_{c-1}^{exp} = 1 - \sum_{j=0}^{c-1} \gamma p_j^{exp} = 1 - \gamma + \gamma \sum_{j=c}^{\infty} p_j^{exp} = 1 - \gamma + \frac{\gamma\rho}{1-\rho} p_{c-1}^{exp}.$$

This implies that $\gamma = 1$ and so $P_{delay}^{app} = \rho p_{c-1}^{exp}/(1-\rho)$. Next, using (4.5.2), the relation (4.5.24) follows.

As a by-product of the above proof, we find

$$p_j^{\mathrm{app}} = p_j^{\mathrm{exp}} \quad \text{for } j = 0, 1, \ldots, c - 1.$$

In other words, the approximate queueing system behaves like an $M/M/c$ queue when not all of the c servers are busy.

Average queue size

Using the technique of generating functions, we can derive approximations to the moments of the queue size from the recursion scheme. Define

$$P_q(z) = \sum_{j=0}^{\infty} p_{c+j}^{\mathrm{app}} z^j, \quad |z| \le 1.$$

It is a matter of simple algebra to derive from (4.5.17) that

$$P_q(z) = \lambda p_{c-1}^{\mathrm{app}} \frac{\alpha(z)}{1 - \lambda \beta(z)}, \tag{4.5.26}$$

where

$$\alpha(z) = \int_0^{\infty} \{1 - B_e(t)\}^{c-1} \{1 - B(t)\} e^{-\lambda(1-z)t} \, dt,$$

$$\beta(z) = \int_0^{\infty} \{1 - B(ct)\} e^{-\lambda(1-z)t} \, dt.$$

Using the representation $L_q = \sum_{j=c}^{\infty} (j-c)p_j$, the derivative $P_q'(1)$ yields an approximation to L_q. By differentiation of (4.5.26) we find after lengthy algebra that

$$L_q^{\mathrm{app}} = \left[(1 - \rho)\gamma_1 \frac{c}{E(S)} + \frac{1}{2}\rho(1 + c_S^2) \right] L_q(\exp), \tag{4.5.27}$$

where $c_S^2 = \sigma^2(S)/E^2(S)$ and

$$\gamma_1 = \int_0^{\infty} \{1 - B_e(t)\}^c \, dt.$$

The quantity $L_q(\exp)$ denotes the average queue size in the $M/M/c$ queue.

In Table 4.5.1 we give for several examples the exact and approximate values for P_{delay} and L_q. We consider the cases of deterministic service ($c_S^2 = 0$), Erlang-2 service ($c_S^2 = 0.5$) and hyperexponential service of order two ($c_S^2 = 2$) with the gamma normalization. In the table we also include the two-moment approximation

$$L_q^{\mathrm{app2}} = (1 - c_S^2)L_q(\det) + c_S^2 L_q(\exp), \tag{4.5.28}$$

where the Cosmetatos approximation

$$L_q^{\mathrm{app}}(\det) = \frac{1}{2} \left[1 + (1 - \rho)(c - 1)\frac{\sqrt{4 + 5c}}{16c\rho} - 2 \right] L_q(\exp)$$

is used for the average queue size $L_q(\det)$ in the $M/D/c$ queue.

Table 4.5.1   Exact and approximate results

		$c_S^2 = 0$		$c_S^2 = 0.5$		$c_S^2 = 2$	
		P_{delay}	L_q	P_{delay}	L_q	P_{delay}	L_q
$c = 2$	exact	0.3233	0.177	0.3308	0.256	0.3363	0.487
$\rho = 0.5$	approx	0.3333	0.194	0.3333	0.260	0.3333	0.479
	approx2	—	0.176	—	0.255	—	0.491
$c = 5$	exact	0.1213	0.077	0.1279	0.104	0.1335	0.181
$\rho = 0.5$	approx	0.1304	0.087	0.1304	0.107	0.1304	0.176
	approx2	—	0.076	—	0.103	—	0.185
$c = 10$	exact	0.0331	0.024	0.0352	0.030	0.0373	0.048
$\rho = 0.5$	approx	0.0361	0.025	0.0361	0.030	0.0361	0.047
	approx2	—	0.023	—	0.030	—	0.049
$c = 2$	exact	0.7019	1.445	0.7087	2.148	0.7141	4.231
$\rho = 0.8$	approx	0.7111	1.517	0.7111	2.169	0.7111	4.196
	approx2	—	1.442	—	2.143	—	4.247
$c = 5$	exact	0.5336	1.156	0.5484	1.693	0.5611	3.250
$\rho = 0.8$	approx	0.5541	1.256	0.5541	1.723	0.5541	3.191
	approx2	—	1.155	—	1.686	—	3.277
$c = 25$	exact	0.1900	0.477	0.2033	0.661	0.2164	1.173
$\rho = 0.8$	approx	0.2091	0.495	0.2091	0.663	0.2091	1.178
	approx2	—	0.477	—	0.657	—	1.196
$c = 50$	exact	0.0776	0.214	0.0840	0.282	0.0908	0.471
$\rho = 0.8$	approx	0.0870	0.207	0.0870	0.277	0.0870	0.488
	approx2	—	0.211	—	0.279	—	0.485

Asymptotic expansions

It is assumed that the probability distribution function $B_c(t) = B(ct)$ satisfies Assumption 4.2.1 in Section 4.2. In other words, the service-time distribution does not have an extremely long tail. Let $B = \sup[s| \int_0^\infty e^{st}\{1 - B(ct)\}\,dt < \infty]$. Then, using the representation (4.5.26) and the general result (C.8) in Appendix C, it is a routine matter to verify that

$$p_j^{\text{app}} \approx \sigma_{\text{app}}\tau^{-j} \quad \text{for } j \text{ large enough,} \tag{4.5.29}$$

where τ is the unique solution to the equation

$$\int_0^\infty e^{-\lambda(1-\tau)t}\{1 - B(ct)\}\,dt = \frac{1}{\lambda} \tag{4.5.30}$$

on the interval $(1, 1 + B/\lambda)$. The constant σ_{app} is given by

$$\sigma_{\text{app}} = \frac{p_{c-1}^{\text{app}}\tau^{c-1}}{\lambda^2} \frac{\int_0^\infty e^{-\lambda(1-\tau)t}\{1 - B_e(t)\}^{c-1}\{1 - B(t)\}\,dt}{\int_0^\infty t e^{-\lambda(1-\tau)t}\{1 - B(ct)\}\,dt}. \tag{4.5.31}$$

We are now in a position to prove that $p_j^{\mathrm{app}}/p_{j-1}^{\mathrm{app}}$ is asymptotically exact. To do so, we invoke a general result for the $GI/G/c$ queue. Denote by $A(t)$ and $B(t)$ the interarrival-time distribution function and the service-time distribution function for this model. For ease let us assume that $A(t)$ and $B(t)$ have densities $a(t)$ and $b(t)$. Let λ denote the long-run average arrival rate of customers. Suppose that $B_c(t) = B(ct)$ satisfies Assumption 4.2.1 in Section 4.2. Then

$$p_j \approx \sigma \tau^{-j} \quad \text{for } j \text{ large enough,} \tag{4.5.32}$$

and

$$1 - W_q(x) \approx \frac{\sigma \delta}{\lambda(\tau - 1)^2 \tau^{c-1}} e^{-\delta x} \quad \text{for } x \text{ large enough,} \tag{4.5.33}$$

where δ is the unique positive solution to the characteristic equation

$$\int_0^\infty e^{-\delta x} a(x)\,\mathrm{d}x \int_0^\infty e^{\delta y/c} b(y)\,\mathrm{d}y = 1 \tag{4.5.34}$$

on the interval $(0, B)$ and the constant τ is given by

$$\tau = \left[\int_0^\infty e^{-\delta x} a(x)\,\mathrm{d}x \right]^{-1}.$$

An explicit expression for the constant σ cannot be given in general. The constant σ can be computed from $\lim_{j \to \infty} p_j \tau^j$. A proof of the above asymptotic expansions for the $GI/G/c$ queue is beyond the scope of this book and can be found in Takahashi (1981). In fact, this reference proves the result only when $A(t)$ and $B(t)$ are phase-type distributions.

Returning to the $M/G/c$ queue, we have that the equation for τ reduces to

$$\tau = 1 + \delta/\lambda,$$

while the characteristic equation (4.5.34) can be rewritten as

$$\lambda \int_0^\infty e^{\delta y} \{1 - B(cy)\}\,\mathrm{d}y = 1.$$

In other words, we arrive at the equation (4.5.30) proving that $p_j^{\mathrm{app}}/p_{j-1}^{\mathrm{app}}$ is asymptotically exact. Moreover, we find that

$$1 - W_q(x) \approx \gamma e^{-\lambda(\tau-1)x} \quad \text{for } x \text{ large enough,} \tag{4.5.35}$$

where an approximation to γ is given by

$$\gamma_{\mathrm{app}} = \frac{\sigma_{\mathrm{app}}}{(\tau - 1)\tau^{c-1}}.$$

Two-moment approximations for the waiting-time percentiles

It is convenient to work with the conditional waiting-time percentiles $\eta(p)$ defined by (4.2.25) rather than with the unconditional waiting-time percentiles $\xi(p)$. The percentiles $\eta(p)$ are defined for all $0 < p < 1$, whereas the unconditional percentiles $\xi(p)$ are defined only for $P_{\text{delay}} < p < 1$. Note the relationship $\eta(p_1) = \xi(p_0)$ when $p_0 = 1 - (1 - p_1)P_{\text{delay}}$.

Just as in the $M/G/1$ case, we suggest the first-order approximation

$$\eta_{app1}(p) = \frac{1}{2}(1 + c_S^2)\eta_{\exp}(p) \qquad (4.5.36)$$

and the second-order approximation

$$\eta_{app2}(p) = (1 - c_S^2)\eta_{\det}(p) + c_S^2\eta_{\exp}(p), \qquad (4.5.37)$$

where $\eta_{\exp}(p)$ and $\eta_{\det}(p)$ are the corresponding percentiles for the $M/M/c$ queue and the $M/D/c$ queue. Both approximations require that the squared coefficient of variation of the service time is not too large (say, $0 \le c_S^2 \le 2$) and the traffic load on the system is not very small. In the multi-server case an appropriate measure for the load on the system is the fraction of time that all servers are busy. This fraction is given by P_{delay}. The second-order approximation performs quite satisfactorily for all parameter values. The simple approximation (4.5.36) is only useful for quick engineering calculations when P_{delay} is not small and p is sufficiently close to 1 (say, $p > 1 - P_{\text{delay}}$). In Table 4.5.2 we give, for several examples, the exact value and the approximate values (4.5.36) and (4.5.37) for the conditional waiting-time percentile. Also, we include in the table the asymptotic value based on the approximation (4.5.35). We consider the cases of Erlang-2 services ($c_S^2 = 0.5$) and hyperexponential services ($c_S^2 = 2$) with the gamma normalization.

Table 4.5.2 Conditional waiting-time percentiles

p		$c_S^2 = 0.5$				$c_S^2 = 2$			
		0.2	0.5	0.9	0.99	0.2	0.5	0.9	0.99
$c = 2$	exact	0.200	0.569	1.72	3.32	0.256	0.930	3.48	7.15
$\rho = 0.5$	app1	0.167	0.520	1.73	3.45	0.335	1.04	3.45	6.91
	app2	0.203	0.580	1.70	3.31	0.264	0.920	3.52	7.20
	asymp	0.282	0.609	1.73	3.33	0.158	0.907	3.47	7.14
$c = 5$	exact	0.082	0.240	0.722	1.37	0.099	0.339	1.32	2.78
$\rho = 0.5$	app1	0.067	0.208	0.691	1.38	0.134	0.416	1.38	2.76
	app2	0.082	0.243	0.725	1.36	0.104	0.346	1.32	2.82
	asymp	0.146	0.277	0.725	1.36	—	0.296	1.32	2.79
$c = 5$	exact	0.193	0.554	1.74	3.42	0.274	0.962	3.43	6.96
$\rho = 0.8$	app1	0.167	0.520	1.73	3.45	0.335	1.04	3.45	6.91
	app2	0.192	0.556	1.73	3.42	0.284	0.967	3.44	6.98
	asymp	0.218	0.562	1.74	3.42	0.232	0.954	3.42	6.96
$c = 25$	exact	0.040	0.118	0.364	0.703	0.052	0.174	0.649	1.35
$\rho = 0.8$	app1	0.033	0.104	0.345	0.691	0.067	0.208	0.691	1.38
	app2	0.040	0.119	0.365	0.701	0.055	0.179	0.651	1.36
	asymp	0.048	0.117	0.353	0.690	0.038	0.182	0.676	1.38

4.5.4 $M/G/\infty$ queue

In this queueing model with Poisson arrivals and general service times a server is immediately available for each arriving customer. The queueing model with infinitely many servers may be a useful approximation to practical queueing systems with a large number of servers. Also this model has interesting applications to inventory systems.

The transient probability $p_j(t)$ of having j customers in the system at time t was computed in Example 1.2.1 in Section 1.2 (identify in this example the ships with customers and the sojourn times of the ships with service times). Letting $t \to \infty$ in formula (1.2.7) for $p_j(t)$ shows that the limiting distribution of the number of customers in the system is given by

$$p_j = \mathrm{e}^{-\lambda E(S)} \frac{[\lambda E(S)]^j}{j!}, \quad j = 0, 1, \ldots,$$

where λ denotes the arrival rate of customers and the generic variable S denotes the service time of a customer. The limiting distribution of the number of customers in the system uses the service-time distribution only through its first moment. This is a very useful insensitivity result.

The following example based on Parikh (1977) illustrates how the $M/G/\infty$ queueing system may be used to estimate measures of system performance in the $M/G/c$ queue with a large number of servers.

Example 4.5.1 A stochastic allocation problem of fleet sizing

Suppose a nationwide company has a large number C of transport vehicle units that are used to ship customer orders. The vehicles must be allocated to a certain number F of fleets. The fleets operate independently of each other and each fleet services its own group of customers. A customer order is shipped singly by one transport vehicle unit. If a customer order finds upon arrival no transport vehicle available, the shipment of the order is delayed until a transport vehicle of the fleet returns. For each fleet the service criterion is the fraction of orders delayed. The goal of the company is to assign the total number of transport vehicles over the various fleets in order to achieve the result that all fleets provide as nearly as possible a uniform level of service. Customer orders for fleet i arrive according to a Poisson process with rate λ_i. The offered load for fleet i is $E_i = \lambda_i E(S_i), i = 1, \ldots, F$. The service time S_i of a customer order for fleet i should be interpreted as the time from shipment of that order until the transport vehicle used is available for another shipment. The service times are quite regular and their coefficients of variation do not exceed 1.

To find the allocation $(c_1, \ldots, c_F)$ with $c_1 + \cdots + c_F = C$ of the transport vehicles to the fleets, we first discuss an approximation of the $M/G/\infty$ queue to the $M/G/c$ queue with c large. Consider the $M/G/c$ queue with an offered load of $E = \lambda E(S)$ and suppose we wish to determine the minimum number of servers in order to achieve a delay probability not exceeding a prespecified value. As pointed out in Section 4.5.3, Erlang's delay probability $P_{\text{delay}}^{\text{exp}}$ for the $M/M/c$ queue is a good approximation to the delay probability P_{delay} in the general $M/G/c$ queue. For the case of a service time with a coefficient of variation smaller than 1, there is much numerical evidence in support of the bound $P_{\text{delay}} \leq P_{\text{delay}}^{\text{exp}}$. A lower bound for P_{delay} in the $M/G/c$ queue is provided by

$$\sum_{j=c}^{\infty} e^{-E} \frac{E^j}{j!} \leq P_{\text{delay}}.$$

This inequality will be intuitively obvious by noting that its left side represents the probability of having at least c customers in the $M/G/\infty$ queue while the right side represents the corresponding probability in the $M/G/c$ queue. A rigorous proof of the inequality can be found in Stoyan (1983). Thus in many practical $M/G/c$ queueing systems with an offered load of E we have for the delay probability P_{delay} the bounds

$$\sum_{j=c}^{\infty} e^{-E} \frac{E^j}{j!} \leq P_{\text{delay}} \leq P_{\text{delay}}^{\text{exp}}.$$

Denote by $c^{\text{exp}}(v)$ and $c^{\infty}(v)$ the smallest values of c for which $P_{\text{delay}}^{\text{exp}} \leq v$ in the $M/M/c$ queue and $\sum_{j=c}^{\infty} e^{-E} E^j/j! \leq v$ in the $M/G/\infty$ queue. In Table 4.5.3 we give $c^{\text{exp}}(v)$ and $c^{\infty}(v)$ for a range of values of v and E. It appears that for v small enough and E sufficiently large the critical levels $c^{\text{exp}}(v)$ and $c^{\infty}(v)$ are very close to each other. The conclusion is that for practical purposes $c^{\infty}(v)$ may be used in the design of the fleet sizes.

The problem of the allocation of the vehicles to the fleets in order to achieve nearly the same service level over the fleets can be solved as follows. Try several values of v until the desired value of v is found for which the sum of the fleet sizes $c_i(v)$, $(i = 1, \ldots, F)$ obtained from the appropriate values $c^{\infty}(v)$ is equal to C. A good approximation to the desired value of v may be computed beforehand by

Table 4.5.3   The critical levels $c^{\text{exp}}(v)$ and $c^{\infty}(v)$

	E	2	5	10	25	50	100	150	200
$v = 0.50$	$c^{\text{exp}}(v)$	3	7	12	28	54	106	157	208
	$c^{\infty}(v)$	3	6	11	26	51	101	151	201
$v = 0.20$	$c^{\text{exp}}(v)$	4	8	14	31	58	111	164	216
	$c^{\infty}(v)$	4	8	14	30	57	109	161	213
$v = 0.10$	$c^{\text{exp}}(v)$	5	9	16	33	61	115	168	221
	$c^{\infty}(v)$	5	9	15	33	60	114	167	219
$v = 0.05$	$c^{\text{exp}}(v)$	6	10	17	35	64	119	173	226
	$c^{\infty}(v)$	6	10	16	34	63	118	171	225
$v = 0.01$	$c^{\text{exp}}(v)$	7	12	19	39	68	125	181	235
	$c^{\infty}(v)$	7	12	19	38	68	125	180	235

using the well-known normal approximation to the Poisson distribution. Denoting by $\Phi(x)$ the standard normal probability distribution function, we have

$$\sum_{j=c}^{\infty} e^{-E} \frac{E^j}{j!} \approx 1 - \Phi\left(\frac{c - E - 0.5}{\sqrt{E}}\right),$$

provided E is not too small. Thus, letting k_v with $1 - \Phi(k_v) = v$ denote the $(1-v)$th percentile of the standard normal distribution, we may approximate $c^{\infty}(v)$ by $E + k_v \sqrt{E} + 0.5$. Since the $c_i(v)$'s must sum to C, we find the approximation

$$k_v \approx \frac{C - \sum_{i=1}^{F} E_i - 0.5F}{\sum_{i=1}^{F} \sqrt{E_i}}.$$

Thus the value of v for which the service level is nearly the same over all fleets can be approximated by

$$v \approx 1 - \Phi\left(\frac{C - \sum_{i=1}^{F} E_i - 0.5F}{\sum_{i=1}^{F} \sqrt{E_i}}\right).$$

As an illustration, consider the numerical data

$$C = 250, \quad F = 5, \quad E_1 = 10, \quad E_2 = 25, \quad E_3 = 35, \quad E_4 = 50 \quad \text{and} \quad E_5 = 75.$$

Using the above approximation for v yields the estimate of 0.039 for the 'uniform' service level. By trying several values of v in the neighbourhood of this estimate, we ultimately find the following assignment:

$$c_1 = 16, \quad c_2 = 34, \quad c_3 = 46, \quad c_4 = 63 \quad \text{and} \quad c_5 = 91$$

for which the respective estimates of the delay probability are 0.057, 0.060, 0.051, 0.051 and 0.048.

4.5.5 The $M^X/G/c$ queue[1]

In the $M^X/G/c$ queue the customers arrive in batches rather than singly. The arrival process of batches is a Poisson process with rate λ. The batch size X has a probability distribution $\{\beta_j, j = 1, 2, \ldots\}$. The service times of the customers are independent of each other and have a general distribution with mean $E(S)$. There are c identical servers. It is assumed that the server utilisation ρ defined by

$$\rho = \frac{\lambda E(X) E(S)}{c}$$

is smaller than 1. The customers from different batches are served in order of arrival and customers from a same batch are served in the same order as their positions in the batch.

---

[1] This section contains specialised material.

A computationally tractable analysis can only be given for the special cases of exponential services and deterministic services. We first analyse these two special cases. Next we discuss a two-moment approximation for the general $M^X/G/c$ queue.

$M^X/M/c$ queue

The process $\{L(t),\ t \geq 0\}$ describing the number of customers present is a continuous-time Markov chain. Equating the rate at which the process leaves the set of states $\{i, i + 1, \ldots\}$ to the rate at which the process enters this set of states, we find for the state probabilities p_j the recursion scheme

$$\min(i, c)\mu p_i = \sum_{k=0}^{i-1} p_k \lambda \sum_{s \geq i-k} \beta_s, \quad i = 1, 2, \ldots, \tag{4.5.38}$$

where $\mu = 1/E(S)$. Starting with $\overline{p}_0 := 1$, we successively compute $\overline{p}_1, \overline{p}_2, \ldots$ and next obtain the desired p_i's by normalization. The computational effort of the recursion scheme can be reduced by using the asymptotic expansion

$$p_j \approx \sigma \tau^{-j} \quad \text{for } j \text{ large enough,} \tag{4.5.39}$$

where τ is the unique solution of the equation

$$\lambda \tau[1 - \beta(\tau)] = c\mu(1 - \tau) \tag{4.5.40}$$

on the interval $(1, R)$ and the constant σ is given by

$$\sigma = \frac{(\tau - 1)\Sigma_{i=0}^{c-1}(c - i)p_i \tau^i/c}{1 - \lambda \tau^2 \beta'(\tau)/(c\mu)}. \tag{4.5.41}$$

Here $\beta(z) = \Sigma_{j=1}^{\infty}\beta_j z^j$ and the number R is the convergence radius of the power series $\beta(z)$. To establish the asymptotic expansion, it should be assumed that $R > 1$. In other words, the batch-size distribution should not have an extremely long tail. The derivation of the asymptotic expansion (4.5.39) is routine. Define the generating function $P(z)$ by $P(z) = \Sigma_{j=0}^{\infty}p_j z^j$, $|z| \leq 1$. It is a matter of simple algebra to derive from (4.5.38) that

$$P(z) = \frac{(1/c)\Sigma_{i=0}^{c-1}(c - i)p_i z^i}{1 - \lambda z\{1 - \beta(z)\}/\{c\mu(1 - z)\}}.$$

Next, by applying the general result (C.8) in Appendix C, we obtain the asymptotic expansion (4.5.39).

From the generating function we also derive by standard arguments that the long-run average queue size is given by

$$L_q = \frac{1}{c(1 - \rho)} \sum_{j=1}^{c-1} j(c - j)p_j + \frac{\rho}{2(1 - \rho)}\left\{\frac{E(X^2)}{E(X)} - 1\right\} + \frac{\rho}{1 - \rho} - c\rho.$$

Next we discuss the computation of the distribution function $W_q(x)$ of the waiting time of a customer. The function $W_q(x)$ is defined in the same way as in Section 4.3.2. To do so, define the probability z_j by

z_j = the long-run fraction of customers who have j other customers in front of them just after arrival.

The delay in queue of a customer who has $j \geq c$ other customers in front of him just after arrival is the sum of $j - c + 1$ independent exponentials with common mean $1/(c\mu)$. Hence this conditional waiting time has an E_{j-c+1} distribution and so

$$1 - W_q(x) = \sum_{j=c}^{\infty} z_j \sum_{k=0}^{j-c} e^{-c\mu x} \frac{(c\mu x)^k}{k!}, \quad x \geq 0.$$

A computationally better representation of $W_q(x)$ is

$$1 - W_q(x) = \sum_{k=0}^{\infty} e^{-c\mu x} \frac{(c\mu x)^k}{k!} \left(1 - \sum_{j=0}^{k+c-1} z_j \right), \quad x \geq 0. \tag{4.5.42}$$

The probabilities z_j are easy to express in terms of the p_j's. To do so, define the probability η_k by

η_k = the long-run fraction of customers who take the kth position in their batch.

Then, using that the long-run fraction of batches finding upon arrival m other customers present equals p_m,

$$z_j = \sum_{m=0}^{j} p_m \eta_{j-m+1}, \quad j = 0, 1, \dots . \tag{4.5.43}$$

Using the expression (4.3.10) for η_k and the relation (4.5.38), it follows that

$$z_j = \frac{1}{E(X)} \frac{\min(j+1, c)}{\lambda} \mu p_{j+1}, \quad j = 0, 1, \dots . \tag{4.5.44}$$

An asymptotic expansion for $W_q(x)$ follows by inserting (4.5.44) and (4.5.39) into (4.5.42). This yields after some algebra

$$1 - W_q(x) \approx \frac{\sigma \tau^{-c}}{\tau - 1} e^{-c\mu(1-1/\tau)x} \quad \text{for } x \text{ large enough}, \tag{4.5.45}$$

where τ and σ are given by (4.5.40) and (4.5.41).

$M^X/D/c$ queue

Suppose that the service time of each customer is a constant D. Denoting by $p_j(t)$ the probability that j customers are present at time t, we find by the same arguments as used in Section 4.5.2 that

$$p_j(t + D) = \sum_{k=0}^{c} p_k(t)r_j(D) + \sum_{k=c+1}^{c+j} p_k(t)r_{j-k+c}(D), \quad j = 0, 1, \ldots,$$

where the compound Poisson probability $r_j(D)$ is defined by

$r_j(D)$ = the probability that exactly j customers arrive during a time interval of length D.

Letting $t \to \infty$, we find the system of linear equations,

$$p_j = r_j(D) \sum_{k=0}^{c} p_k + \sum_{k=c+1}^{c+j} r_{j-k+c}(D)p_k, \quad j = 0, 1, \ldots . \qquad (4.5.46)$$

Just as in the $M/D/c$ case this infinite system of equations can be reduced to a finite system of linear equations by replacing p_j by $p_N \tau^{-(j-N)}, j \geq N$ for an appropriately chosen integer N. This reduction is based on the asymptotic expansion

$$p_j \approx \sigma \tau^{-j} \quad \text{for } j \text{ large enough,} \qquad (4.5.47)$$

where τ is the unique root of the equation

$$\tau^c e^{\lambda D\{1-\beta(\tau)\}} = 1 \qquad (4.5.48)$$

on the interval $(1, R)$ and the constant σ is given by

$$\sigma = [c - \lambda D\tau\beta'(\tau)]^{-1} \sum_{j=0}^{c-1} p_j(\tau^j - \tau^c). \qquad (4.5.49)$$

As before, $\beta(z) = \sum_{j=1}^{\infty} \beta_j z^j$ and the number R denotes the convergence radius of the power series $\beta(z)$. It is assumed that $R > 1$. To prove the asymptotic expansion, we first derive for the generating function $P(z) = \sum_{j=0}^{\infty} p_j z^j$ the expression

$$P(z) = \frac{\sum_{j=0}^{c-1} p_j(z^j - z^c)}{1 - z^c e^{\lambda D\{1-\beta(z)\}}}.$$

This expression follows by multiplying both sides of (4.5.46) by z^j and summing over j and using that the generating function of the compound Poisson probabilities $r_j(D)$ is given by $\exp[-\lambda D\{1 - \beta(z)\}]$; see Section 1.2.2. Next, by applying the general result (C.8) in Appendix C, the asymptotic expansion (4.5.47) follows.

Instead of solving the linear equations (4.5.46), the state probabilities p_j can also be computed by a repeated application of the Fast Fourier Transform method to the generating function $P(z)$; cf. Remark 4.5.1 in Section 4.5.2.

From the generating function $P(z)$ we also derive that the long-run average number of customers in queue is given by

$$L_q = \frac{1}{2c(1-\rho)}\left[(c\rho)^2 - c(c-1) + \sum_{j=0}^{c-1}\{c(c-1) - j(j-1)\}p_j \right.$$

$$\left. + c\rho\left\{ \frac{E(X^2)}{E(X)} - 1 \right\} \right]. \tag{4.5.50}$$

Next we turn to the computation of the distribution function $W_q(x)$ of the waiting time of a customer. The relation (4.5.12) for $W_q(x)$ in the $M/D/c$ queue can be directly generalized to

$$W_q(x) = \sum_{j=1}^{mc+c} \eta_j b_{mc+c-j}(u), \tag{4.5.51}$$

where m, u and $b_k(u)$ are precisely defined as in Section 4.5.2 and η_j denotes the long-run fraction of customers taking the jth position in their batch. The probability η_j is given by (4.3.10) and the probability $b_k(u)$ satisfies the relation (4.5.13) in which the Poisson probabilities are replaced by the compound Poisson probabilities $r_j(u)$. The relation (4.5.51) together with the recursion scheme for the $b_k(u)$'s is not very suited to numerical purposes. In general the numerical instability of the recursion scheme for the $b_k(u)$'s is much more difficult to control in the $M^X/D/c$ queue than in the $M/D/c$ queue. Another complication is that $W_q(x)$ is discontinuous in some of the points $x = jD$ when $\Sigma_{i=1}^c \beta_i < 1$. To explain this, consider a batch finding upon arrival i other customers present with $0 \le i \le c - 1$. Some reflections show that the customers occupying the positions $jc + 1, \ldots, jc + c - i$ in this batch each have a waiting time equal to jD. Thus $W_q(x)$ has at the point jD the mass

$$m_j = \sum_{i=0}^{c-1} p_i \sum_{r=1}^{c-i} \eta_{jc+r}, \quad j = 0, 1, \ldots. \tag{4.5.52}$$

The mass $m_j = 0$ for all $j \ge 1$ only if $\eta_k = 0$ for all $k > c$. The definition of η_k implies that $\eta_k = 0$ for all $k > c$ only if $\Sigma_{i=1}^c \beta_i = 1$. Thus we find that $W_q(x)$ is continuous for all $x > 0$ only if $\Sigma_{i=1}^c \beta_i = 1$, otherwise $W_q(x)$ has a discontinuity in at least one point jD with $j \ge 1$.

In view of the numerical difficulties the exact representation (4.5.51) offers in the batch-arrival case, we propose a simple but useful approximation to $W_q(x)$. This approximation is based on the asymptotic expansion for $W_q(x)$. Let us first give this asymptotic expansion. Using the generating function of the probabilities $1 - b_k(u)$, we derive in the same way as in Section 4.5.2 that

$$1 - W_q(x) \approx \gamma e^{-\lambda\{\beta(\tau)-1\}x} \quad \text{for } x \text{ large enough}, \tag{4.5.53}$$

where

$$\gamma = \frac{\sigma\{\beta(\tau) - 1\}}{(\tau - 1)^2 \tau^{c-1} E(X)}$$

with τ and σ given by (4.5.48) and (4.5.49). The details of the derivation are left to the reader.

The following approximation to $W_q^c(x) = 1 - W_q(x)$ is now suggested:

$$W_{\text{app}}^c(x) = \begin{cases} \alpha e^{-\beta x} - \text{corr}(x) & \text{for } 0 \le x < x_0 \\ \gamma e^{-\lambda\{\beta(\tau)-1\}x} & \text{for } x \ge x_0, \end{cases}$$

where the correction term $\text{corr}(x)$ is defined by

$$\text{corr}(x) = \sum_{j=1}^{[x/D]} m_j$$

with m_j given by (4.5.52). Note that $\text{corr}(x) = 0$ for all $x \ge 0$ when $\Sigma_{i=1}^c \beta_i = 1$. The coefficients α and β and the point x_0 are determined by matching the exact values of $W_q(0)$ and W_q and by requiring continuity of $W_{\text{app}}(x)$ at the crossover point $x = x_0$. It was argued in (4.5.43) that the long-run fraction of customers having j other customers in front of them just after arrival is $\Sigma_{m=0}^j p_m n_{j-m+1}$. Thus, since $\alpha = 1 - W_q(0)$,

$$\alpha = 1 - \sum_{j=0}^{c-1} \sum_{m=0}^{j} p_m n_{j-m+1}.$$

The constants β and x_0 are computed by solving the two equations

$$\int_0^\infty W_{\text{app}}^c(x)\, dx = W_q \quad \text{and} \quad \alpha e^{-\beta x_0} - \sum_{j=1}^{[x_0/D]} m_j = \gamma e^{-\lambda\{\beta(\tau)-1\}x_0},$$

where $L_q = \lambda E(X) W_q$ and L_q is given by (4.5.50). In practice the solving of these two equations offers no numerical difficulties.

In most practical situations one is only interested in the tail of the waiting-time distribution. In those situations it suffices to work with the asymptotic expansion (4.5.53). This asymptotic expansion applies already for moderate values of x provided the traffic load on the system is not very small. The larger the traffic load, the earlier the asymptotic expansion is applicable. An appropriate measure for the traffic load is the probability that all servers are simultaneously busy. This probability is given by $P_B = 1 - \Sigma_{j=0}^{c-1} p_j$. As a rule of thumb, the asymptotic expansion (4.5.53) can be used for practical purposes for $x \ge E(X)D/\sqrt{c}$ when $P_B \ge 0.2$ (say).

$M^X/G/c$ *queue*

An exact and tractable solution for the $M^X/G/c$ queue is in general not possible except for the special cases of deterministic services and exponential services. Using the solutions for these special cases, we can give useful approximations for the general $M^X/G/c$ queue. A practically useful approximation to the average delay in queue per customer is

$$W_q^{\mathrm{app}} = (1 - c_S^2)W_q(\det) + c_S^2 W_q(\exp)$$

provided that c_S^2 is not too large (say, $0 \le c_S^2 \le 2$) and the traffic load is not very small. It was already pointed out in Section 4.3 that the first-order approximation $\frac{1}{2}(1 + c_S^2)W_q(\exp)$ is not applicable in the batch-arrival queue.

A two-moment approximation to the conditional waiting-time percentiles $\eta(p)$ for the delayed customers is provided by

$$\eta_{\mathrm{app}}(p) = (1 - c_S^2)\eta_{\det}(p) + c_S^2\eta_{\exp}(p), \quad 0 < p < 1.$$

However, it turns out that in the batch-arrival case the two-moment approximation to $\eta(p)$ works only for the higher percentiles. Fortunately, higher percentiles are usually the percentiles of interest in practice. Moreover, for p close enough to 1, the percentiles $\eta_{\det}(p)$ and $\eta_{\exp}(p)$ can be computed from the asymptotic expansions (4.5.53) and (4.5.45) for $W_q(x)$. Table 4.5.4 gives for the $M^X/E_2/c$ queue the

Table 4.5.4   The percentiles $\eta(p)$ for the $M^X/E_2/c$ queue

c	ρ	p	Constant batch size				Geometric batch size			
			0.80	0.90	0.95	0.99	0.80	0.90	0.95	0.99
1	0.2	exa	2.927	3.945	4.995	7.458	5.756	8.122	10.49	15.98
		app	2.836	3.901	4.967	7.440	5.745	8.116	10.49	15.99
1	0.5	exa	5.107	7.170	9.231	14.02	9.044	12.84	16.64	25.45
		app	5.089	7.154	9.219	14.01	9.040	12.84	16.64	25.47
2	0.2	exa	1.369	1.897	2.431	3.661	2.989	4.172	5.355	8.101
		app	1.354	1.887	2.419	3.656	2.982	4.167	5.353	8.106
2	0.5	exa	2.531	3.561	4.592	6.985	4.600	6.498	8.395	12.80
		app	2.535	3.567	4.599	6.996	4.601	6.501	8.401	12.81
5	0.2	exa	0.621	0.845	1.063	1.560	1.298	1.773	2.246	3.345
		app	0.640	0.853	1.066	1.560	1.305	1.779	2.253	3.354
5	0.5	exa	1.063	1.476	1.889	2.846	1.898	2.657	3.417	5.179
		app	1.069	1.482	1.895	2.853	1.905	2.665	3.425	5.190
10	0.5	exa	0.553	0.764	0.971	1.451	0.980	1.360	1.740	2.622
		app	0.566	0.772	0.979	1.458	0.991	1.371	1.751	2.634
10	0.7	exa	0.923	1.295	1.667	2.530	1.547	2.181	2.815	4.287
		app	0.930	1.302	1.673	2.536	1.556	2.190	2.824	4.297

exact and approximate values of the conditional waiting-time percentiles $\eta(p)$ both for the case of a constant batch size and the case of a geometrically distributed batch size. In both cases the mean batch size $E(X) = 3$. The normalization $E(S) = 1$ is used for the service time of a customer. The asymptotic expansions (4.5.53) and (4.5.45) are used to compute $\eta_{\text{det}}(p)$ and $\eta_{\text{exp}}(p)$ in the two-moment approximation to $\eta(p)$.

4.6   THE *GI/G/c* QUEUE

It seems obvious that the general *GI/G/c* queue offers enormous difficulties in getting practically useful results. Nevertheless, using specialized techniques for solving large-scale systems of linear equations for structured Markov chains, the continuous-time Markov chain approach has proved to be quite useful for analysing the *GI/G/c* queue when the interarrival time and service time both have phase-type distributions. By a detailed state description involving sufficient information about the number of customers present and the status of both the arrival in progress and the services in progress, it is possible to setup the equilibrium equations for the microstate probabilities. The resulting large-scale system of linear equations possesses a structure enabling the application of special algorithms to solve numerically the equations, provided the number of servers is not too large; see Seelen *et al* (1985). However, this numerical approach is not suited to routine calculations.

In this section we restrict ourselves to the particular models of the *GI/M/c* queue with exponential services and the *GI/D/c* queue with deterministic services. These models allow for a relatively simple algorithmic analysis. The results for these models may serve as a basis for approximations to the complex *GI/G/c* queue. Several performance measures P, such as the average queue length, the average waiting time per customer and the (conditional) waiting-time percentiles, can be approximated by using the familiar interpolation formula

$$P_{\text{app}} = (1 - c_S^2) P_{GI/D/c} + c_S^2 P_{GI/M/c},$$

provided c_S^2 is not too large and the traffic load on the system is not very light. In this formula $P_{GI/D/c}$ and $P_{GI/M/c}$ denote the exact values of the specific performance measure for the special cases of the *GI/D/c* queue and the *GI/M/c* queue with the same mean service time E(S). In Table 4.6.1 we give for the $E_{10}/E_2/c$ queue the exact and approximate values of the average queue size L_q and the average waiting-time percentiles $\eta(0.8)$ and $\eta(0.95)$ for several values of c and ρ. In all examples the normalization $E(S) = 1$ is used. The above linear interpolation formula is in general not to be recommended for the delay probability, particularly not when c_S^2 is close to zero. For example, the delay probability has the respective values 0.0776, 0.3285 and 0.3896 for the $E_{10}/D/5$ queue, the $E_{10}/E_2/5$ queue and the $E_{10}/M/5$ queue, each with $\rho = 0.8$. An interpolation formula like the above one should always be issued together with a caveat against its blind application.

Table 4.6.1 Some numerical result for the $E_{10}/E_2/c$ queue

		$\rho = 0.5$			$\rho = 0.8$			$\rho = 0.9$	
		L_q	$\eta(0.8)$	$\eta(0.95)$	L_q	$\eta(0.8)$	$\eta(0.95)$	L_q	$\eta(0.8)$	$\eta(0.95)$
$c = 1$	exact	0.066	1.21	2.21	0.780	2.59	4.78	2.21	4.99	9.25
	approx	0.082	1.19	2.17	0.813	2.57	4.76	2.25	5.14	9.25
$c = 5$	exact	0.006	0.277	0.499	0.452	0.551	0.993	1.75	1.02	1.87
	approx	0.009	0.243	0.452	0.466	0.530	0.968	1.76	1.02	1.86

The above interpolation formula reflects the empirical finding that measures of system performance are in general much more sensitive to the interarrival-time distribution than to the service-time distribution, in particular when the traffic load is light.

4.6.1 The *GI/M/c* queue

In the *GI/M/c* queue the service times of the customers are exponentially distributed with mean $1/\mu$. The interarrival time A has a general probability distribution function $A(t)$ with density $a(t)$. It is assumed that $\rho = \lambda/(c\mu)$ is smaller than 1, where the arrival rate λ is the inverse of $E(A)$. The time-average probabilities and the customer-average probabilities are denoted by $\{p_j\}$ and $\{\pi_j\}$. In other words,

$$p_j = \text{the long-run fraction of time that } j \text{ customers are present}$$

and

$$\pi_j = \text{the long-run fraction of customers who find upon arrival } j \text{ other} \\ \text{customers present.}$$

There is a simple relation between the p_j's and the π_j's. We have

$$\min(j, c)\mu p_j = \lambda \pi_{j-1} \quad \text{for } j = 1, 2, \dots . \tag{4.6.1}$$

This relation equates the average number of downcrossings from state j to state $j - 1$ per unit time to the average number of upcrossings from state $j - 1$ to state j per unit time. A rigorous proof of the relation (4.6.1) can be found in Example 1.7.1 of Section 1.7.

The probabilities π_j determine the waiting-time distribution function $W_q(x)$. Noting that the conditional waiting time of a customer finding upon arrival $j \geq c$ other customers present is the sum of $j - c + 1$ independent exponentials with mean $1/(c\mu)$ and thus has an Erlang distribution, it follows that

$$1 - W_q(x) = \sum_{j=c}^{\infty} \pi_j \sum_{k=0}^{j-c} e^{-c\mu x} \frac{(c\mu x)^k}{k!}, \quad x \geq 0. \tag{4.6.2}$$

This expression can be reduced to a simple exponential function. To show this, we use that

$$\frac{\pi_{j+1}}{\pi_j} = \eta \quad \text{for all } j \geq c - 1$$

for some constant $0 < \eta < 1$. The proof of this result is a replica of the proof of the corresponding result for the $GI/M/1$ queue dealt with in the continuation of Example 2.1.2 in Section 2.4. Hence we have

$$\pi_j = \eta^{j-c+1}\pi_{c-1} \quad \text{for } j \geq c - 1. \tag{4.6.3}$$

As a by-product of (4.6.1) and (4.6.3), we have

$$p_j = \eta^{j-c}p_c \quad \text{for } j \geq c. \tag{4.6.4}$$

Substituting (4.6.3) into (4.6.2) yields

$$1 - W_q(x) = \frac{\eta}{1-\eta}\pi_{c-1}e^{-c\mu(1-\eta)x}, \quad x \geq 0. \tag{4.6.5}$$

The constant η is the unique solution of the equation

$$\eta = \int_0^\infty e^{-c\mu(1-\eta)t}a(t)\,dt \tag{4.6.6}$$

on the interval $(0, 1)$. To see this, note that $\{\pi_j\}$ is the equilibrium distribution of the embedded Markov chain describing the number of customers present just before an arrival epoch. Substituting (4.6.3) into the balance equations

$$\pi_j = \sum_{k=j-1}^\infty \pi_k \int_0^\infty e^{-c\mu t}\frac{(c\mu t)^{k+1-j}}{(k+1-j)!}a(t)\,dt, \quad j \geq c$$

easily yields the result (4.6.6).

By the relations (4.6.1), (4.6.3) and (4.6.4), the probability distributions $\{p_j\}$ and $\{\pi_j\}$ are completely determined once we have computed $\pi_0, \ldots, \pi_{c-1}$ or $p_0, \ldots, p_c$. Next we show how to find these unknowns for the cases of deterministic, generalized Erlangian and Coxian-2 interarrival times.

Deterministic arrivals

Suppose that there is a constant time D between two consecutive arrivals. Define the embedded Markov chain $\{X_n\}$ by

$$X_n = \text{ the number of customers present just before the } n\text{th arrival.}$$

Denoting the one-step transition probabilities of this Markov chain by p_{ij}, the π_j's are the unique solution to the equations

$$\pi_j = \sum_{k=j-1}^\infty \pi_k p_{kj}, \quad j = 1, 2, \ldots$$

together with the normalizing equation $\Sigma_{j=0}^{\infty} \pi_j = 1$. Substituting (4.6.3) into these equations yields that $\pi_0, \ldots, \pi_{c-1}$ are the unique solution to the finite system of linear equations

$$\pi_j = \sum_{k=j-1}^{c-2} \pi_k p_{kj} + \pi_{c-1} p_{c-1,j}^*, \quad 1 \le j \le c-1,$$

$$\sum_{j=0}^{c-2} \pi_j + \frac{\pi_{c-1}}{1-\eta} = 1,$$

(4.6.7)

where

$$p_{c-1,j}^* = \sum_{k=c-1}^{\infty} \eta^{k-c+1} p_{kj}, \quad 1 \le j \le c-1.$$

The constant η is the unique solution to the equation $\eta = \exp[-c\mu D(1-\eta)]$ on the interval (0,1). It remains to specify the p_{kj}'s for $1 \le j \le c-1$. Since the probability that an exponentially distributed service time is completed within a time D equals $1 - \exp(-\mu D)$, we have

$$p_{kj} = \binom{k+1}{j} e^{-\mu Dj}(1 - e^{-\mu D})^{k+1-j}, \quad 0 \le k \le c-1 \text{ and } 0 \le j \le k+1.$$

The probabilities p_{kj} for $k > c-1$ require a little bit more explanation. We first note that the times between service completions are independent exponentials with common mean $1/(c\mu)$ as long as c or more customers are present. Thus, starting with $k+1 \ge c$ customers present, the time until the $(k+1-c)$th service completion has an E_{k+1-c} distribution. By conditioning on the epoch of this $(k+1-c)$th service completion, we find for any $k \ge c$ that

$$p_{kj} = \int_0^D \binom{c}{j} e^{-\mu(D-x)j}\{1 - e^{-\mu(D-x)}\}^{c-j}(c\mu)^{k+1-c} \frac{x^{k-c}}{(k-c)!} e^{-c\mu x} \, dx$$

$$= \binom{c}{j} e^{-j\mu D} c\mu \int_0^D \frac{(c\mu x)^{k-c}}{(k-c)!} (e^{-\mu x} - e^{-\mu D})^{c-j} \, dx, \quad 0 \le j \le c.$$

This expression is needed to evaluate $p_{c-1,j}^*$. We find

$$p_{c-1,j}^* = p_{c-1,j} + c\mu\eta \binom{c}{j} e^{-j\mu D} \int_0^D e^{c\mu\eta x}(e^{-\mu x} - e^{-\mu D})^{c-j} \, dx$$

for $1 \le j \le c-1$. Numerical integration must be used to calculate $p_{c-1,j}^*$ for $1 \le j \le c-1$. A convenient method is Gauss–Legendre integration. The other coefficients p_{kj} of the linear equations (4.6.7) are simply computed as binomial coefficients. Once the linear equations (4.6.7) have been solved, we can compute the various performance measures.

The analysis for the *D/M/c* can be straightforwardly generalized to the *GI/M/c* queue. However, in general the expression for p_{kj} with $k \ge c$ is quite complicated and leads to a cumbersome and time-consuming calculation of $p^*_{c-1,j}$. Fortunately, a much simpler alternative is available when the interarrival time has a phase-type distribution.

Coxian-2 arrivals

Suppose that the interarrival time has a Coxian-2 distribution with parameters $(b, \lambda_1, \lambda_2)$. In other words, the interarrival time first goes through phase 1 and next it is finished with probability $1 - b$ or goes through a second phase 2 with probability b, where the phases are independent exponentials with respective means $1/\lambda_1$ and $1/\lambda_2$.

The state probabilities p_j, $0 \le j \le c$, can be calculated by using the continuous-time Markov chain approach. Define $X(t)$ as the number of customers present at time t and let $Y(t)$ be the phase of the interarrival time in progress at time t. The process $\{(X(t), Y(t))\}$ is a continuous-time Markov chain with state space $I = \{(n, i) | n = 0, 1, \ldots, i = 1, 2\}$. Denoting the equilibrium probabilities of this Markov chain by p_{ni}, we have $p_n = p_{n1} + p_{n2}$. By equating the rate at which the system leaves the set of states having at least n customers present to the rate at which the system enters this set, we obtain

$$\min(n, c)\mu(p_{n1} + p_{n2}) = \lambda_1(1 - b)p_{n-1,1} + \lambda_2 p_{n-1,2}, \quad n \ge 1. \tag{4.6.8}$$

This system of equations is augmented by the equations

$$[\min(n, c)\mu + \lambda_2]p_{n2} = \min(n + 1, c)\mu p_{n+1,2} + \lambda_1 b p_{n1}, \quad n \ge 0. \tag{4.6.9}$$

These latter equations follow by equating the rate out of state $(n, 2)$ to the rate into this state.

A closer examination of the equations (4.6.8) and (4.6.9) reveals that they cannot be solved recursively starting with $\overline{p}_0 := 1$. However, a recursive computation of $p_0, \ldots, p_c$ is possible by using that

$$\frac{p_{n+1,i}}{p_{ni}} = \eta \quad \text{for } n \ge c \text{ and } i = 1, 2. \tag{4.6.10}$$

The relation (4.6.10) extends the relation $p_{n+1}/p_n = \eta$ for $n \ge c$. A proof of the relation (4.6.10) is not given here. It can be deduced from results in Takahashi (1981) and the general result (2.6.4) in Section 2.6. The constant η can be computed beforehand from the equation (4.6.6). Using the expression for the Coxian-2 density given in Appendix B, this equation becomes

$$\frac{r_1\lambda_1}{c\mu(1 - \eta) + \lambda_1} + \frac{r_2\lambda_2}{c\mu(1 - \eta) + \lambda_2} = \eta, \tag{4.6.11}$$

where $r_1 = 1 - b\lambda_1/(\lambda_1 - \lambda_2)$ and $r_2 = 1 - r_1$. Here it is assumed that $\lambda_1 \ne \lambda_2$. Once η is known, we can express p_{c2} into p_{c1}. Substituting $p_{c+1,i} = \eta p_{ci}$ for

$i = 1, 2$ into (4.6.9) with $n = c$ yields

$$(c\mu + \lambda_2)p_{c2} = c\mu\eta p_{c2} + \lambda_1 b p_{c1}.$$

The following algorithm can now be given.

Algorithm
Step 0. Calculate first η as the unique root of the equation (4.6.11) on (0,1). Let $\overline{p}_{c1} := 1$ and $\overline{p}_{c2} := \lambda_1 b\{c\mu(1 - \eta) + \lambda_2\}^{-1}\overline{p}_{c1}$.
Step 1. For $k = c - 1, \ldots, 0$, use the equation (4.6.8) with $n = k + 1$ and the equation (4.6.9) with $n = k$ to solve for $\overline{p}_{k1}$ and $\overline{p}_{k2}$.
Step 2. Calculate $\overline{p}_n := \overline{p}_{n1} + \overline{p}_{n2}$ for $n = 0, 1, \ldots, c$ and next use relation (4.6.4) to normalize the $\overline{p}_n$'s as

$$p_n := \left[\sum_{j=0}^{c-1} \overline{p}_j + \frac{\overline{p}_c}{1 - \eta}\right]^{-1} \overline{p}_n \quad \text{for } n = 0, 1, \ldots, c.$$

Generalized Erlangian arrivals
Suppose that the interarrival time has the density

$$a(t) = \sum_{i=1}^{m} q_i \alpha^i \frac{t^{i-1}}{(i-1)!} e^{-\alpha t}, \quad t \geq 0.$$

In other words, with probability q_i an interarrival time is the sum of i independent phases each having an exponential distribution with mean $1/\alpha$.

We use again the continuous-time Markov chain approach to compute the probabilities p_j. Define $X(t)$ as the number of customers present at time t and let $Y(t)$ be the number of remaining phases of the interarrival time in progress at time t. The process $\{(X(t), Y(t))\}$ is a continuous-time Markov chain with state space $I = \{(n, i) | n \geq 0, 1 \leq i \leq m\}$. By equating the rate at which the system leaves the set of states having at least n customers present to the rate at which the system enters this set, we find

$$\min(n, c)\mu p_n = \alpha p_{n-1,1}, \quad n \geq 1. \tag{4.6.12}$$

Moreover, by rate out of state (n, i) =rate into state (n, i),

$$[\min(n, c)\mu + \alpha]p_{ni} = \alpha p_{n,i+1} + \min(n + 1, c)\mu p_{n+1,i} + \alpha q_i p_{n-1,1}$$

for $n \geq 0$ and $1 \leq i \leq m$, where $p_{n,m+1} = p_{-1,1} = 0$ by convention. Again a rather simple solution procedure can be given by using that

$$\frac{p_{n+1,i}}{p_{n,i}} = \eta \quad \text{for } n \geq c \text{ and } 1 \leq i \leq m.$$

A proof of this relation will not be given here. The decay factor η is the unique solution to the equation

$$\eta = \sum_{i=1}^{m} q_i \frac{\alpha^i}{[c\mu(1-\eta)+\alpha]^i}$$

on the interval $(0, 1)$. By substitution of (4.6.12) into the balance equation for p_{ni}, we obtain for each $n \geq 0$ that

$$[\min(n,c)\mu + \alpha]p_{ni} = \alpha p_{n,i+1} + \min(n+1,c)\mu p_{n+1,i}$$

$$+ q_i \min(n,c)\mu \sum_{j=1}^{m} p_{nj}, \quad 1 \leq i \leq m. \quad (4.6.13)$$

In particular, using that $p_{c+1,i} = \eta p_{ci}$ for $1 \leq i \leq m$,

$$(c\mu + \alpha)p_{ci} = \alpha p_{c,i+1} + c\mu\eta p_{ci} + q_i c\mu \sum_{j=1}^{m} p_{cj}, \quad 1 \leq i \leq m. \quad (4.6.14)$$

The probabilities $p_0, \ldots, p_c$ can now be computed as follows.

Algorithm
Step 0. Calculate the delay factor η. Let $\overline{p}_{c1} := 1$.
Step 1. Solve the linear equations (4.6.14) with $2 \leq i \leq m$ to obtain $\overline{p}_{ci}$ for $2 \leq i \leq m$.
Step 2. For $k = c - 1, \ldots, 0$, solve the linear equations (4.6.13) with $n = k$ to obtain $\overline{p}_{ki}$ for $1 \leq i \leq m$.
Step 3. Calculate $\overline{p}_n := \sum_{j=1}^{m} \overline{p}_{nj}$ for $n = 0, 1, \ldots, c$ and normalize the $\overline{p}_n$'s as

$$p_n := \left[\sum_{j=0}^{c-1} \overline{p}_j + \frac{\overline{p}_c}{1-\eta} \right]^{-1} \overline{p}_n \quad \text{for } n = 0, 1, \ldots, c.$$

The algorithm requires that a system of linear equations of order m is solved c times. This is computationally feasible provided m is not too large.

As a final remark it is noted that a simpler approach is possible when one is only interested in the waiting-time function $W_q(x)$. This function is completely determined by its mean W_q and the decay factor η. A simple algorithm to compute W_q is given in Takács (1962).

4.6.2 The $GI/D/c$ queue

The multi-server queue with deterministic service times allows for a computationally tractable analysis when the interarrival time has a phase-type distribution. For phase-type arrivals it is possible to extend the analysis given for the $M/D/c$

queue in Section 4.5.2. We first analyse the *GI/D/c* queue with Coxian-2 arrivals. Next we discuss the *GI/D/c* queue with Erlangian arrivals.

Coxian-2 arrivals

Suppose that the interarrival time has a Coxian-2 distribution with parameters $(b, \lambda_1, \lambda_2)$. In other words, the interarrival time first goes through phase 1 and is then finished with probability $1 - b$ or goes through a second phase 2 with probability b, where the phases are independent exponentials with respective means $1/\lambda_1$ and $1/\lambda_2$. Define $X(t)$ as the number of customers present at time t and let $Y(t)$ denote the phase of the interarrival time in progress at time t. For any $t > 0$, let

$$p_{nj}(t) = P\{X(t) = n, Y(t) = j\}, \quad n = 0, 1, \ldots \quad \text{and } j = 1, 2.$$

To find the time-average state probabilities p_n, we need the limiting probabilities

$$p_{nj} = \lim_{n \to \infty} p_{nj}(t), \quad n = 0, 1, \ldots \text{ and } j = 1, 2.$$

Note that $p_n = p_{n1} + p_{n2}$. Just as in the case of the *M/D/c* queue we have that any customer in service at time t is no longer present at time $t + D$, while the customers present at time $t + D$ are exactly those customers waiting in queue at time t or having arrived in $(t, t + D)$. Hence we find for any $t > 0$ that

$$p_{nj}(t + D) = \sum_{k=0}^{c} \sum_{i=1}^{2} p_{ki}(t) a_n^{ij} + \sum_{k=c+1}^{c+n} \sum_{i=1}^{2} p_{ki}(t) a_{n-k+c}^{ij}$$

for $n = 0, 1, \ldots$ and $j = 1, 2$, where

$$a_n^{ij} = P\{n \text{ customers arrive in } (t, t + D) \text{ and } Y(t + D) = j | Y(t) = i\}.$$

The transition probabilities a_n^{ij} are independent of t. Letting $t \to \infty$ in the equations for $p_{nj}(t + D)$, we obtain

$$p_{nj} = \sum_{k=0}^{c} \sum_{i=1}^{2} p_{ki} a_n^{ij} + \sum_{k=c+1}^{c+n} \sum_{i=1}^{2} p_{ki} a_{n-k+c}^{ij}, \quad n \geq 0 \text{ and } i = 1, 2$$

together with the normalization equation $\Sigma_{n=0}^{\infty} \Sigma_{j=1}^{2} p_{nj} = 1$. This infinite system of equations can be reduced to a finite system of linear equations by using the geometric tail approach discussed in Section 2.3.3 of Chapter 2. This powerful approach can be used because of the result

$$\lim_{n \to \infty} \frac{p_{n+1,j}}{p_{n,j}} = \eta \quad \text{for } j = 1, 2$$

where η is given by

$$\eta = e^{-\delta D/c} \tag{4.6.15}$$

and δ is the unique solution of the equation

$$\left\{ \frac{(1-b)\lambda_1}{\delta + \lambda_1} + \frac{b\lambda_1\lambda_2}{(\delta + \lambda_1)(\delta + \lambda_2)} \right\} e^{\delta D/c} = 1 \qquad (4.6.16)$$

on the interval $(0, \infty)$. This result will not be proved here. It is a special case of a general result due to Takahashi (1981); cf. also equation (4.5.34) in Section 4.5.3. The geometric tail approach reduces the infinite system of linear equations to a finite one by replacing p_{nj} by $\eta^{n-N} p_{Nj}$ for $n \geq N$ and $j = 1, 2$ for an appropriately chosen integer N.

It remains to specify the transition probabilities a_n^{ij}. The computation of these probabilities is nontrivial. Using the shorthand notation

$$\phi_{nk} = \int_0^D e^{-\lambda_1 x} \frac{(\lambda_1 x)^n}{n!} e^{-\lambda_2 (D-x)} \frac{[\lambda_2 (D-x)]^k}{k!} \, dx, \qquad n, k = 0, 1, \ldots$$

we have

$$a_n^{11} = (1-b)^n e^{-\lambda_1 D} \frac{(\lambda_1 D)^n}{n!} + \sum_{k=1}^{n} \binom{n}{k} b^k (1-b)^{n-k} \lambda_2 \phi_{n,k-1}, \qquad n \geq 0,$$

$$a_n^{12} = \sum_{k=0}^{n} \binom{n}{k} b^k (1-b)^{n-k} b\lambda_1 \phi_{nk}, \qquad n \geq 0,$$

$$a_n^{21} = \sum_{k=0}^{n-1} \binom{n-1}{k} b^k (1-b)^{n-1-k} \lambda_2 \phi_{n-1,k}, \qquad n \geq 0,$$

$$a_0^{22} = e^{-\lambda_2 D}$$

and

$$a_n^{22} = \sum_{k=0}^{n-1} \binom{n-1}{k} b^k (1-b)^{n-1-k} b\lambda_1 \phi_{n-1,k+1}, \qquad n \geq 1.$$

Here we use the convention $\sum_{k=0}^{-1} = 0$. We only give a derivation of the expression for a_n^{11}. The derivations of the other expressions proceed along the same lines. Let X and Y be independent random variables having exponential distributions with respective means $1/\lambda_1$ and $1/\lambda_2$. An interarrival time is distributed as X with probability $1 - b$ and is distributed as $X + Y$ with probability b. By conditioning on the number of interarrival times that are distributed as $X + Y$, it is readily seen that

$$a_n^{11} = \sum_{k=0}^{n} \binom{n}{k} b^k (1-b)^{n-k} P\{X_1 + \cdots + X_n + Y_1 + \cdots + Y_k$$

$$\leq D < X_1 + \cdots + X_n + Y_1 + \cdots + Y_k + X_{n+1}\}.$$

Next, by conditioning on the Erlang-k density of $Y_1 + \cdots + Y_k$, the desired result easily follows.

The transition probabilities a_n^{ij} are easy to compute once the integrals ϕ_{nk} have been evaluated. The integral representation of ϕ_{nk} multiplied by $\lambda_1 \lambda_2$ can be interpreted as the convolution of an E_{n+1} density and an E_{k+1} density. The convolution of two Erlang densities with the same scale parameters is again an Erlang density. Thus for the case of $\lambda_1 = \lambda_2 = \alpha$,

$$\phi_{nk} = \alpha^{n+k} \frac{D^{n+k+1}}{(n+k+1)!} e^{-\alpha D}, \quad n, k = 0, 1, \ldots .$$

To evaluate the integrals when $\lambda_1 \neq \lambda_2$, we use the recursion relation

$$\phi_{nk} = \frac{\lambda_1 - \lambda_2}{\lambda_1} \phi_{n+1,k} + (1 - \delta_k) \frac{1}{\lambda_1} \frac{(\lambda_1 D)^{n+1}}{(n+1)!} e^{-\lambda_1 D} + \delta_k \frac{\lambda_2}{\lambda_1} \phi_{n+1,k-1}$$

for $n, k = 0, 1, \ldots$, where $\delta_k = 0$ for $k = 0$ and $\delta_k = 1$ for $k \geq 1$. The recursion relation is obtained by writing $x^n \, dx = dx^{n+1}/(n+1)$ in the integral for ϕ_{nk} and doing partial integration. To apply the recursion scheme, we compute the integrals ϕ_{Lk} for $k = 0, \ldots, L+1$ by numerical integration, where L denotes the largest value of n for which a_n^{ij} is used in the geometric tail approach. Next we compute successively for $n = L-1, \ldots, 0$ the quantities $\phi_{nk}, 0 \leq k \leq n+1$, from the recursion relation. An accuracy check for the recursive calculations is provided by the explicit expression

$$\phi_{n0} = \frac{\lambda_1^n}{(\lambda_1 - \lambda_2)^{n+1}} e^{-\lambda_2 D} \left\{ 1 - \sum_{k=0}^{n} e^{-(\lambda_1 - \lambda_2)D} \frac{[(\lambda_1 - \lambda_2)D]^k}{k!} \right\}.$$

Waiting-time probabilities

A theoretical generalization of the exact algorithm for the waiting-time probabilities in the *M/D/c* queue can be formulated. However, this generalized algorithm is not very useful for practical computations. Therefore we resort to a similar approximation as used for the $M^X/D/c$ queue. This approximation is based on the asymptotic expansion of the waiting-time distribution function $1 - W_q(x)$ and matches the exact values of the delay probability and the average delay in queue per customer. By the general result (4.5.33) in Section 4.5.3, we have

$$1 - W_q(x) \approx \frac{\sigma \delta}{\lambda(1/\eta - 1)^2 \eta^{1-c}} e^{-\delta x} \quad \text{for } x \text{ large enough,} \tag{4.6.17}$$

where $\lambda = [1/\lambda_1 + b/\lambda_2]^{-1}$ is the average arrival rate of customers, η and δ are given by (4.6.15) and (4.6.16) and $\sigma = \lim_{n \to \infty} \eta^{-n} p_n$. Once the probabilities p_n have been computed, we calculate σ from

$$\sigma \approx \eta^{-N} p_N,$$

where N denotes the truncation integer used in the geometric tail approach.

It is easy to compute the exact values of the delay probability and the average delay in queue. Denoting by π_n the long-run fraction of customers who see upon arrival n other customers present, we have that the long-run fraction of customers who are delayed is equal to

$$P_{\text{delay}} = 1 - \sum_{n=0}^{c-1} \pi_n.$$

The probability π_n is equal to the average arrival rate of customers seeing n other customers present divided by the average arrival rate λ. Thus

$$\pi_n = \frac{\lambda_1(1-b)p_{n1} + \lambda_2 p_{n2}}{1/(1/\lambda_1 + b/\lambda_2)}, \quad n = 0, 1, \ldots .$$

The average delay in queue follows by applying Little's formula $L_q = \lambda W_q$. Thus using $L_q = \sum_{n=c}^{\infty}(n-c)p_n$, we obtain

$$W_q = (1/\lambda_1 + b/\lambda_2)\left[\sum_{n=c}^{N-1}(n-c)p_n + \{(N-c)(1-\eta) + \eta\}\frac{p_N}{(1-\eta)^2}\right].$$

Next we approximate $W_q^c(x) = 1 - W_q(x)$ by

$$W_{\text{app}}^c(x) = \begin{cases} \alpha e^{-\beta x} & \text{for } x \le x_0, \\ \gamma e^{-\delta x} & \text{for } x > x_0, \end{cases} \tag{4.6.18}$$

where γ is the coefficient of $\exp(-\delta x)$ in (4.6.17) and the constants α, β and x_0 are determined by matching the exact values of $W_q(0)$ and W_q and by requiring continuity of $W_{\text{app}}^c(x)$ at $x = x_0$. Thus $\alpha = P_{\text{delay}}$, while β and x_0 are computed from the two equations,

$$\int_0^\infty W_{\text{app}}^c(x)\,dx = W_q \quad \text{and} \quad \alpha e^{-\beta x_0} = \gamma e^{-\delta x_0}.$$

The simple approximation (4.6.18) is very useful for practical purposes.

Erlangian arrivals

The analysis for the $C_2/D/c$ queue can be extended to the case of generalized Erlangian arrivals. The expression for the transition probabilities a_n^{ij} are simple for the case of a pure Erlangian interarrival-time distribution, but becomes very complicated for the general case. The case of generalized Erlangian arrivals will not be treated. The computation of the (micro)state probabilities for the case of pure Erlangian arrivals is left as an exercise to the reader. The waiting-time probabilities in the $E_k/D/c$ are easiest computed from the corresponding probabilities in the $M/D/kc$ queue. To show this, we first prove the following theorem.

Theorem 4.6.1   *The waiting-time distribution function $W_q(x)$ in the multi-server $GI/D/c$ queue is the same as in the single-server $GI^{(c*)}/D/1$ queue in which the interarrival time is distributed as the sum of c interarrival times in the $GI/D/c$ queue.*

Proof   Since the service times are deterministic, it is no restriction to assign the customers cyclically to the c servers so that server k gets the customers numbered as k, $k + c$, $k + 2c$, ... for any $k = 1, ..., c$. This simple observation proves the theorem.

The theorem has the following important corollary.

Corollary 4.6.2   *The waiting-time distribution function $W_q(x)$ in the $E_k/D/c$ queue is identical to the waiting-time distribution in the $M/D/kc$ queue with the same server utilisation.*

Proof   Recall that an Erlang-m distributed random variable is the sum of m independently and identically distributed exponentials. Thus, by Theorem 4.6.1, the waiting-time distribution functions in the $E_k/D/c$ queue and the $M/D/kc$ queue are both identical to the waiting-time distribution function in the $E_{kc}/D/1$ queue.

An algorithm for the computation of the waiting-time probabilities in the $M/D/kc$ queue is given in Section 4.5.2. This algorithm involves also the computation of the state probabilities. It should be pointed out that the state probabilities for the $E_k/D/c$ queue cannot be deduced from those for the $M/D/kc$ queue. Nevertheless we can conclude that the long-run average queue size in the $E_k/D/c$ is given by

$$L_q = \frac{1}{k} \sum_{j=c}^{\infty} (j - c) p_j,$$

where the p_j's are the state probabilities for the $M/D/kc$ queue. This relation is easily verified by applying twice Little's formula for W_q and using that W_q has the same value for the $E_k/D/c$ queue and the $M/D/kc$ queue.

4.7   MULTI-SERVER QUEUES WITH FINITE-SOURCE INPUT

This section considers a stochastic service system in which service requests are generated by a finite number of sources. There are N identical sources and c servers are available to handle the service requests. It is assumed that $N \geq c$. The service times of the requests are independent of each other and have a general probability distribution function. Each server can handle only one request at a time. A newly generated request finding all servers busy waits in queue until a server becomes available. New requests for service can be generated only by idle sources. A source is said to be idle if the source has no request waiting or being served at the service facility. Any idle source will generate a request in the next Δt time units with probability $\eta \Delta t + o(\Delta t)$ for Δt small, independently of the states of the other sources. In other words, the idle sources act independently of each other and the time until an idle source generates a new request is exponentially distributed with mean $1/\eta$. This time is often called the think time.

We first discuss the special case of exponential services. Then exact algorithms can be given for the state probabilities and the waiting-time probabilities. Next we discuss approximations for the case of general service times.

4.7.1 Exponential service times

Suppose that the service time of a request is exponentially distributed with mean $1/\mu$. Denoting by $X(t)$ the number of service requests present at time t, the process $\{X(t)\}$ is a continuous-time Markov chain with state space $\{0, 1, \ldots, N\}$. By equating the rate at which the process leaves the set of states $\{j, \ldots, N\}$ to the rate at which the process enters this set, we obtain for the equilibrium probabilities p_j the recursion scheme

$$\min(j, c)\mu p_j = (N - j + 1)\eta p_{j-1}, \quad j = 1, \ldots, N.$$

Starting with $\overline{p}_0 := 1$, we recursively compute $\overline{p}_1, \ldots, \overline{p}_N$. Next the desired p_i's are obtained by using the normalizing equation $\Sigma_{j=0}^{N} p_j = 1$.

The time-average probability p_j gives the long-run fraction of time that j customers are present at the service facility. Let us also define the customer-average probability π_j by

$$\pi_j = \text{the long-run fraction of service requests finding upon occurrence}$$
$$j \text{ other requests present } (j = 0, \ldots, N - 1).$$

To find an expression for π_j, note that π_j is equal to the long-run average number of service requests generated per unit time while j other service requests are present divided by the long-run average number of service requests generated per unit time. Using the theory of continuous-time Markov chains (Corollary 2.6.3 in Section 2.6), we have that the long-run average number of service requests generated per unit time while j other requests are present equals $(N - j)\eta p_j$. Thus

$$\pi_j = \frac{(N - j)p_j}{\displaystyle\sum_{k=0}^{N}(N - k)p_k} \quad \text{for } j = 0, \ldots, N - 1. \tag{4.7.1}$$

As a consequence of the definition of π_j, the long-run fraction of delayed requests is given by

$$P_{\text{delay}} = \sum_{j=c}^{N-1} \pi_j.$$

For exponential services we can also compute the waiting-time probability $W_q(x)$ which is defined as the long-run fraction of service requests whose delay in queue is no more than x. Since service completions occur according to a Poisson process with rate $c\mu$ as long as all servers are busy, it follows that

$$W_q(x) = 1 - \sum_{j=c}^{N-1} \pi_j \sum_{k=0}^{j-c} e^{-c\mu x} \frac{(c\mu x)^k}{k!}, \qquad x \geq 0,$$

assuming that service is in order of arrival.

Finally, it is pointed out that the general principle (4.1.2) underlying Little's formula yields

$$\text{the average number of busy servers} = \left[\sum_{j=0}^{N} (N-j)\eta p_j \right] E(S), \qquad (4.7.2)$$

where $E(S) = 1/\mu$ denotes the mean service time.

4.7.2 General service times

It is now assumed that the service time S of a request has a general probability distribution function $B(x)$. An exact analysis is in general not possible except for the case of a single server. Therefore we again resort to an approximate analysis. The same Approximation Assumption as in Section 4.5.3 is made.

Theorem 4.7.1 *Under the Approximation Assumption 4.5.1 of Section 4.5.3, the time-average probabilities p_j are approximated by*

$$p_j^{\text{app}} = \begin{cases} \binom{N}{j} [\eta E(S)]^j p_0^{\text{app}}, & 0 \leq j \leq c-1 \\ (N-c+1)\eta \alpha_{cj} p_{c-1}^{\text{app}} + \sum_{k=c}^{j} (N-k)\eta \beta_{kj} p_k^{\text{app}}, & c \leq j \leq N, \end{cases}$$

where the constants α_{cj} and β_{kj} are given by

$$\alpha_{cj} = \int_0^\infty \{1 - B_e(t)\}^{c-1} \{1 - B(t)\} \phi_{cj}(t) \, dt,$$

$$\beta_{kj} = \int_0^\infty \{1 - B(ct)\} \phi_{kj}(t) \, dt$$

with

$$\phi_{kj}(t) = \binom{N-k}{j-k} (1 - e^{-\eta t})^{j-k} e^{-\eta t(N-j)}, \qquad t > 0 \text{ and } k \geq j \geq c.$$

The function $B_e(t)$ denotes the equilibrium excess distribution of $B(t)$. The proof of Theorem 4.7.1 is similar to that of Theorem 4.5.1 in Section 4.5.3 and will not be given here.

Table 4.7.1   Numerical results for the finite-source model

		$N = 10$		$N = 15$		$N = 20$	
		P_{delay}	L_q	P_{delay}	L_q	P_{delay}	L_q
$c_S^2 = 0$	approx	0.0909	0.052	0.4729	0.613	0.8965	2.945
	exact	0.0857	0.049	0.4581	0.577	0.8937	2.920
$c_S^2 = \frac{1}{2}$	approx	0.0894	0.059	0.4529	0.709	0.8535	3.124
	exact	0.0880	0.059	0.4478	0.703	0.8507	3.119
$c_S^2 = 2$	approx	0.0846	0.082	0.4086	0.918	0.7666	3.474
	exact	0.0863	0.082	0.4153	0.923	0.7728	3.477

The results in Theorem 4.7.1 are exact for both the case of exponential services and the case of $c = 1$ server, since the approximation assumption then holds exactly. The approximation for the time-average probabilities p_j induces an approximation for the customer-average probabilities π_j. The relation (4.7.1) is also valid for the case of general service times. The proof of this relation then uses the property Poisson arrivals see time averages.

Our numerical investigations indicate that the approximation for the p_j's is accurate enough for practical purposes. In Table 4.7.1 we give the exact and approximate values of the delay probability P_{delay} and the average queue size L_q for various service-time distributions and for various values of N. The service-time distributions are the deterministic distribution ($c_S^2 = 0$), the E_2 distribution ($c_S^2 = \frac{1}{2}$) and the H_2 distribution ($c_S^2 = 2$) with the gamma normalization. In the examples we take $c = 5$ and $\eta = 0.4$ and vary N as 10, 15 and 20. Except for the case of $c_S^2 = 0$ the exact values in the table are taken from the tabulations of Seelen *et al* (1985). For the case of deterministic services computer simulation was used to find the exact values, where 500 000 service requests were simulated in each example.

4.8   FINITE-CAPACITY QUEUEING SYSTEMS

This section considers queueing systems having room for only a finite number of customers. Each customer finding upon arrival no waiting place available is rejected. A rejected customer is assumed to have no further influence on the system. In finite-capacity systems the finite waiting room acts as a regulator on the queue size and so no *a priori* assumption on the offered load is needed.

A practical problem of considerable interest is the calculation of the rejection probability. The design of finite buffers is a basic problem in telecommunication and production. We first discuss in Section 4.8.1 the computation of the rejection probability in the $M/G/c/c+N$ queue. In the next Section 4.8.2 we derive a generally applicable heuristic for the rejection probability in finite-capacity queues. This heuristic is based on the solution of the corresponding infinite-capacity queue. The approximation to the rejection probability can be used for both single-arrival queues and batch-arrival queues. In Section 4.8.3 we propose a two-moment approximation for the design problem of finding the smallest buffer size for which the rejection probability remains below a prespecified value.

4.8.1 The $M/G/c/c+N$ queueing system

This queueing model has a Poisson input with rate λ, a general service time S, c identical servers and N waiting positions for customers to await service. An arriving customer who finds all c servers busy and all N waiting places occupied is rejected.

An exact solution of the $M/G/c/c + N$ model is only possible for special cases such as exponential services, a single server or no waiting room. The model with no waiting room ($N = 0$) deserves special attention. This loss model has the famous insensitivity property that the state probabilities require the service time only through its first moment; see Example 2.7.1 of Section 2.7. In general, it is not possible to give a tractable exact solution for the $M/G/c/c+N$ queue. However, the approximate approach given in Section 4.5.3 for the $M/G/c$ queue can be extended to the $M/G/c/c + N$ queue. Before discussing this extension, we give the exact solution of the useful $M/M/c/c + N$ model.

Exponential services

Suppose that the service times are exponentially distributed with mean $1/\mu$. The process $\{X(t)\}$ describing the number of customers present is a continuous-time Markov chain with state space $\{0, 1, \ldots, N + c\}$. By equating the rate at which the process leaves the set of states $\{j, \ldots, N + c\}$ to the rate at which the process enters this set, we obtain for the equilibrium probabilities p_j the recursion scheme

$$\min(j, c)\mu p_j = \lambda p_{j-1} \quad \text{for } j = 1, \ldots, N + c.$$

By the property Poisson arrivals see time averages the time-average probability p_j can also be interpreted as the long-run fraction of customers finding upon arrival j other customers present. In particular,

the long-run fraction of customers rejected $= p_{N+c}$.

Once the p_j's have been computed, it is easy to compute the waiting-time probability $W_q(x)$ which is defined as the long-run fraction of *accepted* customers whose delay in queue is no more than x. Using the property Poisson arrivals see time averages, we have

the long-run fraction of accepted customers finding upon arrival j other

customers present $= \dfrac{p_j}{1 - p_{N+c}} \quad \text{for } 0 \le j \le N + c - 1.$

Since service completions occur according to a Poisson process with rate $c\mu$ as long as all servers are busy, it follows that

$$W_q(x) = 1 - \sum_{j=c}^{N+c-1} \frac{p_j}{1-p_{N+c}} \sum_{k=0}^{j-c} e^{-c\mu x} \frac{(c\mu x)^k}{k!}, \qquad x \geq 0,$$

assuming that service is in order of arrival.

General service times

Suppose that the service time S of a customer has a general probability distribution function $B(x)$ with $B(0) = 0$. The next theorem extends the approximation that was given in Theorem 4.5.1 for the state probabilities in the infinite-capacity $M/G/c$ queue.

Theorem 4.8.1   *Under the Approximation Assumption 4.5.1 given in Section 4.5.3, the state probabilities p_j are approximated by*

$$p_j(\text{app}) = \begin{cases} \dfrac{(c\rho)^j}{j!} p_0(\text{app}), & 0 \leq j \leq c-1, \\[2ex] \lambda p_{c-1}(\text{app})a_{j-c} + \lambda \displaystyle\sum_{k=c}^{j} p_k(\text{app})b_{j-k}, & c \leq j \leq N+c-1 \\[2ex] \rho p_{c-1}(\text{app}) - (1-\rho) \displaystyle\sum_{k=c}^{N+c-1} p_k(\text{app}), & j = N+c, \end{cases}$$

where $\rho = \lambda E(S)/c$ and the constants a_n and b_n are the same as in Theorem 4.5.1.

Proof   The proof of the theorem is a minor modification of the proof of Theorem 4.5.1. The details are left to the reader.

The result of Theorem 4.8.1 is exact for both the case of exponential services and the case of a single server, since for these two special cases the Approximation Assumption holds exactly. Further support to the approximate result of the theorem is provided by the fact that the approximation p_j^{app} is exact for the particular case of no waiting room ($N = 0$). As already pointed out in Example 2.7.1 of Section 2.7, the time-average probabilities p_j in the $M/G/c/c$ model (Erlang's loss model) are given by

$$p_j = \frac{[\lambda E(S)]^j / j!}{\displaystyle\sum_{k=0}^{c} [\lambda E(S)]^k / k!}, \qquad j = 0, 1, \ldots, c.$$

This solution requires the service-time distribution only through its first moment. This insensitivity property is very useful in practice.

Our numerical investigations indicate that the approximation for the state probabilities is accurate enough for practical purposes. In Table 4.8.1 we give the exact and approximate values of the rejection probability P_{rej} ($= p_{N+c}$) for several

Table 4.8.1 Numerical results for P_{rej} in the $M/G/c/c + N$ queue

		$\rho = 0.5$		$\rho = 0.8$		$\rho = 1.5$	
		$N = 6$	$N = 10$	$N = 6$	$N = 10$	$N = 6$	$N = 10$
$c_S^2 = 0$	approx	0.0286	0.00036	0.1221	0.0179	0.3858	0.3348
	exact	0.0287	0.00035	0.1224	0.0175	0.3870	0.3352
$c_S^2 = {}^1\!/_2$	approx	0.0311	0.0010	0.1306	0.0308	0.3975	0.3395
	exact	0.0314	0.0010	0.1318	0.0314	0.4000	0.3400
$c_S^2 = 2$	approx	0.0370	0.0046	0.1450	0.0603	0.4114	0.3555
	exact	0.0366	0.0044	0.1435	0.0587	0.4092	0.3537

examples. In all examples we take $c = 5$ severs. Deterministic services ($c_S^2 = 0$), E_2 services ($c_S^2 = {}^1\!/_2$) and H_2 services ($c_S^2 = 2$) with the gamma normalization are considered. For the latter two services the exact values of P_{rej} are taken from the tabulations of Seelen $et\ al$ (1985). For deterministic services computer simulation was used to find the exact value of P_{rej}, where in each example 500 000 customer arrivals were simulated, and even 2000 000 customer arrivals in the example with $\rho = 0.5$ and $N = 10$.

It is interesting to point out that the numerical results in Table 4.8.1 support the long-outstanding conjecture for the $GI/G/c/c + N$ queue that $P_{\text{rej}} \to 1 - 1/\rho$ as $N \to \infty$ when $\rho > 1$.

The approximation for P_{rej} is computationally quite demanding particularly when N gets large. In the event that P_{rej} must be computed for several values of N, the computational effort can be reduced by using the relation

$$p_j(\text{app}) = \gamma p_j^{(\infty)}(\text{app}), \quad j = 0, 1, \ldots, N + c - 1 \qquad (4.8.1)$$

with

$$\gamma = \left\{ 1 - \rho \sum_{k=N+c}^{\infty} p_k^{(\infty)}(\text{app}) \right\}^{-1}.$$

In this relation $p_j^{(\infty)}(\text{app})$ denotes the approximation given in Theorem 4.5.1 for the state probability $p_j^{(\infty)}$ in the infinite-capacity $M/G/c$ queue. It is required that $\rho < 1$, since otherwise the probabilities $p_j^{(\infty)}$ are not defined. The relation (4.8.1) is readily verified from the results in Theorems 4.5.1 and 4.8.1.

The proportionality relation (4.8.1) is in general not satisfied when the exact values of p_j and $p_j^{(\infty)}$ are taken instead of the approximate values. Nevertheless the proportionality relation is extremely useful. In the next section it will be shown that the proportionality relation forms the basis for a generally useful heuristic for the rejection probability.

4.8.2 Heuristic for the rejection probability

The proportionality relation (4.8.1) will be the basis for a generally applicable approximation to the rejection probability for both single-arrival queues and batch-arrival queues. We first state conditions under which the proportionality relation (4.8.1) holds exactly. In the following $\{p_j\}$ and $\{p_j^{(\infty)}\}$ denote the time-average state probabilities for the finite-capacity model and for the infinite-capacity model. To ensure the existence of the probabilities $p_j^{(\infty)}$, it is assumed that the server utilisation ρ is smaller than 1.

Theorem 4.8.2 *For both the M/G/c/c + N queue with c = 1 server and the M/G/c/c + N queue with exponential services,*

$$p_j = \gamma p_j^{(\infty)} \quad for \ j = 0, 1, \ldots, N + c - 1,$$

where

$$\gamma = \left\{ 1 - \rho \sum_{j=N+c}^{\infty} p_j^{(\infty)} \right\}^{-1}$$

Proof   The proof is based on the theory of regenerative processes. The process describing the number of customers present is a regenerative stochastic process in both the finite-capacity model and the infinite-capacity model. For both models, let a cycle be defined as the time elapsed between two consecutive arrivals that find the system empty. For the finite-capacity model, we define the random variables

T = the length of one cycle,

T_j = the amount of time that j customers are present during one cycle $(j = 0, 1, \ldots, N + c)$,

N_j = the number of service completion epochs at which j customers are left behind in one cycle $(j = 0, 1, \ldots, N + c - 1)$.

Also, for the finite-capacity model, let

μ_j = the expected amount of time from a service completion epoch at which j customers are left behind until the next service completion epoch $(j = 0, 1, \ldots, N + c - 1)$.

The corresponding quantities for the infinite-capacity model are denoted by T^{∞}, T_j^{∞}, N_j^{∞} and μ_j^{∞}. The following relations are obvious:

$$E(T) = \mu_0 + \sum_{j=1}^{N+c-1} E(N_j)\mu_j \tag{4.8.2}$$

and

$$E(T^{\infty}) = \mu_0^{\infty} + \sum_{j=1}^{\infty} E(N_j^{\infty})\mu_j^{\infty}. \tag{4.8.3}$$

For both the $M/G/c/c + N$ queue with $c = 1$ and the $M/G/c/c + N$ queue with exponential services, we have

$$\mu_j^\infty = \begin{cases} \mu_j & \text{for } 0 \leq j \leq N + c - 1 \\ \mu/c & \text{for } j \geq c, \end{cases} \tag{4.8.4}$$

where μ denotes the mean service time of a customer. This result uses the memoryless property of the exponential services in the multi-server case. By the theory of regenerative processes (see Section 1.3 in Chapter 1),

$$p_j = \frac{E(T_j)}{E(T)} \qquad \text{for } 0 \leq j \leq N + c$$

and $\tag{4.8.5}$

$$p_j^{(\infty)} = \frac{E(T_j^\infty)}{E(T^\infty)} \qquad \text{for } j \geq 0.$$

By the same arguments as used to derive relation (4.2.5) in the proof of Theorem 4.2.1, we find

$$E(N_j) = \lambda E(T_j), \quad 0 \leq j \leq N + c - 1$$

and

$$E(N_j^\infty) = \lambda E(T_j^\infty), \quad j \geq 0. \tag{4.8.6}$$

These relations express that the expected number of downcrossings from state $j+1$ to state j in one cycle equals the expected number of upcrossings from state j to state $j + 1$ in one cycle. Next we make another crucial observation. Using the lack of memory of the Poisson arrival process, some reflections show that the distribution of N_j for $j \leq N + c - 1$ does not depend on the limit on the queue size. In other words, N_j has the same distribution as N_j^∞ for $0 \leq j \leq N + c - 1$. Thus

$$E(N_j) = E(N_j^\infty) \quad \text{for } 0 \leq j \leq N + c - 1. \tag{4.8.7}$$

Together the relations (4.8.6) and (4.8.7) yield $E(T_j) = E(T_j^\infty)$ for $0 \leq j \leq N + c - 1$. Next, by applying (4.8.5), we find

$$p_j = \gamma p_j^{(\infty)} \quad \text{for } 0 \leq j \leq N + c - 1 \tag{4.8.8}$$

with

$$\gamma = \frac{E(T^\infty)}{E(T)}. \tag{4.8.9}$$

To obtain an explicit expression for γ, divide both sides of relation (4.8.2) by $E(T^\infty)$ and use the relations (4.8.4)–(4.8.7) to find

$$\frac{1}{\gamma} = \frac{\mu_0^\infty}{E(T^\infty)} + \sum_{j=1}^{N+c-1} \lambda p_j^{(\infty)} \mu_j^\infty. \tag{4.8.10}$$

By dividing both sides of (4.8.3) by $E(T^\infty)$ and using that $\mu_j^\infty = \mu/c$ for $j \geq c$, we find

$$1 = \frac{\mu_0^\infty}{E(T^\infty)} + \sum_{j=1}^{N+c-1} \lambda p_j^{(\infty)} \mu_j^\infty + \frac{\lambda\mu}{c} \sum_{j=N+c}^{\infty} p_j^{(\infty)}. \qquad (4.8.11)$$

The relations (4.8.10) and (4.8.11) yield the desired expression for γ.

$M^X/G/c/c + N$ queue with batch arrivals

Theorem 4.8.2 can be extended to the batch-arrival $M^X/G/c/c + N$ queue. In this model batches of customers arrive according to a Poisson process with rate λ and the batch size X has a discrete probability distribution $\{\beta_j, \ j \geq 1\}$ with mean β. Denoting by μ the mean service time of a customer, it is assumed that the server utilisation $\rho = \lambda\beta\mu/c$ is smaller than 1. For finite-buffer queues with batch arrivals we must distinguish between the following two cases:

(a) *Partial rejection*: An arriving batch whose size exceeds the remaining capacity of the buffer is partially rejected by turning away only those customers in excess of the remaining capacity.

(b) *Complete rejection*: An arriving batch whose size exceeds the remaining capacity of the buffer is rejected in its whole.

The emphasis of the discussion will be on the case of partial rejection.

Theorem 4.8.3   *Under partial rejection, it holds for both the $M^X/G/c/c + N$ with $c = 1$ server and the $M^X/G/c/c + N$ queue with exponential services that*

$$p_j = \gamma p_j^{(\infty)} \quad for \ j = 0, 1, \ldots, N+c-1,$$

where

$$\gamma = \left\{ 1 - \rho \sum_{j=N+c}^{\infty} q_j^{(\infty)} \right\}^{-1}$$

with

$$q_j^{(\infty)} = \frac{1}{\beta} \sum_{k=0}^{j} p_k^{(\infty)} \sum_{s=j-k+1}^{\infty} \beta_s, \quad j = 0, 1, \ldots .$$

Proof   The proof of Theorem 4.8.2 can be easily extended to the batch-arrival case. The relations (4.8.2)–(4.8.5) remain the same. The relations in (4.8.6) should be modified as

$$E(N_j) = \lambda \sum_{k=0}^{j} E(T_k) \sum_{s=j-k+1}^{\infty} \beta_s, \quad 0 \leq j \leq N+c-1 \qquad (4.8.12)$$

and

$$E(N_j^\infty) = \lambda \sum_{k=0}^{j} E(T_k^\infty) \sum_{s=j-k+1}^{\infty} \beta_s, \quad j \geq 0. \tag{4.8.13}$$

These two relations express that the expected number of state transitions out of the set of states having $j + 1$ or more customers present equals the expected number of state transitions into this set in one cycle.

Relation (4.8.7) remains valid for the batch-arrival model with partial rejection. To see this, note that the number of customers in the system is at the boundary level $N + c$ just after partial rejection of a batch. In the infinite-capacity system such a batch would have raised the number in the system above the level $N + c$. However, since the arrival process is memoryless, we have for any $0 \leq j \leq N + c - 1$ that the number of downcrossings from $j + 1$ to j in one cycle does not depend on excursions above the level $N + c$.

Using induction, it follows from the relations (4.8.7), (4.8.12) and (4.8.13) that $E(T_j) = E(T_j^\infty)$ for $0 \leq j \leq N + c - 1$. Next, by applying (4.8.5), we again obtain the proportionality relation (4.8.8) where γ is given by (4.8.9). To find an expression for γ, we use (4.8.4), (4.8.7) and (4.8.13) to obtain from (4.8.2) and (4.8.3) that

$$\frac{1}{\gamma} = \frac{\mu_0^\infty}{E(T^\infty)} + \sum_{j=1}^{N+c-1} \left[\lambda \sum_{k=0}^{j} p_k^{(\infty)} \sum_{s=j-k+1}^{\infty} \beta_s \right] \mu_j^\infty \tag{4.8.14}$$

and

$$1 = \frac{\mu_0^\infty}{E(T^\infty)} + \sum_{j=1}^{N+c-1} \left[\lambda \sum_{k=0}^{j} p_k^{(\infty)} \sum_{s=j-k+1}^{\infty} \beta_s \right] \mu_j^\infty + \frac{\lambda\mu}{c} \sum_{j=N+c}^{\infty} \sum_{k=0}^{j} p_k^{(\infty)} \sum_{s=j-k+1}^{\infty} \beta_s. \tag{4.8.15}$$

Since $\rho = \lambda\beta\mu/c$, we next obtain from (4.8.14) and (4.8.15) the desired expression for γ.

It is left to the reader to verify that the probability $q_j^{(\infty)}$ in Theorem 4.8.3 can be interpreted as the long-run fraction of service completions at which j customers are left behind in the infinite-capacity model. (*Hint*: this fraction is the ratio of $E(N_j^\infty)$ and $\lambda\beta E(T^\infty)$). Since the probabilities $q_j^{(\infty)}$ sum up to 1, the infinite sum $\sum_{j \geq N+c}^{\infty} q_j^{(\infty)}$ can be replaced by $1 - \sum_{j=0}^{N+c-1} q_j^{(\infty)}$.

Define now the rejection probability P_{rej} by

$$P_{\text{rej}} = \text{the long-run fraction of customers rejected.}$$

An exact expression for P_{rej} can be given under the conditions of the Theorems 4.8.2 and 4.8.3. Since Theorem 4.8.2 is a special case of Theorem 4.8.3 with $X = 1$, it suffices to give P_{rej} for the situation described in Theorem 4.8.3.

Corollary 4.8.4   *Under partial rejection, it holds for both the $M^X/G/c/c+N$ queue with $c = 1$ server and the $M^X/G/c/c + N$ queue with exponential services that*

$$P_{\text{rej}} = \frac{(1 - \rho)(\gamma - 1)}{\rho},$$

where γ is given in Theorem 4.8.3.

Proof   By Little's formula for the average number of busy servers in the finite-capacity model, we have

$$\sum_{j=0}^{N+c} \min(j, c) p_j = \lambda \beta (1 - P_{\text{rej}}) \mu. \tag{4.8.16}$$

Similarly, for the infinite-capacity model,

$$\sum_{j=0}^{\infty} \min(j, c) p_j^{(\infty)} = \lambda \beta \mu. \tag{4.8.17}$$

Hence, by (4.8.16),

$$P_{\text{rej}} = 1 - \frac{1}{\lambda\beta\mu} \sum_{j=0}^{N+c-1} \min(j, c) p_j - \frac{c}{\lambda\beta\mu} p_{N+c}$$

$$= 1 - \frac{1}{\lambda\beta\mu} \sum_{j=0}^{N+c-1} \min(j, c) p_j - \frac{c}{\lambda\beta\mu} \left(1 - \sum_{j=0}^{N+c-1} p_j \right).$$

Substituting the proportionality relation $p_j = \gamma p_j^{(\infty)}$, $0 \le j \le N + c - 1$ into this equation yields

$$P_{\text{rej}} = 1 - \frac{c}{\lambda\beta\mu} - \frac{\gamma}{\lambda\beta\mu} \sum_{j=0}^{N+c-1} \min(j, c) p_j^{(\infty)} + \frac{c\gamma}{\lambda\beta\mu} \sum_{j=0}^{N+c-1} p_j^{(\infty)}$$

$$= 1 - \frac{c}{\lambda\beta\mu} + \frac{c\gamma}{\lambda\beta\mu} - \frac{\gamma}{\lambda\beta\mu} \sum_{j=0}^{N+c-1} \min(j, c) p_j^{(\infty)} - \frac{\gamma}{\lambda\beta\mu} \sum_{j=N+c}^{\infty} c p_j^{(\infty)}.$$

Next, by using (4.8.17), we find

$$P_{\text{rej}} = 1 - \frac{c}{\lambda\beta\mu} + \frac{c\gamma}{\lambda\beta\mu} - \frac{\gamma}{\lambda\beta\mu} \lambda\beta\mu.$$

This yields the desired result.

We now come to a heuristic for the rejection probability in the $GI^X/G/c/c + N$ queue with partial rejection. In this model batches of customers arrive according to a renewal process with a mean interarrival time of $1/\lambda$, the batch size X has a discrete probability distribution $\{\beta_j, \ j \geq 1\}$ with mean β and the service time of a customer has mean μ. It is assumed that $\rho = \lambda\beta\mu/c$ is smaller than 1. To give the heuristic, define for the infinite-capacity $GI^X/G/c$ queue the customer-average probabilities $\pi_j^{(\infty)}$ by

π_j^∞ = the long-run fraction of batches that find upon arrival j other customers in the system $(j = 0, 1, \ldots)$.

Note that $\pi_j^{(\infty)} = p_j^{(\infty)}$ for all $j \geq 0$ when the arrival process of batches is a Poisson process, where the probabilities $p_j^{(\infty)}$ are the time-average state probabilities. The result of Corollary 4.8.4 suggests now the following heuristic for the rejection probability P_{rej}.

Heuristic   *An approximation to the rejection probability in the $GI^X/G/c/c + N$ queue with partial rejection is*

$$P_{\text{rej}}^{\text{app}} = \frac{(1 - \rho) \displaystyle\sum_{j=N+c}^{\infty} r_j^{(\infty)}}{1 - \rho \displaystyle\sum_{j=N+c}^{\infty} r_j^{(\infty)}} \tag{4.8.18}$$

where

$$r_j^{(\infty)} = \frac{1}{\beta} \sum_{k=0}^{j} \pi_k^{(\infty)} \sum_{s=j-k+1}^{\infty} \beta_s, \quad j = 0, 1, \ldots .$$

The heuristic is exact when the arrival process of batches is a Poisson process. Since the probabilities $r_j^{(\infty)}$ sum up to 1, the infinite sum $\Sigma_{j \geq N+c} r_j^{(\infty)}$ can be replaced by $1 - \Sigma_{j=0}^{N+c-1} r_j^{(\infty)}$. Note that for the case of single arrivals $r_j^{(\infty)} = \pi_j^{(\infty)}$ for all $j \geq 0$.

Our numerical investigations indicate that the heuristic performs quite well for practical purposes. In the Tables 4.8.2 and 4.8.3 we give some numerical results. In the examples we have tested the heuristic for the particular case of deterministic input. The deterministic arrival process with its full information about future arrivals is diametrically opposed to the Poisson arrival process with its memoryless property. In Table 4.8.2 we give the approximate and exact values of P_{rej} for several cases of the $D/E_k/1/1 + N$ queue, while Table 4.8.3 deals with the multi-server $D/M/c/c + N$ queue. The numerical result shows that the heuristic performs remarkably well. In all examples the heuristic value of P_{rej} is of the same order of magnitude as the exact value. This is what is typically needed when a heuristic is used for dimensioning purposes. Tables 4.8.2 and 4.8.3 deal with the case of single arrivals. However, the performance of the heuristic is of a similar quality for the case of batch arrivals with partial rejection.

Table 4.8.2 Numerical results for the $D/E_k/1/1 + N$ queue

		$\rho = 0.8$			$\rho = 0.95$	
		$N = 0$	$N = 3$	$N = 8$	$N = 0$	$N = 10$	$N = 40$
E_4	approx	1.39×10^{-1}	3.74×10^{-4}	3.47×10^{-8}	2.07×10^{-1}	6.88×10^{-4}	2.75×10^{-9}
	exact	2.16×10^{-1}	5.50×10^{-4}	5.09×10^{-8}	2.99×10^{-1}	7.69×10^{-4}	3.07×10^{-9}
E_2	approx	1.95×10^{-1}	7.22×10^{-3}	6.76×10^{-5}	2.61×10^{-1}	6.22×10^{-3}	1.11×10^{-5}
	exact	2.50×10^{-1}	8.63×10^{-3}	8.05×10^{-5}	3.20×10^{-1}	6.54×10^{-3}	1.17×10^{-5}
E_1	approx	2.53×10^{-1}	3.57×10^{-2}	3.10×10^{-3}	3.14×10^{-1}	2.30×10^{-2}	7.28×10^{-4}
	exact	2.87×10^{-1}	3.86×10^{-2}	3.33×10^{-3}	3.49×10^{-1}	2.36×10^{-2}	7.41×10^{-4}

Table 4.8.3 Numerical results for the $D/M/c/c + N$ queue

		$\rho = 0.8$			$\rho = 0.95$	
		$N = 0$	$N = 10$	$N = 25$	$N = 0$	$N = 50$	$N = 75$
$c = 5$	approx	1.02×10^{-2}	6.99×10^{-4}	6.59×10^{-7}	1.48×10^{-1}	1.39×10^{-6}	6.83×10^{-9}
	exact	1.11×10^{-2}	7.49×10^{-4}	7.06×10^{-7}	1.59×10^{-1}	1.44×10^{-6}	6.74×10^{-9}
$c = 25$	approx	1.59×10^{-3}	1.44×10^{-4}	1.37×10^{-7}	4.98×10^{-2}	7.54×10^{-7}	3.53×10^{-9}
	exact	1.71×10^{-3}	1.55×10^{-4}	1.46×10^{-7}	5.23×10^{-2}	7.80×10^{-7}	3.65×10^{-9}
$c = 100$	approx	2.16×10^{-4}	2.08×10^{-6}	1.97×10^{-9}	9.60×10^{-3}	1.94×10^{-7}	9.07×10^{-10}
	exact	2.32×10^{-4}	2.23×10^{-6}	2.11×10^{-9}	9.96×10^{-3}	2.00×10^{-7}	9.39×10^{-10}

Complete rejection

For the batch-arrival $GI^x/G/c/c + N$ queue with complete rejection, the following heuristic can be used for the rejection probability

$$
P_{\text{rej}}^{\text{app}} = \frac{(1 - \rho)\left(1 - \sum_{j=0}^{N+c-1} u_j^{(\infty)}\right)}{1 - \rho\left(1 - \sum_{j=0}^{N+c-1} u_j^{(\infty)}\right)},
$$

where

$$
u_j^{(\infty)} = \frac{1}{\beta} \sum_{k=0}^{j} \pi_k^{(\infty)} \sum_{s=j-k+1}^{N+c-k} \beta_s, \quad j = 0, 1, \ldots .
$$

The heuristic can be shown to be exact for both the $M^X/G/1/1 + N$ queue and the $M^X/M/c/c + N$ queue with constant batch size $X = Q$ satisfying $Q \leq N + 1$, see also Exercise 4.13. In Table 4.8.4 we give some numerical results for P_{rej} in the $M^X/G/1/1 + N$ queue with complete rejection. For the batch size we consider both the two-point distribution $P\{X = 1\} = P\{X = 7\} = \frac{1}{2}$ and the geometric

Table 4.8.4   The $M^X/G/1/N + 1$ queue with complete rejection

		X: geometric			X: two-point		
		$N = 0$	$N = 50$	$N = 250$	$N = 0$	$N = 50$	$N = 250$
$c_S^2 = 0.1$	approx	8.99×10^{-1}	1.40×10^{-2}	1.62×10^{-7}	8.86×10^{-1}	8.88×10^{-3}	1.29×10^{-8}
	exact	9.40×10^{-1}	1.59×10^{-2}	1.82×10^{-7}	8.86×10^{-1}	9.01×10^{-3}	1.31×10^{-8}
$c_S^2 = 10$	approx	8.99×10^{-1}	6.09×10^{-2}	2.64×10^{-4}	8.86×10^{-1}	5.58×10^{-2}	1.79×10^{-4}
	exact	9.42×10^{-1}	6.11×10^{-2}	2.64×10^{-4}	8.86×10^{-1}	5.55×10^{-2}	1.79×10^{-4}

distribution $P\{X = j\} = (1/4)(3/4)^j$, $j \geq 1$. In both cases the mean batch size $\beta = 4$. The service-time distributions are the E_{10} distribution ($c_S^2 = 0.1$) and the H_2 distribution ($c_S^2 = 10$) with the gamma normalization. The offered load ρ is taken equal to 0.8. The results in Table 4.8.4 indicate that the approximation performs quite well for practical purposes.

Minimal buffer size

In practical applications one often wishes to dimension the buffer size in order to achieve a prespecified value of the rejection probability. For any $0 < \alpha < 1$, let us define $N(\alpha)$ as

$N(\alpha) = $ the minimal buffer size for which the rejection probability does not exceed the value α.

In many situations $N(\alpha)$ increases logarithmically in α as α gets smaller. This behaviour is also reflected in the heuristic for P_{rej}. To explain this, let us consider the $GI^X/G/c/c + N$ queue with partial rejection. Under general conditions it will hold that

$$r_j^{(\infty)} \approx \xi \eta^j \quad \text{for } j \text{ large enough}$$

for some $\xi > 0$ and $0 < \eta < 1$. In other words, the probabilities $r_j^{(\infty)}$ exhibit a geometric tail behaviour. Then the heuristic for P_{rej} shows that the corresponding approximation $N_{\text{app}}(\alpha)$ to $N(\alpha)$ behaves like

$$N_{\text{app}}(\alpha) \approx \frac{1}{\ln(\eta)} \ln \left[\frac{\alpha(1 - \eta)}{\xi(1 - \rho + \alpha\rho)} \right] - c \quad \text{for } \alpha \text{ small enough.} \quad (4.8.19)$$

This insightful relation may be very useful for dimensioning purposes.

4.8.3 Two-moment approximation for the minimal buffer size

The practical applicability of the heuristic for P_{rej} stands or falls with the computation of the state probabilities $\pi_j^{(\infty)}$. In some queueing models it is computationally feasible to calculate these probabilities using embedded Markov chain analysis or continuous-time Markov chain analysis. In these cases the computational effort can be reduced when the state probabilities exhibit a geometric tail behaviour; cf. Section 2.3.3. However, in many queueing models the exact computation of the state probabilities $\pi_j^{(\infty)}$ is not practically feasible. This is for instance the case in the $M^X/G/c$ queue with general service times. In such situations one might try to approximate the exact solution of the complex model through the exact solutions of simpler related models. In this chapter we have already seen several examples of such two-moment approximations. The rejection probability itself is not directly amenable to a two-moment approximation, but indirectly a two-moment approximation is possible through the 'percentile' $N(\alpha)$. Recall that $N(\alpha)$ is defined as the minimal buffer size for which the rejection probability does not exceed a prespecified value α.

A simple and useful two-moment approximation to $N(\alpha)$ can be given for the $M^X/G/c/c + N$ queue. Denoting by c_S^2 the squared coefficient of the service time of a customer, this approximation is given by

$$N_{\text{app}}(\alpha) = (1 - c_S^2)N_{\text{det}}(\alpha) + c_S^2 N_{\text{exp}}(\alpha), \qquad (4.8.20)$$

where $N_{\text{det}}(\alpha)$ and $N_{\text{exp}}(\alpha)$ are the (approximate) values of $N(\alpha)$ for the $M^X/D/c/c + N$ queue and the $M^X/M/c/c + N$ queue. The buffer sizes $N_{\text{det}}(\alpha)$ and $N_{\text{exp}}(\alpha)$ are computed by using the heuristic for P_{rej} for the particular cases of deterministic services and exponential services. Relatively simple algorithms can be given to compute the state probabilities in the $M^X/M/c$ queue and the $M^X/D/c$ queue, see the Sections 4.5.1, 4.5.2 and 4.5.5. The two-moment approximation (4.8.20) is only recommended when c_S^2 is not too large (say, $0 \le c_S^2 \le 2$).

In Table 4.8.5 we illustrate the performance of the two-moment approximation (4.8.20) for the particular case of the $M/G/c/c + N$ queue where the number of

Table 4.8.5　The minimal buffer size in the $M/G/c/c + N$ queue

		$\rho = 0.5$					$\rho = 0.8$				
$c = 1$	α	10^{-2}	10^{-4}	10^{-6}	10^{-8}	10^{-10}	10^{-2}	10^{-4}	10^{-6}	10^{-8}	10^{-10}
$c_S^2 = \frac{1}{2}$	exact	4	9	15	20	25	10	26	41	57	73
	approx	4	10	15	20	25	10	26	41	57	73
$c_S^2 = 2$	exact	7	16	26	35	45	19	49	80	110	141
	approx	7	17	25	36	46	19	50	80	111	140
$c = 10$	α	10^{-2}	10^{-4}	10^{-6}	10^{-8}	10^{-10}	10^{-2}	10^{-4}	10^{-6}	10^{-8}	10^{-10}
$c_S^2 = \frac{1}{2}$	exact	1	7	12	17	23	8	24	39	55	71
	approx	1	7	13	17	23	8	24	39	55	71
$c_S^2 = 2$	exact	1	10	20	29	39	14	44	74	105	135
	approx	1	10	20	29	39	14	45	75	106	135

servers has the two values $c = 1$ and $c = 10$. For both E_2 services ($c_S^2 = \frac{1}{2}$) and H_2 services ($c_S^2 = 2$) with the gamma normalization, the approximate and exact values of $N(\alpha)$ are given for $\alpha = 10^{-2}, 10^{-4}, 10^{-6}, 10^{-8}$ and 10^{-10}.

Here we have rounded up any fractional value resulting from the interpolation formula (4.8.20). The results in the table also nicely demonstrate that $N(\alpha)$ increases logarithmically in α as α increases.

EXERCISES

4.1 Consider the $M/G/1$ queue with exceptional first service. This model differs from the standard $M/G/1$ queue only in the service times of the customers reactivating the server after an idle period. Those customers have special service times with distribution function $B_0(t)$, while the other customers have ordinary service times with distribution function $B(t)$. Use the regenerative approach to verify that the state probabilities can be computed from the recursion scheme (4.2.1) in which $\lambda p_0 a_{j-1}$ is replaced by $\lambda p_0 \bar{a}_{j-1}$, where $\bar{a}_n$ is obtained by replacing $B(t)$ by $B_0(t)$ in the integral representation for a_n. Also, argue that p_0 satisfies $1 - p_0 = \lambda[p_0 \mu_0 + (1 - p_0)\mu_1]$, where μ_1 and μ_0 denote the means of the ordinary service times and the special service times.

4.2 Consider the $M/G/1$ queue with server vacations. In this variant of the $M/G/1$ queue a server vacation begins when the server becomes idle. During a server vacation the server performs other work and is not available for providing service. The length V of a server vacation has a general probability distribution function $V(x)$ with probability density $v(x)$. If upon return from a vacation the server finds the system empty, a new vacation period begins; otherwise, the server starts servicing the customers present. Denote by p_{0j} (p_{1j}) the time-average probability that j customers are in the system and the server is on vacation (available for service). Use the regenerative approach to verify the recursion scheme

$$p_{0j} = \frac{1-\rho}{E(V)} \int_0^\infty e^{-\lambda t} \frac{(\lambda t)^j}{j!} \{1 - V(t)\}\, dt, \quad j \geq 0,$$

$$p_{1j} = \frac{1-\rho}{E(V)} \sum_{k=1}^j v_k a_{j-k} + \lambda \sum_{k=1}^j (p_{0k} + p_{1k})a_{j-k}, \quad j \geq 1,$$

where a_n is given in Theorem 4.2.1 and v_k is the probability of k arrivals during a single vacation period. (*Hint*: take as cycle the time elapsed between two consecutive epochs at which either the server becomes idle or finds upon return from vacation an empty system.)

4.3 Suppose an $M/G/1$ queueing system in which the service time of a customer depends on the queue size at the moment the customer enters service. The service time has a probability distribution function $B_1(x)$ when R or less customers are present upon the moment the customer enters service; otherwise, the service time has probability distribution function $B_2(x)$. Denote by p_{1j} (p_{2j}) the time-average probability that j customers are in the system and service according to B_1 (B_2) is provided. Use the regenerative approach to verify the recursion scheme

$$p_{1j} = \lambda p_0 a_{j-1}^{(1)} + \lambda \sum_{k=1}^{\min(j,R)} p_{1k} a_{j-k}^{(1)}, \quad j = 1, 2, \ldots,$$

$$p_{2j} = \lambda \sum_{k=R+1}^{j} (p_{1k} + p_{2k})a_{j-k}^{(2)}, \quad j \geq R,$$

where $a_n^{(i)}$ is the same as the constant a_n in Theorem 4.2.1 except that $B(t)$ is replaced by $B_i(t), i = 1, 2$. Also, argue that $1 - p_0 = \lambda\{\mu_1\Sigma_{j=0}^{R}p_j + \mu_2(1 - \Sigma_{j=0}^{R}p_j)\}$, where μ_i is the mean of the distribution B_i. (*Hint*: use that the long-run fraction of service completions at which j customers are left behind equals the long-run fraction of customers finding upon arrival j other customers present.)

4.4 Consider the $M^X/G/1$ queue with Coxian-2 services. Use continuous-time Markov chain analysis to develop a recursion scheme for the calculation of the state probabilities.

4.5 Consider the $M^X/G/\infty$ queue. Extend the analysis in Example 1.2.2 in Section 1.2 to prove that the mean m and the variance v of the limiting distribution of the number of busy servers are given by

$$m = \lambda E(X)E(S)$$

$$v = \lambda E(X)E(S) + \lambda E[X(X-1)] \int_0^\infty \{1 - B(x)\}^2 \, dx,$$

where λ is the average arrival rate of batches, the random variable X denotes the batch size and $B(x)$ is the probability distribution of the service time S of a customer.

4.6 Consider the $M/G/c$ queue with service in order of arrival. Let the generic random variable L_∞ be distributed according to $P\{L_\infty = 0\} = \Sigma_{k=0}^{c}p_k$ and $P\{L_\infty = j\} = p_{j+c}$ for $j \geq 1$. In other words, L_∞ is distributed according the limiting distribution of the number in queue. Let the generic random variable W_∞ be distributed according to $P\{W_\infty \leq x\} = W_q(x), x \geq 0$. Prove that

$$E[L_\infty(L_\infty - 1)\ldots(L_\infty - k + 1)] = \lambda^k E(W_\infty^k) \quad \text{for all } k = 1, 2, \ldots.$$

(*Hint*: define the random variable L_n as the number of customers left behind in *queue* when the nth arrival enters service and define the random variable D_n as the delay in queue of the nth arrival. Argue first that

$$P\{L_n = j\} = \int_0^\infty e^{-\lambda x} \frac{(\lambda x)^j}{j!} d_n(x) \, dx, \quad j = 1, 2, \ldots,$$

where $d_n(x)$ is the density of $P\{D_n \leq x\}$ for $x > 0$. Next argue that the limiting distribution of L_n as $n \to \infty$ is equal to the distribution of L_∞ by observing that the long-run fraction of customers finding upon arrival j other customers in queue must be equal to the long-run fraction of customers leaving j other customers behind in queue when entering service.) This problem is based on Haji and Newell (1971).

4.7 Consider the $M/D/c$ queue with service in order of arrival. Let the generic random variables L_∞ and W_∞ be defined as in Exercise 4.6. Verify that

$$E(W_\infty^2) = \frac{1}{3c\lambda^2(1-\rho)} \left[(c\rho)^3 - c(c-1)(c-2) + \sum_{j=0}^{c-1}\{c(c-1)(c-2) \right.$$

$$\left. - j(j-1)(j-2)\}p_j \right] - \frac{(c-1-c\rho^2)}{\lambda^2(1-\rho)} E(L_\infty)$$

where $E(L_\infty)$ is given by formula (4.5.11) in Section 4.5.2.

4.8 Consider the $M/G/c$ queue with service in order of arrival. In Exercise 4.6 the reader was asked to argue that

$$p_{j+c} = \int_0^\infty e^{-\lambda x} \frac{(\lambda x)^j}{j!} w_q(x) \, dx, \quad j = 1, 2, \ldots,$$

where $w_q(x)$ is the density of $W_q(x)$ for $x > 0$. Denote by $V(x)$ the long-run fraction of *delayed* customers whose delay in queue is no more than x. Use the above relation to show that the conditional waiting-time probability $V(x)$ can be approximated by $V_{app}(x)$ satisfying the integral equation

$$V_{app}(x) = (1 - \rho)\{1 - (1 - B_e(x))^c\} + \lambda \int_0^x V_{app}(x - y)\{1 - B(cy)\} \, dy, \quad x \geq 0,$$

where $B_e(x)$ is the equilibrium excess distribution function associated with the service-time distribution function $B(x)$. Assuming that the service-time distribution has no extremely long tail, use the same argument as in Section 1.8 to prove that

$$1 - V_{app}(x) \approx \frac{e^{-\delta x} \int_0^\infty e^{\delta y} [1 - \rho B_e(cy) - (1 - \rho)\{1 - (1 - B_e(y))^c\}] \, dy}{\lambda \int_0^\infty y e^{\delta y} \{1 - B(cy)\} \, dy}$$

for x large enough, where $\delta > 0$ is the solution to $\lambda \int_0^\infty e^{\delta t}\{1 - B(ct)\} \, dt = 1$.

4.9 Consider the stochastic service system with slotted service from Example 2.4.2 in Section 2.4. Suppose now that the system has an infinite-capacity buffer. Denote by $\overline{\pi}_j^{(\infty)}$ the long-run fraction of time slots starting with j messages present in the buffer just prior to transmission. Prove that

$$\overline{\pi}_j^{(\infty)} \approx \gamma \tau^{-j} \quad \text{for } j \text{ large enough}$$

for some constants $\gamma > 0$ and $\tau > 1$, where τ is the unique solution to the equation $z^c - e^{-\lambda(1-z)} = 0$ on $(1, \infty)$. Adjust the heuristic (4.8.18) in terms of the $\overline{\pi}_j^{(\infty)}$.

4.10 Use an up- and downcrossings argument to verify that the probabilities $r_j^{(\infty)}$ used in the heuristic (4.8.18) for the rejection probability can be interpreted as the long-run fraction of service completions at which j customers are left behind in the $GI^X/G/c$ queue.

4.11 Verify for the partial-rejection case that the results of Theorem 4.8.3 and Corollary 4.8.4 also hold for the $M^X/G/1/1+N$ queue in which the first customer in each busy period receives exceptional service.

4.12 Consider a communication system with a single transmission channel at which batches of packets arrive according to a Poisson process with rate λ, where the batch size has a general distribution. The packets are temporarily stored in a finite buffer to await transmission. Rejection occurs for those packets of an arriving batch which are in excess of the remaining buffer capacity. The transmission time of each packet is a constant slot length of one time unit. The beginnings of the time slots provide the only opportunity to start the transmission of a packet. The transmission channel is subject to random service interruptions. The transmission of a packet is successful with probability f; otherwise, the transmission has to be retried in the next time slot. Show that this system can be modelled as an $M^X/G/1/1 + N$

queue with exceptional first service so that the result of Exercise 4.11 can be applied. (This problem is based on Tijms and Van Ommeren (1989).)

4.13 Consider the batch-arrival $M^X/G/c/c + N$ queue with complete rejection. Assuming that the batch size X is a constant Q with $Q \leq N + 1$, prove the proportionality relation $p_j = \gamma p_j^{(\infty)}$, $j = 0, 1, \ldots, N + c - Q$ holds for both the single-server case with general services and the multi-server case with exponential services.

4.14 Consider a finite-capacity stackyard at which batches of containers arrive according to a Poisson process with rate $\lambda = 2$ and the batch size is geometrically distributed with mean 10. A batch finding upon arrival not enough space is partially accepted. The holding times of the various containers at the stackyard are independent of each other and have a mean of $m = 15$. What is the required capacity of the stackyard such that the fraction of containers rejected is less than $\nu = 0.01$? Use an approximate analysis to answer this question when the time that a container is stored at the stackyard has (a) a uniform distribution at (10,20), (b) an exponential distribution and (c) an H_2 distribution with the gamma normalization and a squared coefficient of variation of 4. (*Hint*: use the heuristic for the rejection probability in the $M^X/G/c/c$ queue. Since the required stackyard will be typically large, it seems reasonable to approximate the state probabilities $\pi_j^{(\infty)}$ in the $M^X/G/c$ queue by those in the $M^X/G/\infty$ queue. Fit a negative binomial distribution to the state probabilities in the $M^X/G/\infty$ queue by using the results in Exercise 4.5.)

4.15 Let us consider a multi-server queueing system with server breakdowns. A busy server is subject to breakdowns occurring according to a Poisson process with rate $\eta = 1/4$; a server cannot break down whenever idle. Each service interruption caused by a breakdown takes a fixed time $\tau = 1/2$. If a service is interrupted by a breakdown, then upon completion of the interruption the service is continued from the point at which it was interrupted. Customers arrive according to a Poisson process with rate $\lambda = 1.5$, the service time of a customer has mean $E(S) = 1$ and a squared coefficient of variation $c_S^2 = 1/4$. The number of servers is $c = 2$. Calculate an approximation to the average time spent by a customer in the system.

4.16 Customers with items to repair arrive at a repair facility according to a Poisson process with rate λ. The repair time of an item has a uniform distribution on $[a, b]$. There are ample repair facilities so that each defective item immediately enters repair. If the repair time of an item takes longer than τ time units with τ a given number between a and b, the customer gets after a time τ a loaner for his defective item until the item returns from repair. A sufficiently large supply of loaners is available. Identify an appropriate queueing model in order to show that the average number of loaners which are out equals $1/2 \lambda (b - \tau)^2/(b - a)$. (This problem and the next one are based on Karmarkar and Kubat (1983).)

4.17 Suppose that a repair facility has a supply of N serviceable units that can be temporarily loaned to customers bringing items for repair. The units for repair arrive according to a Poisson process with rate λ. The repair time of each unit is exponentially distributed with mean $1/\mu$. There are ample repair facilities so that each defective item immediately enters repair. A customer bringing an item for repair gets a loaner, when available; otherwise the customer waits until either a loaner becomes free or the repair of his item is completed, whichever occurs first. The loaners are provided to waiting customers on a first-come-first-served basis; a loaner is returned by a customer upon completion of repair of his item. Determine an explicit expression for the average amount of time a customer has to wait until a loaner is provided or the repair of his item is completed. (*Hint*: denote by $E(W_j)$ the conditional expected waiting time of a customer finding upon arrival $N + j$ units in repair. Use a recursion equation to verify that $E(W_j) = (j + 1)/[\mu(N + j + 1)]$, $j \geq 0$.)

4.18 (a) Consider the $(S - 1, S)$ inventory model in which the demand process is a Poisson process with rate λ. Demands occurring when the system is out of stock are backordered. Under the $(S - 1, S)$ policy a replenishment order for one unit is placed each time one unit is demanded. The replenishment lead times are independent random variables having a common probability distribution with mean τ. Use results for the $M/G/\infty$ queue to prove that

$$\text{the average stock on hand } = \sum_{j=0}^{S}(S - j)e^{-\lambda t}\frac{(\lambda \tau)^j}{j!}.$$

(b) Consider the (R, S) inventory model with limited order sizes. In this model the inventory position is reviewed every R periods. At each review the inventory position is ordered up to the level S provided that the order size does not exceed the constant Q; otherwise, the replenishment order is of size Q. It is assumed that $Q > R\mu_1$, where μ_1 is the mean demand per period. Prove for the backlog model that the process describing the inventory position just after review can be reduced to the waiting-time process in the $D/G/1$ queue with $D = Q$. (This problem is based on De Kok (1989).)

4.19 Consider the $D/G/\infty$ queue. Denote by π_j the long-run fraction of customers finding upon arrival j other customers present. Prove that the first two moments of the probability distribution $\{\pi_j\}$ are given by

$$m_1 = \sum_{k=1}^{\infty}\{1 - B(kD)\} \quad \text{and} \quad m_2 = m_1 + 2\sum_{i=1}^{\infty}\sum_{j=i+1}^{\infty}\{1 - B(iD)\}\{1 - B(jD)\},$$

where $B(x)$ denotes the probability distribution function of the service time. Explain why it seems reasonable to fit a normal distribution to these two moments when the traffic load is not light.

4.20 Consider the $E_r/D/\infty$ queue. Denote by p_j the long-run fraction of time that j customers are present. Verify that $p_n = \sum_{i=1}^{r}p_{ni}$ with

$$p_{oi} = \frac{1}{r}\sum_{j=1}^{i}e^{-\alpha D}\frac{(\alpha D)^{j-i}}{(j - i)!}, \quad 1 \leq i \leq r$$

and

$$p_{ni} = \frac{1}{r}\sum_{j=1}^{r}e^{-\alpha D}\frac{(\alpha D)^{r-j+1+(n-1)r+i-1}}{(r - j + 1 + (n - 1)r + i - 1)!}, \quad n \geq 1 \text{ and } 1 \leq i \leq r,$$

where α is the scale parameter of the Erlang-r distributed interarrival time. Also, argue that the long-run fraction of customers finding upon arrival n other customers present is given by rp_{nr}.

BIBLIOGRAPHIC NOTES

The queueing theory literature is voluminous. A good account of the basic theory is provided by the books by Cooper (1981), Cox and Smith (1961) and Takács (1962). The books by Hayes (1984) and Kleinrock (1976) also discuss queueing networks and their applications to computer science. A book emphasizing the (approximate) analysis of the transient behaviour of queues is Newell (1971). A book on discrete-time queues is Bruncel and Kim (1993).

A thorough treatment of most of the background material in Section 4.1 can be found in the book by Wolff (1989). The regenerative approach used in Sections 4.2 and 4.3 to analyse single-server queues with Poisson input has its origin in the paper of Hordijk and Tijms (1976). This versatile approach was used in Tijms, *et al.* (1981) and Tijms and Van Hoorn (1982) to give an approximate analysis of multi-server queues with state-dependent Poisson input; see also Van Hoorn (1984). The discussion in Sections 4.5.3, 4.7 and 4.8.1 is based on these references. The approximations obtained by the regenerative approach for the $M/G/c$ queue are extended in Van Hoorn and Seelen (1986) to the $GI/G/c$ queue with phase-type arrivals. The two-moment approximation given in Section 4.5.3 for the average queue size in the $M/G/c$ queue goes back to Cosmetatos (1976) and Page (1972); see also the paper by Cox (1955). Another useful approximation to the average queue size in the $M/G/c$ queue is given by Boxma, *et al.* (1979). The analysis of the $M^X/G/c$ queue in Section 4.5.5 is based on Eikeboom and Tijms (1987). The analysis of the $C_2/D/c$ queue in Section 4.6.2 is taken from Van Hoorn (1986). Section 4.8.2 dealing with the heuristic for the rejection probability in finite-buffer queues with batch arrivals and partial rejection is based on Tijms (1992). Bounds for the rejection probability are discussed in Miyazawa and Tijms (1993) and the modification of the heuristic for the batch-arrival queue with complete rejection is taken from Gouweleeuw (1994). The two-moment approximation for the minimal buffer size in Section 4.8.2 is motivated by the paper of De Kok and Tijms (1985). This paper deals with finite-buffer queues with Poisson arrivals and continuous input.

REFERENCES

Ackroyd, M.H. (1980). 'Computing the waiting - time distribution for the $G/G/1$ queue by signal processing methods', *IEEE Trans. Commun.*, **28**, 52–58.

Boxma, O.J., Cohen, J.W. and Huffels, N. (1979). 'Approximations of the mean waiting time in an $M/G/s$ queueing system', *Operat. Res.*, **27**, 1115–1127.

Bruncel, H. and Kim, B.G. (1993). *Discrete-time Models for Communication Systems Including ATM*, Kluwer, Boston.

Chaudry, M.L. and Templeton, J.G.C. (1983). *A First Course in Bulk Queues*, Wiley, New York.

Cohen, J.W. (1982). *The Single-Server Queue*, 2nd edn., North-Holland, Amsterdam.

Cooper, R.B. (1981). *Introduction to Queueing Theory*, Edward Arnold, London.

Cosmetatos, G.P. (1976). 'Some approximate equilibrium results for the multi-server queue $M/G/r$', *Operat. Res. Quart.*, **27**, 615–620.

Cox, D.R. (1955). 'The statistical analysis of congestion', *J. Roy. Statist. Soc. A.*, **118**, 324–335.

Cox, D.R. and Smith, W.L. (1961). *Queues*, Chapman and Hall, London.

Crommelin, C.D. (1932). 'Delay probability formulas when the holding times are constant', *P.O. Elect. Engr. J.*, **25**, 41–50.

De Kok, A.G. (1989). 'A moment-iterating method for approximating the waiting-time characteristics of the $GI/G/1$ queue', *Prob. Engineering Inform. Sci.*, **3**, 273-288.

De Kok, A.G. and Tijms, H.C. (1985). 'A two-moment approximation for a buffer design problem requiring a small rejection probability', *Perform. Evaluation*, **7**, 77-84.

Eikeboom, A.M. and Tijms, H.C. (1987). 'Waiting-time percentiles in the multi-server $M^X/G/c$ queue with batch arrivals', *Prob. Engineering Inform. Sci.*, **1**, 75-96.

Gouweleeuw, F.N. (1994). 'The loss probability in finite-buffer queues with batch arrivals and complete rejection', *Prob. Engineering Inform. Sci.*, **8**, 221-227.

Haji, R. and Newell, G.F. (1971). 'A relation between stationary queue and waiting-time distribution', *J. Appl. Prob.*, **8**, 617-620.

Hayes, J.F. (1984). *Modelling and Analysis of Computer Communications Networks*, Plenum Press, New York.

Hordijk, A. and Tijms, H.C. (1976). 'A simple proof of the equivalence of the limiting distributions of the continuous time and the embedded process of the queue size in the $M/G/1$ queue', *Statistica Neerlandica*, **30**, 97-100.

Karmarkar, U.S. and Kubat, P. (1983), 'The value of loaners in product support', *IEE Trans.*, **15**, 5-11.

Kleinrock, L. (1976). *Queueing Systems*, Vol. 2: *Computer Applications*, Wiley, New York.

Krämer, W. and Langenbach-Belz, M. (1976). 'Approximate formulas for the delay in the queueing system $GI/G/1$', in: *Proceedings of 8th International Teletraffic Congress (Melbourne)*, paper 235, pp 1-8. North-Holland, Amsterdam

Miyazawa, M. and Tijms, H.C. (1993). 'Comparison of two approximations for the loss probabilities in finite-buffer queues', *Prob. Engineering Inform. Sci.*, **7**, 19-27.

Neuts, M.F. (1981). *Matrix-Geometric Solutions in Stochastic Models — an Algorithmic Approach*, The John Hopkins University Press, Baltimore, MD.

Newell, G.F. (1971). *Applications of Queueing Theory*, Chapman and Hall, London.

Page, E. (1972). *Queueing Theory in O.R.*, Butterworths, London.

Parikh, S.C. (1977), 'On a fleet sizing and allocation problem', *Management Sci.*, **23**, 972-977.

Seelen, L.P., Tijms, H.C. and Van Hoorn, M.H. (1985). *Tables for Multi-Server Queues*, North-Holland, Amsterdam.

Stoyan, D. (1983). *Comparison Methods for Queues and Other Stochastic Models*, Wiley, New York.

Takács, L. (1962). *Introduction to the Theory of Queues*, Oxford University Press, Oxford.

Takahashi, Y. (1981). 'Asymptotic exponentiality of the tail of the waiting time distribution in a $Ph/Ph/c$ queue', *Adv. Appl. Prob.*, **13**, 619-630.

Tijms, H.C. (1992). 'Heuristics for finite-buffer queues', *Prob. Engineering Inform. Sci.*, **6**, 277-285.

Tijms, H.C. and Van Hoorn, M.H. (1982). 'Computational methods for single-server and multi-server queues with Markovian input and general service times', in: *Applied Probability-Computer Sciences, The Interface*, eds. R.L. Disney and T.J. Ott, Vol. II, pp. 71-102, Birkhauser, Boston, MA.

Tijms, H.C., Van Hoorn, M.H. and Federgruen, A. (1981). 'Approximations for the steady-state probabilities in the $M/G/c$ queue', *Adv. Appl. Prob.*, **13**, 186-206.

Tijms, H.C. and Van Ommeren, J.W. (1989). 'Asymptotic analysis for buffer behavior in communication systems', *Prob. Engineering Inform. Sci.*, **3**, 1-12.

Tran-Gia, P. (1986). 'Discrete-time analysis for the interdeparture distribution of $GI/G/1$ queues', in: *Teletraffic Analysis and Computer Performance Evaluation*, eds. O.J. Boxma, J.W. Cohen and H.C. Tijms, pp. 341-357, North-Holland, Amsterdam.

Van Hoorn, M.H. (1984). *Algorithms and Approximations for Queueing Systems*, CWI Tract No. 8, CWI, Amsterdam.

Van Hoorn, M.H. (1986). 'Numerical analysis of multi-server queues with deterministic service and special phase-type arrivals', *Zeitschrift für Operat. Res.* A, **30**, 15–28.

Van Hoorn, M.H. and Seelen, L.P. (1986). 'Approximations for the $GI/G/c$ queue', *J. Appl. Prob.*, **23**, 484–494.

Van Ommeren, J.W. (1988). 'Exponential expansion for the tail of the waiting-time probabilities in the single-server queue with batch arrivals', *Adv. Appl. Prob.*, **20**, 880–895.

Whitt, W. (1992). 'Understanding the efficiency of multi-server service systems', *Management Sci.*, **38**, 708–723.

Wolff, R.W. (1989). *Stochastic Modelling and the Theory of Queues*, Prentice-Hall, Englewood Cliffs, NJ.

Appendices

APPENDIX A. USEFUL TOOLS IN APPLIED PROBABILITY

This appendix summarizes some basic tools that can be found in most introductory texts on probability. We give those basic tools that are utilized in this book.

Computation of probabilities and expectations by conditioning

In many applied probability problems it is only possible to compute certain probabilities and expectations by using appropriate conditioning arguments. Conditional probabilities are based on extra information and are therefore easier to compute. We summarize a number of useful results involving conditional probabilities and conditional expectations.

First we state the law of total probability for the computation of some event A. Suppose that $B_1, B_2, \ldots, B_n$ are mutually exclusive events such that event A can only occur if one of the events B_k occurs. The *law of total probability* states that

$$P(A) = \sum_{k=1}^{n} P(A|B_k)P(B_k). \tag{A.1}$$

More generally, the probability $P(A)$ can be computed as the weighted average of conditional probabilities having as weights the probabilities of the possible outcomes of some random variable Y. Thus, by conditioning on a random variable Y, we can compute $P(A)$ from

$$P(A) = \sum_{y} P(A|Y = y)P(Y = y) \tag{A.2}$$

when Y has a discrete distribution, while $P(A)$ can be computed from

$$P(A) = \int_{-\infty}^{\infty} P(A|Y = y)f(y)\,dy \tag{A.3}$$

when Y has a probability density $f(y)$. The relation (A.2) is a special case of (A.1), while the relation (A.3) can be considered as a 'continuous' version of (A.2) by noting that the probability density $f(y)$ allows for the interpretation $f(y)\Delta y \approx P(y < Y \le y + \Delta y)$ for Δy small.

Similarly, the computation of an expectation can often be considerably facilitated by conditioning on the outcome of some appropriate random variable. The *law of total expectation* states that, for any two random variables X and Y,

$$E(X) = \sum_y E(X|Y = y)P(Y = y) \tag{A.4}$$

when Y has a discrete distribution and

$$E(X) = \int_{-\infty}^{\infty} E(X|Y = y)f(y)\,dy \tag{A.5}$$

when Y has a continuous distribution with probability density $f(y)$. Here it is assumed that the relevant expectations exist.

As an illustration, we give the following problem. In a series of independent games between two teams, each team has probability $1/2$ of winning a particular game. The series terminates as soon as either team has won $m\,(= 3)$ consecutive games. What are the mean and the standard deviation of the length of the series? To answer this question, define the random variable X_i by

$X_i = $ the number of games to follow after one of the teams has won
the last i games.

The goal is to calculate $E(X_0)$ and $\sigma(X_0)$. To find $E(X_i)$, condition on the outcome of the next game. Let the random variable $Y = 1$ if the team ahead wins the next game and let $Y = 0$ otherwise. From $E(X_i) = E(X_i|Y = 0)P\{Y = 0\} + E(X_i|Y = 1)P\{Y = 1\}$, we obtain

$$E(X_i) = \frac{1}{2}E(1 + X_1) + \frac{1}{2}E(1 + X_{i+1})$$

$$= 1 + \frac{1}{2}E(X_1) + \frac{1}{2}E(X_{i+1}), \quad i = 0, 1, \ldots, m - 1,$$

with the convention $E(X_m) = 0$. Thus we obtain a system of linear equations in the unknowns $E(X_i)$. Similarly, we find

$$E(X_i^2) = \frac{1}{2}E(1 + X_1)^2 + \frac{1}{2}E(1 + X_{i+1})^2$$

$$= 1 + E(X_1) + E(X_{i+1}) + \frac{1}{2}E(X_1^2) + \frac{1}{2}E(X_{i+1}^2), \quad i = 0, 1, \ldots, m - 1,$$

with the convention $E(X_m^2) = 0$. This is again a system of linear equations in the unknowns $E(X_i^2)$ once the $E(X_i)$'s have been computed. The numerical answers are $E(X_0) = 7$ and $\sigma(X_0) = 4.7$ when $m = 3$.

Convolution

Let X and Y be two non-negative, independent random variables with respective probability distribution functions $F(x)$ and $G(y)$. It is assumed that $G(y)$ has a probability density $g(y)$. A basic problem is to find the probability distribution function of the sum $X + Y$. By conditioning on Y, we find from (A.3) that

$$P(X + Y \leq z) = \int_0^\infty P\{X + Y \leq z | Y = y\} g(y) \, dy$$

$$= \int_0^z P\{X \leq z - y\} g(y) \, dy, \quad z \geq 0,$$

where the second equality uses that X is non-negative and is independent of Y. This gives the convolution formula

$$P\{X + Y \leq z\} = \int_0^z F(z - y) g(y) \, dy, \quad z \geq 0. \tag{A.6}$$

As an application of the convolution formula, we give the following important result. Suppose that $X_1, X_2, \ldots, X_n$ are independent and identically distributed random variables with probability distribution function

$$P\{X_k \leq x\} = 1 - e^{-\lambda x}, \quad x \geq 0.$$

Then, for any $n \geq 1$,

$$P\{X_1 + \cdots + X_n \leq x\} = 1 - \sum_{j=0}^{n-1} e^{-\lambda x} \frac{(\lambda x)^j}{j!}, \quad x \geq 0. \tag{A.7}$$

In other words, the sum of n independent random variables having a common exponential distribution is Erlang-n distributed. The proof of (A.7) is by induction on n. Obviously, (A.7) holds for $n = 1$. Assume that (A.7) has been verified for $n = 1, \ldots, m - 1$. Then, by applying the convolution formula (A.6) with $X = X_1 + \cdots + X_{m-1}$ and $Y = X_m$, the relation (A.7) readily follows for $n = m$.

Moments of a non-negative random variable

Let N be a non-negative, integer-valued random variable. A useful representation for the expectation of N is

$$E(N) = \sum_{k=0}^\infty P\{N > k\}. \tag{A.8}$$

To verify this result, write $\sum_{k=0}^\infty P\{N > k\} = \sum_{k=0}^\infty \sum_{j=k+1}^\infty P\{N = j\}$ and interchange the order of summation. The relation (A.8) can be generalized. For any non-negative random variable X with probability distribution function $F(x)$, we have

$$E(X) = \int_0^\infty [1 - F(x)] \, dx. \tag{A.9}$$

A probabilistic proof of (A.9) is as follows. Imagine that X is the lifetime of a machine. Define the indicator variable $I(t)$ by $I(t) = 1$ if the machine is still working at time t and by $I(t) = 0$ otherwise. Then, using that $E[I(t)] = P\{I(t) = 1\}$ and $P\{I(t) = 1\} = P\{X > t\}$, it follows that

$$E(X) = E\left[\int_0^\infty I(t)\,dt\right] = \int_0^\infty E[I(t)]\,dt = \int_0^\infty P\{X > t\}\,dt$$

which proves (A.9). The result (A.9) can be extended to

$$E(X^k) = k\int_0^\infty x^{k-1}[1 - F(x)]\,dx, \quad k = 1, 2, \ldots . \tag{A.10}$$

To see this, note that (A.9) implies

$$E(X^k) = \int_0^\infty P\{X^k > t\} = \int_0^\infty P\{X > t^{1/k}\}\,dt$$

and next use the change of variable $t = x^k$.

Mean and variance of a random sum of stochastic variables

Let $X_1, X_2, \ldots$ be a sequence of independent and identically distributed random variables whose first two moments are finite. Also, let N be a non-negative and integer-valued random variable having finite first two moments. If the random variable N is independent of the random variables $X_1, X_2, \ldots$, then

$$E\left(\sum_{k=1}^N X_k\right) = E(N)E(X_1), \tag{A.11}$$

$$\text{var}\left(\sum_{k=1}^N X_k\right) = E(N)\text{var}(X_1) + \text{var}(N)E^2(X_1), \tag{A.12}$$

where $E^2(X_1)$ denotes the squared expectation of X_1. The proof uses the law of total expectation. By conditioning on N, we find

$$E\left(\sum_{k=1}^N X_k\right) = \sum_{n=0}^\infty E\left(\sum_{k=1}^N X_k | N = n\right) P\{N = n\}$$

$$= \sum_{n=0}^\infty E\left(\sum_{k=1}^n X_k\right) P\{N = n\}$$

$$= \sum_{n=0}^\infty nE(X_1)P\{N = n\},$$

which verifies (A.11). Note that the second equality uses that the random variables

$X_1, \ldots, X_n$ are independent of the event $\{N = n\}$. In a similar way, we find

$$E\left[\left(\sum_{k=1}^{N} X_k\right)^2\right] = \sum_{n=0}^{\infty} E\left[\left(\sum_{k=1}^{N} X_k\right)^2 \mid N = n\right] P\{N = n\}$$

$$= \sum_{n=0}^{\infty} E\left[\sum_{k=1}^{n} X_k^2 + 2\sum_{i<j} X_i X_j\right] P\{N = n\}$$

$$= \sum_{n=0}^{\infty} [nE(X_1^2) + n(n-1)E^2(X_1)] P\{N = n\}$$

$$= E(N)E(X_1^2) + E[N(N-1)]E^2(X_1). \tag{A.13}$$

Using $\sigma^2(X) = E(X^2) - E^2(X)$, the result (A.12) now follows from (A.11) and (A.13).

Wald's equation

The result (A.11) remains valid when the assumption that the random variable N is independent of the sequence $X_1, X_2, \ldots$ is somewhat weakened. Suppose that the following conditions are satisfied:

(i) $X_1, X_2, \ldots$ is a sequence of independent and identically distributed random variables with finite mean,

(ii) N is a non-negative, integer-valued random variable with $E(N) < \infty$,

(iii) the event $\{N = n\}$ is independent of $X_{n+1}, X_{n+2}, \ldots$ for each $n \geq 1$.

Then we have that

$$E\left(\sum_{k=1}^{N} X_k\right) = E(X_1)E(N). \tag{A.14}$$

This equation is known as Wald's equation. It is a very useful result in applied probability. To prove (A.14), let us assume for ease that the X_i's are non-negative. The following trick is used. For $n = 1, 2, \ldots$, we define the random variable I_k by

$$I_k = \begin{cases} 1 & \text{if } N \geq k, \\ 0 & \text{if } N < k. \end{cases}$$

Then $\sum_{k=1}^{N} X_k = \sum_{k=1}^{\infty} X_k I_k$ and so

$$E\left(\sum_{k=1}^{N} X_k\right) = E\left(\sum_{k=1}^{\infty} X_k I_k\right) = \sum_{k=1}^{\infty} E(X_k I_k),$$

where the interchange of expectation and summation is justified by the non-negativity of the random variables involved. The random variable I_k can take on only the two values 0 and 1. The outcome of I_k is completely determined by the event $\{N \leq k - 1\}$. This event can depend on $X_1, \ldots, X_{k-1}$, but not on $X_k, X_{k+1}, \ldots$. This implies that I_k is independent of X_k. Consequently, $E(X_k I_k) = E(X_k)E(I_k)$ for all $k \geq 1$. Using that $E(I_k) = P\{I_k = 1\}$ and $P\{I_k = 1\} = P\{N \geq k\}$, it now follows that

$$E\left(\sum_{k=1}^N X_k\right) = \sum_{k=1}^\infty E(X_1)P\{N \geq k\}.$$

Together this relation and (A.8) verify (A.14).

Reliability application

To illustrate the above concepts consider the following reliability problem. A computer system has a built-in redundancy in the form of a standby unit to support an operating unit. The two units are identical. When the operating unit fails, its tasks are taken over immediately by the standby unit if available. The failed unit immediately enters repair. The system goes down when the operating unit fails while the other unit is still in repair. The lifetime of an operating unit has a probability density $g(x)$ with finite mean μ_L. The repair time of a failed unit has the probability distribution function $H(x)$. The lifetimes and repair times are mutually independent. Supposing that both units are as good as new at time 0, we wish to find the expected time until the system goes down for the first time.

To solve this problem, denote by L_0 the lifetime of the operating unit installed at time 0 and denote by $L_1, L_2, \ldots$ the lifetimes of the subsequent operating units. Also, let $R_1, R_2, \ldots$ be the repair times associated with successive failures. Then the time until the first system breakdown is distributed as $L_0 + L_1 + \cdots + L_N$ where N denotes the first n for which $R_n > L_n$. The random variables $L_1, \ldots, L_N$ and N are mutually dependent but the event $\{N = n\}$ is independent of $L_{n+1}, L_{n+2}, \ldots$ for each n. Thus, by using Wald's equation,

$$E(\text{time until first system breakdown}) = E(L_0) + E(L_1)E(N)$$

$$= \{1 + E(N)\}\mu_L.$$

To calculate $E(N)$, let $p = P\{R_1 > L_1\}$ and note that N has the geometric distribution $P\{N = n\} = (1 - p)^{n-1}p$, $n = 1, 2, \ldots$. Hence $E(N) = 1/p$. By conditioning on L_1 and using the law of total probability, we find

$$p = \int_0^\infty \{1 - H(x)\}g(x)\,\mathrm{d}x.$$

A few words on numerical integration for an infinite integral: in general, one divides the integration interval $(0, \infty)$ into a number of subintervals, where Gauss–Legendre or Romberg integration is used for the finite integrals and the remaining infinite interval is transformed into the interval $(0, 1)$; see Press *et al*

(1986). One should be very careful in choosing the subintervals when the integrand is very peaked or has long tails.

Convergence theorems

To conclude this appendix, we state a number of basic convergence theorems that will be used in this book. These theorems can be found in any textbook on real analysis, see e.g. Rudin (1964).

Theorem A.1   *Let a_{nm}, $n, m = 0, 1, \ldots$ be real numbers. If all the numbers a_{nm} are non-negative or if $\Sigma_{n=0}^{\infty} \Sigma_{m=0}^{\infty} |a_{nm}| < \infty$, then*

$$\sum_{n=0}^{\infty} \sum_{m=0}^{\infty} a_{nm} = \sum_{m=0}^{\infty} \sum_{n=0}^{\infty} a_{nm}.$$

This theorem is a special case of what is known as Fubini's Theorem in analysis.

Theorem A.2   *Let $\{p_m, \ m = 0, 1, \ldots\}$ be a sequence of non-negative numbers. Suppose that the numbers a_{nm}, $n, m = 0, 1, \ldots$ are such that*

$$\lim_{n \to \infty} a_{nm} = a_m$$

exists for all $m = 0, 1, \ldots$.

(a) If all numbers a_{nm} are non-negative, then

$$\lim_{n \to \infty} \inf \sum_{m=0}^{\infty} a_{nm} p_m \geq \sum_{m=0}^{\infty} a_m p_m.$$

(b) If there is a finite constant $M > 0$ such that $|a_{nm}| \leq M$ for all n, m and if $\Sigma_{m=0}^{\infty} p_m < \infty$, then

$$\lim_{n \to \infty} \sum_{m=0}^{\infty} a_{nm} p_m = \sum_{m=0}^{\infty} a_m p_m.$$

The first part of the theorem is a special case of Fatou's lemma and the second part of the theorem is a special case of the bounded convergence theorem.

The above theorems can be stated in greater generality. For example, a more general version of the bounded convergence theorem is as follows. Let $\{X_n\}$ be a sequence of random variables that converge with probability 1 to a random variable X. Then

$$\lim_{n \to \infty} E(X_n) = E(X)$$

provided that $|X_n| \leq Y$, $n \geq 1$, for some random variable Y with $E(Y) < \infty$. Recall that convergence with probability 1 means that

$$P\{\omega \in \Omega : \lim_{n \to \infty} X_n(\omega) = X(\omega)\} = 1$$

where Ω is the common sample space of the random variables X_n, $n \geq 1$, and the random variable X. Often one uses the term of 'almost sure convergence' instead of the term 'convergence with probability one'.

Finally, we mention the important concept of the Cesaro limit. A sequence $\{a_n, \ n \geq 1\}$ of real numbers is said to have a Cesaro limit if $\lim_{n\to\infty}(1/n)\sum_{k=1}^{n} a_k$ exists. A sequence $\{a_n\}$ may have a Cesaro limit while the ordinary limit does not exist. For example, suppose that $a_n = 1$ for n even and $a_n = 0$ for n odd. Then $\lim_{n\to\infty} a_n$ does not exist, while $\lim_{n\to\infty}(1/n)\sum_{k=1}^{n} a_k = 1/2$. However, if the ordinary limit exists then the Cesaro limit exists as well and is equal to the ordinary limit.

APPENDIX B. USEFUL PROBABILITY DISTRIBUTION FUNCTIONS

This appendix discusses a number of important distributions which have been found useful for describing random variables in inventory, reliability and queueing applications. These distributions include the gamma, lognormal and Weibull distributions.

We first introduce some general concepts. Let X be a positive random variable with the probability distribution function $F(t)$ having finite mean $E(X)$ and finite standard deviation $\sigma(X)$. The random variable X may represent the lifetime of some item or the time to complete some task. The *coefficient of variation* of the positive random variable X is defined by

$$c_X = \frac{\sigma(X)}{E(X)}.$$

This is a dimensionless quantity. In applications one often works with the squared coefficient of variation c_X^2 rather than with c_X. The (squared) coefficient of variation is a measure of the variability of the random variable X. For example, the deterministic distribution has $c_X^2 = 0$, the exponential distribution has $c_X^2 = 1$ and the Erlang-k distribution has the intermediate value $c_X^2 = 1/k$.

Assuming that the probability distribution function $F(t)$ of the random variable X has a density $f(t)$, the *failure (or hazard) rate function* $r(t)$ of X is defined by

$$r(t) = \frac{f(t)}{1 - F(t)}$$

for those values of t for which $F(t) < 1$. The failure rate has a useful probabilistic interpretation. Think of the random variable X as the lifetime of some item. Since $F(t + \Delta t) - F(t) \approx f(t)\Delta t$, it follows that

$$P\{X \in (t, t + \Delta t]|X > t\} \approx \frac{f(t)\Delta t}{1 - F(t)} \quad \text{for } \Delta t \text{ small.}$$

This shows that $r(t)\Delta t$ gives approximately the probability that an item of age t will fail in the next time Δt. Noting that $r(t) = -L'(t)$ with $L(t) = \ln[1 -$

$F(t)$], it follows that the failure rate function uniquely determines the corresponding probability distribution function by

$$1 - F(t) = \exp\left\{-\int_0^t r(x)\,dx\right\}, \quad t \ge 0.$$

An immediate consequence of this relation is that the special case of a constant failure rate $r(x) = \lambda$ for all x corresponds to the probability distribution function $F(x) = 1 - e^{-\lambda x}$, $x \ge 0$. Hence, the exponential distribution has a constant failure rate. In other words, an item in use is as good as a new one when the lifetime is exponential. Other important cases are the case of an increasing failure rate ('the older, the worse') and the case of a decreasing failure rate ('the older, the better'). A random variable with increasing (decreasing) failure rate has the property that its coefficient of variation is smaller (larger) than or equal to 1. The failure rate is a concept that enables us to discriminate between distribution functions on physical considerations.

We now discuss the exponential, gamma, lognormal and Weibull distributions for a positive random variable X and list without proof relevant properties of these distributions.

The exponential distribution

The density $f(t)$ is given by

$$f(t) = \lambda e^{-\lambda t}, \quad t \ge 0,$$

where the scale parameter λ is positive. The probability distribution function $F(t)$ is given by

$$F(t) = 1 - e^{-\lambda t}, \quad t \ge 0.$$

The mean and the squared coefficient of variation of the exponential distribution are given by

$$E(X) = \frac{1}{\lambda} \quad \text{and} \quad c_X^2 = 1.$$

The exponential distribution is of extreme importance in applied probability. The main reason for this is its memoryless property. To state this property, it is convenient to think of the exponentially distributed random variable X as the lifetime of a light bulb. The memoryless property is that, for any $t \ge 0$,

$$P\{X > t + x \mid X > t\} = e^{-\lambda x}, \quad x \ge 0. \tag{B.1}$$

In other words, the residual life of the bulb has the same exponential distribution as the original lifetime regardless of how long the bulb has already been in use. The relation (B.1) follows directly from

$$P\{X > t + x \mid X > t\} = \frac{P\{X > t + x, X > t\}}{P\{X > t\}} = \frac{e^{-\lambda(t+x)}}{e^{-\lambda t}},$$

showing that $P\{X > t + x|X > t\} = P\{X > x\}$. The exponential distribution is the only continuous distribution having the memoryless property.

Saying that a bulb in use is as good as a new one is equivalent to saying that the bulb has a constant failure rate. In other words, the probability that a bulb of age t will fail within the next Δt time units is given by

$$P\{X \in (t, t + \Delta t)|X > t\} = \lambda \Delta t + o(\Delta t) \quad \text{for each } t > 0$$

as $\Delta t \to 0$. Here $o(h)$ is the standard notation for some unspecified function $g(h)$ having the property that $g(h)/h \to 0$ as $h \to 0$, that is $o(h)$ is negligibly small compared with h as $h \to 0$. The failure rate representation follows from (B.1) and the expansion $1 - e^{-h} = h + o(h)$ as $h \to 0$.

Before giving an illustration of the memoryless property, we mention the following useful fact for the exponential distribution. If X_1 and X_2 are two independent exponentials with respective means $1/\lambda_1$ and $1/\lambda_2$, then

$$P\{\min(X_1, X_2) \le t\} = 1 - e^{-(\lambda_1 + \lambda_2)t}, \quad t \ge 0 \tag{B.2}$$

and

$$P\{X_1 < X_2\} = \frac{\lambda_1}{\lambda_1 + \lambda_2} \tag{B.3}$$

In other words, the minimum of the two independent exponentials X_1 and X_2 is exponentially distributed with mean $1/(\lambda_1 + \lambda_2)$ and with probability $\lambda_1/(\lambda_1 + \lambda_2)$ the random variable X_1 is smaller than X_2.

As an illustration of the memoryless property, let us consider the following problem. Every time unit an item arrives at a conveyor with a single work station. The work station can process only one item at a time and has no buffer to store temporarily other items. An item finding upon arrival the work station busy is lost. The processing times of the items are independent random variables having a common exponential distribution with mean $1/\mu$. What is the long-run fraction of items that are lost? Denote by $t = 0, 1, \ldots$ the epochs at which the items arrive at the work station. Let the indicator variable I_k be 1 if the item arriving at $t = k$ is lost and let $I_k = 0$ otherwise. For the item arriving at time $t = n + 1$ it makes no difference whether the work station starts working on a new item at $t = n$ or the work station continues working on an old item at $t = n$. By the memoryless property of the exponential distribution, the time to complete the processing of the item present at time $t = n$ is exponentially distributed with mean $1/\mu$ in both situations. Thus the random variables $I_1, I_2, \ldots$ are independent and identically distributed with

$$P\{I_k = 1\} = P\{\text{processing time of an item is larger than } 1\} = e^{-\mu}.$$

The long-run fraction of items that are lost is given by $\lim_{n \to \infty}(1/n)\sum_{k=1}^{n} I_k$. By the strong law of large numbers this limit is equal to $E(I_1)$ with probability 1. Hence the long-run fraction of items that are lost is equal to $e^{-\mu}$ with probability 1.

The gamma distribution

The density $f(t)$ is given by

$$f(t) = \frac{\lambda^\alpha t^{\alpha-1}}{\Gamma(\alpha)} e^{-\lambda t}, \quad t \geq 0,$$

where the shape parameter α and the scale parameter λ are both positive. Here $\Gamma(\alpha)$ is the complete gamma function defined by

$$\Gamma(\alpha) = \int_0^\infty e^{-t} t^{\alpha-1} \, dt, \quad \alpha > 0.$$

This function has the property $\Gamma(\alpha + 1) = \alpha\Gamma(\alpha)$ for any $\alpha > 0$. The probability distribution function $F(t)$ may be written as

$$F(t) = \frac{1}{\Gamma(\alpha)} \int_0^{\lambda t} e^{-u} u^{\alpha-1} \, du, \quad t \geq 0.$$

The latter integral is known as the incomplete gamma function. It is important to note that efficient numerical procedures are available for the computation of the incomplete gamma function; see Press *et al* (1986).

If the shape parameter α is a positive integer k, the gamma distribution is the well-known Erlang-k (E_k) distribution with

$$f(t) = \lambda^k \frac{t^{k-1}}{(k-1)!} e^{-\lambda t} \quad \text{and} \quad F(t) = 1 - \sum_{j=0}^{k-1} e^{-\lambda t} \frac{(\lambda t)^j}{j!}, \quad t \geq 0.$$

The Erlang-k distribution has a very useful interpretation. A random variable with an Erlang-k distribution can be represented as the sum of k independent random variables having a common exponential distribution.

The mean and the squared coefficient of variation of the gamma distribution are given by

$$E(X) = \frac{\alpha}{\lambda} \quad \text{and} \quad c_X^2 = \frac{1}{\alpha}.$$

These equations show that a unique gamma distribution can be fitted to each positive random variable with given first two moments. To characterize the shape and the failure rate of the gamma density, we distinguish between the cases $c_X^2 < 1$ ($\alpha > 1$) and $c_X^2 \geq 1$ ($\alpha \leq 1$). The gamma density is always *unimodal*; that is, the density has only one maximum. For the case of $c_X^2 < 1$ the density first increases to the maximum at $t = (\alpha - 1)/\lambda > 0$ and next decreases to zero as $t \to \infty$, whereas for the case of $c_X^2 \geq 1$ the density has its maximum at $t = 0$ and thus decreases from $t = 0$ on. The failure rate function is increasing from zero to λ if $c_X^2 < 1$ and is decreasing from infinity to zero if $c_X^2 > 1$. The exponential distribution ($c_X^2 = 1$) has a constant failure rate λ and is a natural boundary between the cases $c_X^2 < 1$ and $c_X^2 > 1$.

The lognormal distribution

The density $f(t)$ is given by

$$f(t) = \frac{1}{\alpha t \sqrt{2\pi}} \exp[-(\ln(t) - \lambda)^2/2\alpha^2], \quad t > 0,$$

where the shape parameter α is positive and the scale parameter λ may assume each real value. The probability distribution function $F(t)$ equals

$$F(t) = \Phi\left(\frac{\ln(t) - \lambda}{\alpha}\right), \quad t > 0,$$

where $\Phi(x) = (1/\sqrt{2\pi}) \int_{-\infty}^{x} \exp(-u^2/2)\,du$ is the standard normal probability distribution function. The mean and the squared coefficient of variation of the lognormal distribution are given by

$$E(X) = \exp(\lambda + \alpha^2/2) \quad \text{and} \quad c_X^2 = \exp(\alpha^2) - 1.$$

Thus a unique lognormal distribution can be fitted to each positive random variable with given first two moments. The lognormal density is always unimodal with a maximum at $t = \exp(\lambda - \alpha^2)$. The failure rate function first increases and next decreases to zero as $t \to \infty$ and thus the failure rate is only decreasing in the long-life range.

The Weibull distribution

The density $f(t)$ is given by

$$f(t) = \alpha\lambda(\lambda t)^{\alpha-1} \exp[-(\lambda t)^\alpha], \quad t > 0,$$

with the shape parameter $\alpha > 0$ and scale parameter $\lambda > 0$. The probability distribution function $F(t)$ equals

$$F(t) = 1 - \exp[-(\lambda t)^\alpha], \quad t \geq 0.$$

The mean and the squared coefficient of variation of the Weibull density are given by

$$E(X) = \frac{1}{\lambda}\Gamma\left(1 + \frac{1}{\alpha}\right) \quad \text{and} \quad c_X^2 = \frac{\Gamma(1 + 2/\alpha)}{\{\Gamma(1 + 1/\alpha)\}^2} - 1.$$

A unique Weibull distribution can be fitted to each positive random variable with given first two moments. For that purpose a nonlinear equation in α must be solved numerically. The Weibull density is always unimodal with a maximum at $t = \lambda^{-1}(1 - 1/\alpha)^{1/\alpha}$ if $c_X^2 < 1$ ($\alpha > 1$), and at $t = 0$ if $c_X^2 \geq 1$ ($\alpha \leq 1$). The failure rate function is increasing from 0 to infinity if $c_X^2 < 1$ and is decreasing from infinity to zero if $c_X^2 > 1$.

The gamma and Weibull densities are similar in shape, and for $c_X^2 < 1$ the lognormal density takes on shapes similar to the gamma and Weibull densities. For the case of $c_X^2 \geq 1$ the gamma and Weibull densities have their maximum value at $t = 0$ so that most outcomes will be small and very large outcomes occur only occassionally. The lognormal density tends to zero as $t \to 0$ faster than any power of t, and thus the lognormal distribution will typically produce fewer small outcomes than the other two distributions. This latter fact explains the popular use of the lognormal distribution in actuarial studies. The differences between the gamma, Weibull and lognormal densities become most significant in their tail behaviour. The densities for large t go down like $\exp[-\lambda t]$, $\exp[-(\lambda t)^\alpha]$ and $\exp[-(\ln(t) - \lambda)^2/2\alpha^2]$. Thus for a given mean and coefficient of variation the lognormal density has always the longest tail. The gamma density has the second longest tail only if $\alpha > 1$; that is, only if its coefficient of variation is less than one. In Figure B.1 we illustrate these facts by drawing the gamma, Weibull and lognormal densities for $c_X^2 = 0.25$, where $E(X)$ is taken to be 1.

To conclude this appendix, we discuss several useful generalizations of exponential and Erlangian distributions. In many queueing and inventory

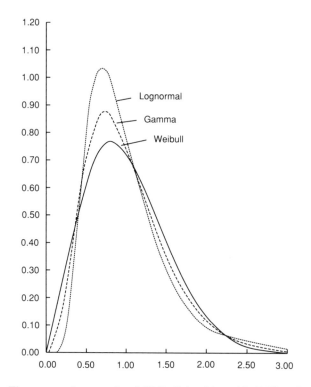

Figure B.1   The gamma, lognormal and Weibull densities with $E(X) = 1$ and $c_X^2 = 0.25$

applications there is a very substantial (numerical) advantage in using the generalized distributions rather than other distributions.

Generalized Erlangian distributions

An Erlang-k (E_k) distributed random variable can be represented as the sum of k independent exponentially distributed random variables with the same means. A generalized Erlangian distribution is one built out of a random sum of exponentially distributed components. A particularly convenient distribution arises when these components have the same means. In fact, such a distribution can be used to approximate arbitrarily closely any distribution having its mass on the positive half-axis. We discuss two special cases of mixtures of Erlangian distributions with the same scale parameters. First, we consider the $E_{k-1,k}$ distribution which is defined as a mixture of E_{k-1} and E_k distributions with the same scale parameters. The probability density of an $E_{k-1,k}$ distribution has the following form:

$$f(t) = p\mu^{k-1}\frac{t^{k-2}}{(k-2)!}e^{-\mu t} + (1-p)\mu^k\frac{t^{k-1}}{(k-1)!}e^{-\mu t}, \quad t \geq 0,$$

where $0 \leq p \leq 1$. In words, a random variable having this density is with probability p (respectively $1 - p$) distributed as the sum of $k - 1$ (respectively k) independent exponentials with common mean $1/\mu$. By choosing the parameters p and μ as

$$p = \frac{1}{1+c_X^2}[kc_X^2 - \{k(1+c_X^2) - k^2c_X^2\}^{1/2}] \quad \text{and} \quad \mu = \frac{k-p}{E(X)}$$

the associated $E_{k-1,k}$ distribution fits the first two moments of a positive random variable X provided that

$$\frac{1}{k} \leq c_X^2 \leq \frac{1}{k-1}.$$

We note that only coefficients of variation between 0 and 1 can be achieved by mixtures of the $E_{k-1,k}$ type. Also, it is noteworthy that the $E_{k-1,k}$ density can be shown to have an increasing failure rate.

Next we consider the $E_{1,k}$ distribution which is defined as a mixture of E_1 and E_k distributions with the same scale parameters. The density of the $E_{1,k}$ distribution has the form

$$f(t) = p\mu e^{-\mu t} + (1-p)\mu^k\frac{t^{k-1}}{(k-1)!}e^{-\mu t}, \quad t \geq 0,$$

where $0 \leq p \leq 1$. By choosing

$$p = \frac{2kc_X^2 + k - 2 - (k^2 + 4 - 4kc_X^2)^{1/2}}{2(k-1)(1+c_X^2)} \quad \text{and} \quad \mu = \frac{p + k(1-p)}{E(X)}$$

the associated $E_{1,k}$ distribution fits the first two moments of a positive random variable X provided that

$$\frac{1}{k} \leq c_X^2 \leq \frac{k^2 + 4}{4k}$$

Thus the $E_{1,k}$ distribution can also achieve coefficients of variation greater than 1.

For use in applications the $E_{k-1,k}$ density is in general more suited than the $E_{1,k}$ density since the $E_{k-1,k}$ density is always unimodal and has a similar shape to the frequently occurring gamma density. The $E_{1,k}$ density may be useful in sensitivity analysis. For both theoretical and practical purposes it is often easier to work with mixtures of Erlangian distributions than with gamma distributions, since mixtures of Erlangian distributions with the same scale parameters allow for the probabilistic interpretation that they represent a random sum of independent exponentials with the same means.

Hyperexponential distribution

A commonly used representation of a positive random variable with a coefficient of variation greater than 1 is a mixture of two exponentials with different means. The distribution of such a mixture is called a *hyperexponential* (H_2) distribution of order two. The density of the H_2 distribution has the form

$$f(t) = p_1 \mu_1 e^{-\mu_1 t} + p_2 \mu_2 e^{-\mu_2 t}, \quad t \geq 0,$$

where $0 \leq p_1, p_2 \leq 1$. Note that always $p_1 + p_2 = 1$, since the density $f(t)$ represents a probability mass of 1. In words, a random variable having the H_2 density is distributed with probability p_1 (respectively p_2) as an exponential variable with mean $1/\mu_1$ (respectively $1/\mu_2$). The hyperexponential density has always a coefficient of variation of at least 1 and is unimodal with a maximum at $t = 0$. The failure rate function of the hyperexponential distribution is decreasing. The H_2 density has three parameters and is therefore not uniquely determined by its first two moments. For a two-moment fit the H_2 density with *balanced means* is often used; that is, the normalization $p_1/\mu_1 = p_2/\mu_2$ is used. The parameters of the H_2 density having balanced means and fitting the first two moments of a positive random variable X with $c_X^2 \geq 1$ are

$$p_1 = \frac{1}{2} \left(1 + \sqrt{\frac{c_X^2 - 1}{c_X^2 + 1}} \right), \quad p_2 = 1 - p_1,$$

$$\mu_1 = \frac{2p_1}{E(X)}, \quad \mu_2 = \frac{2p_2}{E(X)}.$$

In a somewhat more general context we shall give below another normalization we believe to be more natural. A three-moment fit by an H_2 density is not always possible, but it is unique whenever it exists. An H_2 density can only be fitted to the first three moments m_1, m_2 and m_3 of a positive random variable X with $c_X^2 > 1$ when the requirement $m_1 m_3 \geq \tfrac{3}{2} m_2^2$ is satisfied; see Whitt (1982). Then the parameters of the three-moment fit are given by

$$\mu_{1,2} = \frac{1}{2} \left\{ a_1 \pm \sqrt{a_1^2 - 4a_2} \right\}, \quad p_1 = \frac{\mu_1(1 - \mu_2 m_1)}{\mu_1 - \mu_2}, \quad p_2 = 1 - p_1,$$

where $a_2 = (6m_1 - 3m_2/m_1)/(3m_2^2/2m_1 - m_3)$ and $a_1 = 1/m_1 + \tfrac{1}{2}m_2 a_2/m_1$. The requirement $m_1 m_3 \geq \tfrac{3}{2} m_2^2$ holds both for a gamma distributed and for a lognormal distributed random variable X with $c_X^2 > 1$.

Coxian-2 distribution

The hyperexponential density requires that the weights p_1 and p_2 are non-negative. However, in order that $p_1 \mu_1 \exp(-\mu_1 t) + p_2 \mu_2 \exp(-\mu_2 t)$ represents a probability density it is not necessary to require that p_1 and p_2 are both non-negative. The class of H_2 distributions can be shown to be a subclass of the class of so-called *Coxian-2* (C_2) distributions. A random variable X is said to be Coxian-2 distributed if X can be represented as

$$X = \begin{cases} X_1 + X_2 & \text{with probability } b, \\ X_1 & \text{with probability } 1 - b, \end{cases}$$

where X_1 and X_2 are independent random variables having exponential distributions with respective means $1/\mu_1$ and $1/\mu_2$. In words, the lifetime X first goes through an exponential phase X_1 and next it goes through a second exponential phase X_2 with probability b or it goes out with probability $1 - b$; see Figure B.2. It can be assumed without loss of generality that

$$\mu_1 \geq \mu_2.$$

A Coxian-2 distribution having parameters (b, μ_1, μ_2) with $\mu_1 < \mu_2$ can be shown to have the same probability density as the Coxian-2 distribution having parameters (b^*, μ_1^*, μ_2^*) with $\mu_1^* = \mu_2, \mu_2^* = \mu_1$ and $b^* = 1 - (1 - b)\mu_1/\mu_2$. In the discussion below it is assumed that the Coxian-2 distribution has $\mu_1 \geq \mu_2$.

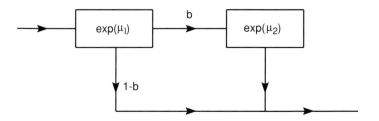

Figure B.2　The Coxian distribution with two phases

The probability density of the Coxian-2 distributed random variable X is given by

$$f(t) = \begin{cases} p_1\mu_1 e^{-\mu_1 t} + (1 - p_1)\mu_2 e^{-\mu_2 t} & \text{if } \mu_1 \neq \mu_2, \\ p_1\mu_1 e^{-\mu_1 t} + (1 - p_1)\mu_1^2 t e^{-\mu_1 t} & \text{if } \mu_1 = \mu_2, \end{cases}$$

where $p_1 = 1 - b\mu_1/(\mu_1 - \mu_2)$ if $\mu_1 \neq \mu_2$ and $p_1 = 1 - b$ if $\mu_1 = \mu_2$. Thus the class of H_2 densities is contained in the class of Coxian-2 densities. Note that the H_2 distribution allows for two different but equivalent probabilistic interpretations. The H_2 distribution can be interpreted both in terms of exponential phases in parallel and in terms of exponential phases in series.

A Coxian-2 density has always a unimodal shape. Moreover, it holds for Coxian-2 distributed random variable X that

$$c_X^2 \geq \frac{1}{2},$$

where $c_X^2 \geq 1$ only if the density is of the form $p_1\mu_1 \exp(-\mu_1 t) + p_2\mu_2 \exp(-\mu_2 t)$ such that p_1 and p_2 are both non-negative. The Coxian-2 density has three parameters (b, μ_1, μ_2). Hence an infinite number of Coxian-2 densities can in principle be used for a two-moment fit to a random variable X with $c_X^2 > 1/2$ (the E_2 density is the only possible choice when $c_X^2 = 1/2$). A particularly useful choice for a two-moment match is the Coxian-2 density with parameters

$$\mu_1 = \frac{2}{E(X)}\left(1 + \sqrt{\frac{c_X^2 - \frac{1}{2}}{c_X^2 + 1}}\right), \quad \mu_2 = \frac{4}{E(X)} - \mu_1, \quad b = \frac{\mu_2}{\mu_1}\{\mu_1 E(X) - 1\}.$$

This particular Coxian-2 density has the remarkable property that its third moment is also the same as that of the gamma density with mean $E(X)$ and squared coefficient of variation c_X^2. The unique Coxian-2 density having this property will therefore be called the Coxian-2 density with the *gamma normalization*. This normalization is a natural one in many applications.

APPENDIX C. LAPLACE TRANSFORMS AND GENERATING FUNCTIONS

This appendix first gives a brief outline of some results from Laplace transform theory that are useful in applied probability problems. Next we discuss in some detail how an asymptotic estimate of the tail of a discrete probability distribution can sometimes be obtained from the generating function of the probability distribution.

Laplace transform

Suppose that $f(x)$ is a continuous real-valued function in $x \geq 0$ such that $|f(x)| \leq Ae^{Bx}$, $x \geq 0$, for some constants A and B. The *Laplace transform* of $f(x)$ is defined by the integral

$$f^*(s) = \int_0^\infty e^{-sx} f(x)\, dx$$

as a function of the complex variable s with $Re(s) > B$. The integral always exists when $Re(s) > B$. The following results can easily be verified from the definition of $f^*(s)$. Here it is assumed that the following various integrals exist.

(a) If the function $f(x) = ag(x) + bh(x)$ is a linear combination of the functions $g(x)$ and $h(x)$ with Laplace transforms $g^*(s)$ and $h^*(s)$, then

$$f^*(s) = ag^*(s) + bh^*(s). \tag{C.1}$$

(b) If $F(x) = \int_0^x f(y)\, dy$, then

$$\int_0^\infty e^{-sx} F(x)\, dx = \frac{f^*(s)}{s}. \tag{C.2}$$

If $f(x)$ has a derivative $f'(x)$ then

$$\int_0^\infty e^{-sx} f'(x)\, dx = sf^*(s) - f(0). \tag{C.3}$$

(c) If the function $f(x)$ is given by a convolution

$$f(x) = \int_0^x g(x - y)h(y)\, dy$$

of two functions $g(x)$ and $h(x)$ with Laplace transforms $g^*(s)$ and $h^*(s)$, then

$$f^*(s) = g^*(s)h^*(s). \tag{C.4}$$

(d) If the function $f(x)$ is a probability density of a non-negative random variable X, then

$$f^*(s) = E(e^{-sX}) \tag{C.5}$$

and

$$E(X^k) = (-1^k) \lim_{s \to 0} \frac{d^k f^*(s)}{ds^k}, \quad k = 1, 2, \ldots .$$

(e) Assuming that both limits exist,

$$\lim_{x \to \infty} f(x) = \lim_{s \to 0} sf^*(s). \tag{C.6}$$

In specific applications requiring the determination of some unknown function $f(x)$ it is often possible to obtain the Laplace transform $f^*(s)$ of $f(x)$. A very useful result is that the function $f(x)$ is uniquely determined by its Laplace transform $f^*(s)$. In principle the function $f(x)$ can be obtained by inversion of its Laplace transform. Extensive tables are available for the inverse of basic forms of $f^*(s)$; see, for example, Abramowitz and Stegun (1965). An inversion formula that is sometimes helpful in applications is the Heaviside formula. Suppose that

$$f^*(s) = \frac{P(s)}{Q(s)},$$

where $P(s)$ and $Q(s)$ are polynomials in s such that the degree of $P(s)$ is smaller than that of $Q(s)$. It is no restriction to assume that $P(s)$ and $Q(s)$ have no zeros in common. Let $s_1, \ldots, s_k$ be the distinct zeros of $Q(s)$ in the complex plane. For ease of presentation, assume that each root s_j is simple (i.e. has multiplicity one). Then it is known from algebra that $P(s)/Q(s)$ admits the partial fraction expansion

$$\frac{P(s)}{Q(s)} = \frac{r_1}{s - s_1} + \frac{r_2}{s - s_2} + \cdots + \frac{r_k}{s - s_k},$$

where $r_j = \lim_{s \to s_j} (s - s_j)P(s)/Q(s)$ and so $r_j = P(s_j)/Q'(s_j), 1 \le j \le k$. The inverse of the Laplace transform $f^*(s) = P(s)/Q(s)$ is now given by

$$f(x) = \sum_{j=1}^{k} \frac{P(s_j)}{Q'(s_j)} e^{s_j x},$$

as can be verified by taking the Laplace transform of both sides of this equation. This result is readily extended to the case in which some of the roots of $Q(s) = 0$ are not simple. For example, the inverse of the Laplace transform

$$f^*(s) = \frac{P(s)}{(s - a)^m},$$

where $P(s)$ is a polynomial of degree lower than m, is given by

$$f(x) = e^{ax} \sum_{j=1}^{m} \frac{P^{(m-j)}(a)x^{j-1}}{(m - j)!(j - 1)!}. \tag{C.7}$$

Here $P^n(a)$ denotes the nth derivative of $P(x)$ at $x = a$.

It is usually not possible to give an explicit expression for the inverse of a given Laplace transform. In those situations one might try to obtain the unknown function $f(x)$ by numerical inversion of its Laplace transform $f^*(s)$. The numerical inversion of a Laplace transform is a tricky problem in numerical analysis, but a method that works satisfactorily in many cases has been proposed by Abate and Whitt (1992).

Illustration

To illustrate the above results, consider the following integro-differential equation in the unknown function $q(x)$:

$$q'(x) = -\frac{\lambda}{\sigma}\{1 - B(x)\} + \frac{\lambda}{\sigma}q(x) - \frac{\lambda}{\sigma}\int_0^x q(x - y)b(y)\,dy, \quad x > 0.$$

Here $B(x)$ is the probability distribution function of a positive random variable with probability density $b(x)$ and finite mean m, while λ and σ are positive constants such that $\lambda m/\sigma < 1$. The integro-differential equation appears in Section 1.8 and the function $q(x)$ represents amongst others the complementary waiting-time distribution in the standard single-server queueing system with Poisson input. From physical considerations we know that $q(x)$ is decreasing in x and has limit 0 as $x \to \infty$.

Denoting by $q^*(s)$ and $b^*(s)$ the Laplace transforms of $q(x)$ and $b(x)$, it follows by using (C.1)–(C.4) that for all s with $Re(s) > 0$,

$$sq^*(s) - q(0) = -\frac{\lambda}{\sigma}\left\{\frac{1}{s} - \frac{b^*(s)}{s}\right\} + \frac{\lambda}{\sigma}q^*(s) - \frac{\lambda}{\sigma}q^*(s)b^*(s).$$

The unknown $q(0)$ follows by letting $s \to 0$ in both sides of this equation and applying (C.6) with f replaced by q. Noting that $\lim_{s\to 0}\{1 - b^*(s)\}/s = m$ (use L'Hospital's rule), we obtain $q(\infty) - q(0) = -\lambda m/\sigma$. Thus, by $q(\infty) = 0$,

$$q(0) = \frac{\lambda m}{\sigma}.$$

Suppose now that $b(x)$ is the probability density of a Coxian-2 distributed random variable X. In other words, $X = X_1 + X_2$ with probability b and $X = X_1$ with probability $1 - b$ for two independent exponentials X_1 and X_2 with respective means $1/\mu_1$ and $1/\mu_2$. Using (C.5) and noting that

$$E(e^{-sX}) = bE\left(e^{-s(X_1+X_2)}\right) + (1 - b)E\left(e^{-sX_1}\right),$$

it follows that

$$b^*(s) = \frac{b\mu_1\mu_2}{(s + \mu_1)(s + \mu_2)} + \frac{(1 - b)\mu_1}{s + \mu_1}.$$

Next, using that $m = 1/\mu_1 + b/\mu_2$, we obtain after some algebra

$$q^*(s) = \frac{\lambda(\sigma\mu_1\mu_2)^{-1}\left[(b\mu_1 + \mu_2)s + b\mu_1(\mu_1 + \mu_2) + \mu_2^2\right]}{(s + \mu_1)(s + \mu_2) - (\lambda/\sigma)(s + b\mu_1 + \mu_2)}.$$

Thus, using the inversion formula (C.7), we obtain

$$q(x) = A_1e^{-b_1x} + A_2e^{-b_2x}, \quad x \geq 0.$$

Explicit expressions for the coefficients A_i and b_i are easily derived. However, they are omitted since our only purpose is to establish that $q(x)$ is the sum of two exponential functions when $b(x)$ is a Coxian-2 density.

Asymptotic expansions from the generating function

The generating function is in fact a special case of the Laplace transform. Let $\{p_n, n = 0, 1, \ldots\}$ be a discrete probability distribution. The generating function (or z-transform) of this distribution is defined by

$$P(z) = \sum_{n=0}^{\infty} p_n z^n, \quad |z| \le 1,$$

with z being a complex variable. Note that $P(z)$ can be interpreted as a Laplace transform by taking $z = e^{-s}$. In many applications the p_n's are unknown probabilities, but an explicit expression for the generating function $P(z)$ is obtained by some reasoning. Under rather weak regularity conditions an asymptotic estimate for the probability p_n with n large can be derived from the generating function $P(z)$. As in the Fast Fourier Transform method which provides a numerical tool for recovering all the p_j's from $P(z)$, we need help from complex numbers. To be specific, let us assume that $P(z)$ can be represented as

$$P(z) = \frac{N(z)}{D(z)},$$

where $N(z)$ and $D(z)$ are analytic functions whose domains of definition can be extended to a region $|z| < R$ in the complex plane for some $R > 1$. It is no restriction to assume that $N(z)$ and $D(z)$ have no common zeros; otherwise, cancel out common zeros. Let us further assume that the following regularity conditions are satisfied:

C1 The equation $D(z) = 0$ has a *real* root z_0 on the interval $(1, R)$.

C2 The function $D(z)$ has no zeros in the domain $1 < |z| < z_0$ of the complex plane.

C3 The zero $z = z_0$ of $D(z)$ is of multiplicity 1 and is the only zero of $D(z)$ on the circle $|z| = z_0$.

Theorem C.1   *Under the conditions C1–C3,*

$$p_j \approx \gamma_0 z_0^{-j} \quad \text{for } j \text{ large enough,} \tag{C.8}$$

where the constant γ_0 is given by

$$\gamma_0 = -\frac{1}{z_0} \frac{N(z_0)}{D'(z_0)}. \tag{C.9}$$

Here $D'(z_0)$ denotes the derivative of $D(x)$ at $x = z_0$.

Proof   We first mention the following basic facts from complex function theory. The most important fact is that a function $f(z)$ is analytic at a point $z = a$ if and only if $f(z)$ can be expanded in a power series $f(z) = \sum_{n=0}^{\infty} a_n (z-a)^n$ in $|z-a| < \rho$ for some $\rho > 0$; see e.g. Courant (1964). The coefficient a_n of the Taylor series is the nth derivative of $f(z)$ at $z = a$ divided by $n!$. The analytic function $f(z)$ is said to have a zero of multiplicity k in $z = a$ if $a_0 = \ldots = a_{k-1} = 0$ and $a_k \ne 0$. Another basic result is the following. The Taylor series $\sum_{n=0}^{\infty} a_n (z-a)^n$ of a function

$f(z)$ at the point $z = a$ coincides with the function $f(z)$ in the interior of the largest circle whose interior lies wholly within the domain where $f(z)$ is analytic.

The proof of (C.8) now proceeds as follows. The conditions C1–C3 imply that there is a circle around $z = 0$ with radius R_0 larger than z_0 such that $P(z)$ is analytic in $|z| < R_0$ except for the isolated point $z = z_0$. Since $D(z)$ has a zero of multiplicity 1 at $z = z_0$, it follows from the Taylor series that $D(z) = (z - z_0)\phi(z)$ in $|z| < R_0$, where $\phi(z)$ is an analytic function with $\phi(z_0) \neq 0$. Thus we can write $P(z)$ as $P(z) = H(z)/(z - z_0)$ for some analytic function $H(z)$ in $|z| < R_0$ with $H(z_0) \neq 0$. Using a Taylor expansion $H(z) = H(z_0) + (z - z_0)U(z)$, we next find that $P(z)$ can be represented as

$$P(z) = \frac{r_0}{z - z_0} + U(z) \tag{C.10}$$

in $|z| < R_0$, $z \neq z_0$. Here $U(z)$ is an analytic function in the domain $|z| < R_0$ and the residue r_0 is given by

$$r_0 = \lim_{z \to z_0} (z - z_0)P(z) = N(z_0)/D'(z_0).$$

The remainder of the proof is simple. Since $U(z)$ is analytic for $|z| < R_0$ we have the power series representation $U(z) = \sum_{j=0}^{\infty} u_j z^j$ for $|z| < R_0$. Let R_1 be any number with $z_0 < R_1 < R_0$. Then, for some constant b, $|u_j| \leq bR_1^{-j}$ for all $j \geq 0$. This follows from the fact that the series $\sum_{j=0}^{\infty} u_j z^j$ is convergent for $z = R_1$ and so the sequence $\{u_j R_1^j\}$ is bounded. Using the power series representation of $U(z)$ and the fact that the power series representation $P(z) = \sum_{j=0}^{\infty} p_j z^j$ extends to $|z| < z_0$, it follows from (C.10) that

$$\sum_{j=0}^{\infty} p_j z^j = \frac{-r_0}{z_0} \sum_{j=0}^{\infty} (z/z_0)^j + \sum_{j=0}^{\infty} u_j z^j, \quad |z| < z_0.$$

Equating coefficients yields

$$p_j = -r_0 z_0^{-j-1} + u_j, \quad j \geq 0.$$

Since $|u_j| \leq bR_1^{-j}$, $j \geq 0$, for some constant b and since $R_1 > z_0$, the coefficient u_j tends faster to zero than z_0^{-j}. This completes the verification of the asymptotic expansion (C.8).

The asymptotic expansion (C.8) is very useful for both theoretical and computational purposes. It appears that in many applications the asymptotic expansion for p_j can already be used for relatively small values of j.

Illustration

To illustrate the asymptotic expansion (C.8), consider the *M/G/*1 queue in which customers arrive according to a Poisson process with rate λ and the service time of a customer has a probability distribution function $B(x)$. It is assumed that the offered load $\rho = \lambda\mu$ is smaller than 1, where μ is the mean service time. Denote by $\{p_j, \ j = 0, 1, \ldots\}$ the limiting distribution of the number of customers present at an arbitrary epoch. In Section 4.2 of Chapter 4 it was shown that the generating function $P(z) = \sum_{n=0}^{\infty} p_n z^n$, $|z| \le 1$, is given by

$$P(z) = (1 - \rho)\frac{1 - \lambda(1 - z)A(z)}{1 - \lambda A(z)}$$

where $A(z) = \sum_{n=0}^{\infty} a_n z^n$, $|z| \le 1$, with $a_n = (1/n!) \int_0^{\infty}\{1 - B(t)\}e^{-\lambda t}(\lambda t)^n \, dt$. Note that $A(z)$ can be written as

$$A(z) = \int_0^{\infty} e^{-\lambda(1-z)t}\{1 - B(t)\} \, dt. \tag{C.11}$$

The generating function $P(z)$ can indeed be represented by the ratio of two functions $N(z)$ and $D(z)$ with $N(z) = (1 - \rho)[1 - \lambda(1 - z)A(z)]$ and $D(z) = 1 - \lambda A(z)$. The domain of definition of $N(z)$ and $D(z)$ can be extended outside the unit circle when the following assumptions are made:

(i) $\int_0^{\infty} e^{st}\{1 - B(t)\} \, dt < \infty$ for some $s > 0$,

(ii) $\lim_{s \to B} \int_0^{\infty} e^{st}\{1 - B(t)\} \, dt = \infty$,   where $B = \sup\{s | \int_0^{\infty} e^{st}\{1 - B(t)\} \, dt < \infty\}$.

The assumption (i) requires that the service-time distribution has no extremely long tail. This assumption is satisfied in most applications. The other assumption is a technical one. Let $R > 1$ be defined by

$$R = 1 + \frac{B}{\lambda}.$$

It follows from the representation (C.11) that the function $A(z)$ has an analytic continuation to the region $|z| < R$. This implies that $N(z)$ and $D(z)$ are analytic functions for $|z| < R$. It now holds that

$$p_j \approx \frac{(1 - \rho)}{\lambda^2}\left[\int_0^{\infty} te^{-\lambda(1-z_0)t}\{1 - B(t)\} \, dt\right]^{-1} z_0^{-j} \quad \text{for } j \text{ large enough} \tag{C.12}$$

where z_0 is the unique root of the equation

$$\int_0^{\infty} e^{-\lambda(1-z)t}\{1 - B(t)\} \, dt = \frac{1}{\lambda}$$

on the interval $(1, R)$.

For completeness we give a proof of (C.12). This proof is representative for other applications as well. The reader who is only interested in the result (C.12) may skip the proof. To verify that the conditions C1–C3 are satisfied, we first note that $A(x)$ is strictly increasing for $1 < x < R$. Further, using the assumption of $\lambda\mu < 1$ and the assumption (ii), it follows that $A(1) = \mu < 1/\lambda$ and $A(x) \to \infty$ as $x \to R$. Thus we can conclude that there is a unique $z_0 \in (1, R)$ such that the graph of $A(x)$ and the line $y = 1/\lambda$ intersect at $x = z_0$. Hence $1 - \lambda A(x)$ has a unique zero z_0 on the interval $(1, R)$. As a by-product of the proof, we find that

$$A(x) - 1/\lambda < 0 \quad \text{for } 1 < x < z_0$$

and

$$A(x) - 1/\lambda > 0 \quad \text{for } z_0 < x < R.$$

Using the mean-value theorem $f(c + d) = f(c) + df'(c) + \tfrac{1}{2}d^2 f''(c + \theta d)$ for some $0 \le \theta \le 1$, it follows that $f'(c) \ne 0$ when $f(x)$ is strictly convex on (a, b) and changes sign at $x = c$ with $c \in (a, b)$. The function $A(x) - 1/\lambda$ is strictly convex on $(1, R)$, since the second derivative $A''(x)$ is positive. Also $A(x) - 1/\lambda$ changes sign at $x = z_0$. Thus the derivative of $A(x) - 1/\lambda$ at $x = z_0$ is not equal to 0, implying that the zero z_0 of $1 - \lambda A(x)$ is of multiplicity 1. To verify that $1 - \lambda A(z)$ has no zero for $1 < |z| < z_0$, we use the basic result that the power series representation $A(z) = \sum_{n=0}^{\infty} a_n z^n$ extends to $|z| < R$. Since the a_n's are reals and positive, it now follows that $|A(z)| \le A(|z|)$ for $|z| < R$. This inequality and the inequality $A(x) - 1/\lambda < 0$ for $1 < x < z_0$ imply that $|A(z)| < 1/\lambda$ for all z in the domain $1 < |z| < z_0$, showing that $1 - \lambda A(z)$ has no zero in this domain. It remains to verify that $1 - \lambda A(z)$ has z_0 as the only zero on the circle $|z| = z_0$. Thus we must prove that $x = z_0$ and $y = 0$ for any complex number $z = x + \mathrm{i}y$ with $A(z) = 1/\lambda$ and $|z| = z_0$. To do so, note that $|z| = (x^2 + y^2)^{1/2}$ and so $(x^2 + y^2)^{1/2} = z_0$ implying that $x \le z_0$. Next, using that $|e^{\mathrm{i}u}| = 1$ for any real u, we obtain from (C.11) that

$$\frac{1}{\lambda} = |A(z)| \le \int_0^\infty |e^{-\lambda(1-z)t}| \{1 - B(t)\} \, \mathrm{d}t$$

$$= \int_0^\infty e^{-\lambda(1-z_0)t} \{1 - B(t)\} \, \mathrm{d}t = \frac{1}{\lambda}.$$

This implies that x must be equal to z_0. Also, by $(x^2 + y^2)^{1/2} = z_0$, we find $y = 0$. This completes the proof.

APPENDIX D. NUMERICAL SOLUTION OF MARKOV CHAIN EQUATIONS

In Markov chain applications one often has to solve a finite system of linear equations of the form

$$x_i = \sum_{j=1}^N p_{ji} x_j, \quad i = 1, \ldots, N, \tag{D.1}$$

$$\sum_{i=1}^{N} x_i = 1. \tag{D.2}$$

Under a mild regularity condition posed on the underlying Markov chain (see Chapter 2), this system of linear equations has a unique solution. Moreover, any solution to the balance equations (D.1) is uniquely determined up to a multiplicative constant. This constant is found by using the normalizing equation (D.2).

In general there are two methods to solve the Markov chain equations: (a) direct methods; (b) iterative methods.

A convenient direct method is a Gaussian elimination method such as the Gauss–Jordan method. This reliable method is recommended as long as the dimension N of the system of linear equations does not exceed 200 (say). The computational effort of Gaussian elimination is proportional to N^3. Ready-to-use codes for Gaussian elimination methods are widely available; see e.g. the source book by Press *et al* (1986). A Gaussian elimination method requires that the whole coefficient matrix is stored, since this matrix must be updated at each step of the algorithm. This explains why a Gaussian elimination method suffers from computer memory problems when N gets large. Another direct method for solving Markov chain equations is the probabilistic method proposed in Grassmann *et al* (1985). However, this method has the same drawbacks as Gaussian elimination when N gets large. In solving (D.1) and (D.2) by a direct method one of the balance equations (D.1) is omitted in order to obtain a square system of linear equations.

Iterative method of ·successive overrelaxation

Iterative methods have to be used when the size of the system of linear equations gets large. In specific applications an iterative method can usually avoid computer memory problems by exploiting the structure of the application. An iterative method works with the original matrix of coefficients. In applications these coefficients are usually composed from a few constants. Then only these constants have to be stored in memory.

The iterative method of successive overrelaxation is a suitable method for solving the linear equations of large Markov chains. The well-known Gauss–Seidel method is a special case of the method of successive overrelaxation. The iterative methods generate a sequence of vectors $x^{(0)} \to x^{(1)} \to x^{(2)} \to \dots$ converging towards a solution of the balance equation (D.1). The normalization is done at the end of the calculations. To apply successive overrelaxation, we first rewrite the balance equations (D.1) into the form

$$x_i = \sum_{\substack{j=1 \\ j \neq i}}^{N} a_{ij} x_{ij}, \quad i = 1, \dots, N,$$

where

$$a_{ij} = \frac{p_{ji}}{1 - p_{ii}}, \quad i, j = 1, \dots, N, \ j \neq i.$$

The standard successive overrelaxation methods uses a fixed relaxation factor ω for speeding up the convergence. The method starts with an initial approximation vector $x^{(0)} \neq 0$. In the kth iteration of the algorithm an approximation vector $x^{(k)}$ is found by a recursive computation of the components $x_i^{(k)}$ for $i = 1, \ldots, N$ such that the calculation of the new estimate $x_i^{(k)}$ uses both the new estimates $x_j^{(k)}$ for $j < i$ and the old estimates $x_j^{(k-1)}$ for $j > i$. The steps of the algorithm are now as follows.

Step 0. Choose a nonzero vector $x^{(0)}$. Let $k := 1$.
Step 1. Calculate successively for $i = 1, \ldots, N$ the component $x_i^{(k)}$ from

$$x_i^{(k)} = (1 - \omega)x_i^{(k-1)} + \omega \left(\sum_{j=1}^{i-1} a_{ij}x_j^{(k)} + \sum_{j=i+1}^{N} a_{ij}x_j^{(k-1)} \right).$$

Step 2. If the stopping criterion

$$\sum_{i=1}^{N} \left| x_i^{(k)} - x_i^{(k-1)} \right| \leq \varepsilon \sum_{i=1}^{N} \left| x_i^{(k)} \right|,$$

is satisfied with $\varepsilon > 0$ a prespecified accuracy number, then go to step 3. Otherwise $k := k + 1$ and go to step 1.
Step 3. Calculate the solution to (D.1) and (D.2) from

$$x_i^* = \frac{x_i^{(k)}}{\sum_{j=1}^{N} x_j^{(k)}}, \qquad 1 \leq i \leq N.$$

The specification of the tolerance number ε depends typically on the particular problem considered and the accuracy required in the final answers. In addition to the stopping criterion, it may be helpful to use an extra accuracy check for the equilibrium probabilities of the underlying Markov chain. An extra accuracy check may prevent a decision upon a premature termination of the algorithm when the tolerance number ε is not chosen sufficiently small. Notice that the normalizing equation (D.2) is used only at the very end of the algorithm. In applying successive overrelaxation it is highly recommended that all of the balance equations (D.1) are used rather than omitting one redundant equation and substituting the normalizing equation (D.2) for it.

The convergence speed of the successive overrelaxation method may dramatically depend on the choice of the relaxation factor ω, and even worse the method may diverge for some choices of ω. A suitable value of ω has to be determined experimentally. Usually $1 \leq \omega \leq 2$. The choice $\omega = 1.2$ is often recommended. The 'optimal' value of the relaxation factor ω depends on the structure of the particular problem considered. It is pointed out that the iteration method with $\omega = 1$ is the well-known Gauss–Seidel method. This method is convergent in all practical cases. The ordering of the states may also have a considerable effect on the convergence speed of the successive overrelaxation algorithm. In general one should order the states such that the upper diagonal part of the matrix of coefficients is as sparse as possible. In specific applications the transition structure of the Markov chain often suggests an appropriate ordering of the states.

To conclude this section, we briefly mention an alternative iterative method that has the advantage of providing at each iteration lower and upper bounds on the quantities to be calculated. In many applications the state probabilities of the underlying Markov chain are calculated with the only purpose of obtaining a single performance measure (e.g. the rejection probability or the average queue size in a queueing application). In such situations the problem can be reduced to calculating the average cost per unit time in a single Markov chain on which an appropriate cost structure is superimposed. A convenient method to calculate the average cost is the value-iteration algorithm from Section 3.4, where the data transformation from Section 3.5 should be applied first when the underlying Markov chain has a continuous-time parameter. This value-iteration method can be extended to calculate simultaneously all of the state probabilities of the Markov chain such that lower and upper bounds on the state probabilities are provided at each iteration; see Van der Wal and Schweitzer (1987).

REFERENCES

Abate, J., and Whitt, W. (1992). 'The Fourier-series method for inverting transforms of probability distributions', *Queueing Systems*, **10**, 5–88.

Abramowitz, M., and Stegun, I. (1965). *Handbook of Mathematical Functions*, Dover, New York.

Courant, R. (1964). *Differential and Integral Calculus*, Vol. II, Blackie & Son, Glasgow.

Grassmann, W., Taksar, M. and Heyman, D.P. (1985). 'Regenerative analysis and steady-state distributions for Markov Chains', *Operat. Res.*, **33**, 1107–1116.

Press, W.H., Flannery, B.P., Teukolsky, S.A. and Vetterling, W.T. (1986). *Numerical Recipes*, Cambridge University Press, Cambridge.

Rudin, W. (1964). *Principles of Mathematical Analysis*, 2nd edn., McGraw-Hill, New York.

Van der Wal, J. and Schweitzer, P.J. (1987). 'Iterative bounds on the equilibrium distribution of a finite Markov chain', *Prob. Engineering and Inform. Sci.*, **1**, 117–131.

Whitt, W. (1982). 'Approximating a point process by a renewal process, I: two basic methods', *Operat. Res.*, **30**, 125–147.

Index

Learning Resources
Centre